Managing Emerging Technologies for Socio-Economic Impact

SCIENCE, INNOVATION, TECHNOLOGY AND ENTREPRENEURSHIP

Series Editors: Elias G. Carayannis, *Professor of Science, Technology, Innovation and Entrepreneurship, School of Business, George Washington University, USA* and Aris Kaloudis, *NIFU STEP Studies in Innovation, Research and Education, Oslo, Norway*

There is ample and growing evidence that intangible resources such as knowledge, know-how and social capital will prove to be the coal, oil, and diamonds of the twenty-first century for developed, developing, and emerging economies alike. Moreover, there are strong indications and emerging trends that there are qualitative and quantitative differences between the drivers of economic growth in the twentieth and twenty-first centuries.

This new era is punctuated by:

- Development of a service-based economy, with activities demanding intellectual content becoming more pervasive and decisive.
- Increased emphasis on higher education and life-long learning to make effective use of the rapidly expanding knowledge base.
- Massive investments in research and development, training, education, software, branding, marketing, logistics, and similar services.
- Intensification of competition between enterprises and nations based on new product design, marketing methods and organizational forms.
- Continual restructuring of economies to cope with constant change.

This valuable new series concentrates on these important areas by focusing on the key pillars of science, technology, innovation, and entrepreneurship.

Titles in the series include:

Diversity in the Knowledge Economy and Society
Heterogeneity, Innovation and Entrepreneurship
Edited by Elias G. Carayannis, Aris Kaloudis and Åge Mariussen

Crisis, Innovation and Sustainable Development
The Ecological Opportunity
Edited by Blandine Laperche, Nadine Levratto and Dimitri Uzunidis

The Innovation Union in Europe
A Socio-Economic Perspective on EU Integration
Edited by Elias G. Carayannis and George M. Korres

Managing Emerging Technologies for Socio-Economic Impact
Edited by Dimitris G. Assimakopoulos, Ilan Oshri and Krsto Pandza

Managing Emerging Technologies for Socio-Economic Impact

Edited by

Dimitris G. Assimakopoulos

Professor of Technology and Innovation Management, Grenoble Ecole de Management, LINC Lab, France

Ilan Oshri

Professor of Technology and Globalisation, Loughborough University, School of Business and Economics, UK

Krsto Pandza

Professor of Strategy and Innovation, Leeds University Business School, UK

SCIENCE, INNOVATION, TECHNOLOGY AND ENTREPRENEURSHIP

Edward Elgar
PUBLISHING

Cheltenham, UK • Northampton, MA, USA

Published by
Edward Elgar Publishing Limited
The Lypiatts
15 Lansdown Road
Cheltenham
Glos GL50 2JA
UK

Edward Elgar Publishing, Inc.
William Pratt House
9 Dewey Court
Northampton
Massachusetts 01060
USA

A catalogue record for this book
is available from the British Library

Library of Congress Control Number: 2014957079

This book is available electronically in the **Elgar**online
Business subject collection
DOI 10.4337/9781782547884

MIX
Paper from
responsible sources
FSC® C013604

ISBN 978 1 78254 787 7 (cased)
ISBN 978 1 78254 788 4 (eBook)

Typeset by Servis Filmsetting Ltd, Stockport, Cheshire
Printed and bound in Great Britain by CPI Group (UK) Ltd, Croydon CR0 4YY

Contents

Contributors

Andreas Alexiou has a background in economics and information systems management (Warwick Business School, 2010). He is currently a PhD researcher in strategy for the Rotterdam School of Management, Erasmus University, the Netherlands, and his research interests involve the antecedents of organizational resilience in regard to disruptive technological change as well as the role of soft factors – such as emotions and energy – in shaping organizational outcomes such as learning and dynamic capabilities. At the same time, he is exploring the role of new technologies such as virtual worlds and serious games in facilitating learning and training in organizations.

Dimitris G. Assimakopoulos is a professor of technology management and the founder and director of the Laboratory on Learning and Innovation in Networks and Communities (LINC Lab) at Grenoble Ecole de Management, France. He also serves as the president of the European Doctoral Programmes Association in Management and Business Administration (EDAMBA). He holds an HDR in economics from Grenoble University and a PhD from Sheffield University, UK. He has also twice been a visiting scholar in economic sociology at Stanford University, USA. His research focuses on the origins and evolution of technological communities in the context of emerging technologies and knowledge-based innovation, in a variety of settings, such as semiconductors in Silicon Valley, software development in China, and particle therapy worldwide. Journal articles have appeared in *Environment and Planning B*, *International Journal of Technology Management*, *International Small Business Journal*, *Sloan Management Review*, *Organizational Dynamics*, *R&D Management* and *Science and Public Policy*.

Andrea Carafa is a Marie Curie researcher in science, technology, and innovation networks that address societal challenges, based at the LINC Lab, Grenoble Ecole de Management, France. He collaborated with the GSI and CERN, and was a visiting researcher at the EC DG Research and Bocconi University (Milan, Italy). He holds degrees from ESADE Business School (Barcelona, Spain), Aarhus University (Denmark) and LUISS University (Rome, Italy), and received training in several other countries.

Diogo Cotta is currently a Marie Curie fellow and a PhD student in operations management at IE Business School, Spain. Diogo has a bachelors degree in economics from the Technical University of Lisbon, Portugal, and a masters degree in history and philosophy of science from the New University of Lisbon. His research focuses on organizational structures and processes that are able to deal with novelty. Most recently he has been working on organization-wide systems of transactive memory.

Joze P. Damijan is a professor at the University of Ljubljana (Faculty of Economics), and affiliated to LICOS, Centre for Institutional Economics, and to VIVES, Centre for Regional Economic Policy at KU Leuven, Belgium. His academic interest is focused on international trade and globalization and their impacts on firm performance and macroeconomic development. His main topics of research are international trade and trade policy, corporate governance, corruption and development, economic geography, foreign direct investment, firm innovation, transfer of technology and factors enhancing competitiveness and firm productivity growth. Recent publications include papers in *World Economy*, *Review of World Economics*, *Journal of Comparative Economics*, *World Development*, *Economics of Transition*, *Economic Letters*, as well as chapters in books published by Edward Elgar and Palgrave Macmillan, for example. He serves as an editor of the *Economic and Business Review* and is associated with a number of editorial boards.

Evgenia Dolgova is a Marie Curie postdoctoral research fellow at Leeds University Business School, UK, which she joined after her PhD programme at Rotterdam School of Business, Erasmus University, the Netherlands. Evgenia specializes in dynamic social network analysis, applying bibliographic methods and semantic network analysis to study how emergent technologies spread in society. Taking nanotechnology and massive open online courses as an example, she investigates the processes that lead to adoption or rejection of new technologies. She also holds a masters degree in psychology from the Ludwig-Maximillians-University of Munich, Germany.

Renata Dombrovski is a PhD student and a Marie Curie researcher at the Faculty of Economics, University of Ljubljana, Slovenia. She is employed under the EU FP7 Project 'Management of Emerging Technologies for Economic Impact'. Renata graduated from the University of Rijeka (Croatia) with an MSc degree in public finances. Her research is focused on econometric examination of the effectiveness of the European Union renewable energy policies in promoting clean technologies.

Davy van Doren has worked in various research positions, with an academic background in evolutionary biology, environmental sciences (doctorate, Leiden University, the Netherlands, 2006) and sustainable development (masters, Utrecht University, the Netherlands, 2010). After projects related to marine biodiversity (Naturalis Biodiversity Center, the Netherlands), ecosystem conservation (Royal Boskalis Westminister, the Netherlands), biofuel supply chain management (Port of Rotterdam, the Netherlands) and climate change communication (Utrecht University, the Netherlands), van Doren currently holds a position at the Fraunhofer Institute for Systems and Innovation Research (Germany). His current research centres on technological and non-technological drivers of advanced biotechnology development, including issues related to public policy, collaborative behaviour and responsible research and innovation (RRI). Within this research, van Doren has a particular focus on the emerging technology synthetic biology.

Marija Drenkovska holds an MSc (2011) from the Faculty of Economics, University of Ljubljana, Slovenia, and a BA in Economics (2005) from the Faculty of Economics, Sts Cyril and Methodius University – Skopje, Macedonia. She is a Fellow Marie Curie researcher at the Academic Unit for Economics, Faculty of Economics, University of Ljubljana. Her main research interests include development economics, measurement of innovative activities in technological follower countries and their impact on growth. Marija has co-authored several chapters in edited books. Currently her work is focused on her PhD dissertation thesis.

Pradeep Durgam works at the Department of Information Service & Economy at Aalto University School of Business as a Doctoral Researcher. He graduated from IE Business School, Spain with an MBA degree specializing in corporate strategy. He has over 10 years of experience in CRM (Strategy and Technology) focusing on consumer products and the retail sector. From 2003 to 2010, he worked in companies like Siemens and Capgemini Consulting, amongst others. He was a part of the ManETEI project as a Marie Curie Fellow from 2010 until 2014 and considers ManETEI to be an excellent environment that gave him the opportunity to collaborate with academic and industry champions, as well as allowing him to identify and explain the multifaceted occurrence of emerging technologies in the field of CRM. He has since gone a step further and included his ManETEI research in his PhD thesis.

Abdelghani Es-Sajjade's research interests include organizational design and interdependence, technology and innovation management, and social identity in organizations. He is intrigued by the notion of collective agency, and by the fact that most theory on organizational design focuses on environmental or firm-level aspects in explaining the evolvement of organizational design, in lieu of group-level aspects. He is a graduate from the Erasmus University Rotterdam School of Management, the Netherlands, and currently works as a research fellow for the Leeds University Business School, UK. As a former practitioner in technology-led companies he engages closely with industry to develop research topics that are relevant to the theory and practice of management and organization.

Bei Gao is a PhD candidate at University College Dublin, Michael Smurfit Graduate Business School (Ireland). She joined the ManETEI (Management of Emerging Technologies for Economic Impact) project as a Marie Curie research fellow in 2010. She studied social science and business administration at Lund University (Sweden), gaining a masters degree in sustainable business leadership. Her undergraduate study was in the field of materials science and engineering, and she graduated from the Harbin Institute of Technology (China). Her research interests are in tools for emergent technology assessment, concerning policy influence, social innovation processes, the role of communities and emergent industrial architectures. Her current work is concentrated on mapping the national innovation networks of renewable energy and policy analysis, referring to the research areas on complex innovation systems, evolutionary economics, social network analysis and policy modeling.

Saeed Khanagha is an assistant professor of strategic management at Nijmegen School of Management, the Netherlands. He completed his PhD at Rotterdam School of Management, Erasmus University, the Netherlands. During his PhD studies, he focused on exploring the intersection of innovation management, corporate strategy and organization studies. In particular, he is researching the process technology and business model change in the established organizations through a mixture of qualitative and quantitative approaches. His research has been published in academic journals such as *European Management Review* and *R&D Management*, and presented in Academy of Management annual meetings and other prestigious conferences.

Juan Pablo Madiedo M. is a Marie Curie fellow and PhD candidate in operations management at IE Business School, Spain. He holds an MSc in engineering and an MBA from Universidad de los Andes, Colombia. Previously he held a position as professor and researcher at Universidad de los Andes School of Management, Colombia. His research focuses on innovation and business analytics management. Since 2004 he has worked in operations and quality management in the oil and gas and health care industries.

Natalia Maya is currently working for Bayer Technology Services (BTS) in Germany and her main field of interest is technology management. Previously she worked in projects related to the use of information and communication technologies for development in Colombia. She studied electronics engineering and telecommunications at the University of Cauca (Colombia), where she earned a masters degree as well. She also holds a masters degree in technology and policy from MIT, USA.

Ilan Oshri is a professor of globalization and technology, and director of the Centre for Global Sourcing and Services, Loughborough School of Business and Economics, UK. Oshri received his PhD in technological innovations from Warwick Business School, UK (2003). Ilan then held assistant professor and associate professor positions at Rotterdam School of Management in the Netherlands. Ilan was the founder of the Case Development Centre and also the co-founder of the Global Sourcing Workshop. Ilan is an associate at the LSE Outsourcing Unit, UK, a regular speaker at industry events and a contributor to various professional magazines. Since 2003 Ilan has built a research programme that brings together technology management and globalization aspects. His early years as an academic were devoted to understanding component sharing in both co-located and globally distributed teams. In recent years, the mainstream of his work has revolved around outsourcing and offshoring of information technology and business processes. He is particularly interested in understanding how multinationals make decisions about their outsourcing and offshoring strategies, how they set up structures and mechanisms to execute outsourcing and offshoring strategies, and how value and innovation is delivered under sourcing terms. Ilan uses multiple theoretical lenses to study the above, including the knowledge-based view of the firm and contingency theory.

Darius Pacauskas is a PhD candidate in information and service economy department, Aalto School of Business, Finland. His research interest belongs to the area of value co-creation, where he has two main aims: to understand how value co-creation activities support user-innovators' creativity; and how user-innovators' connection with the product or service which they are developing adds value for the companies. He is eager to enrich theoretical and practical knowledge of the value co-creation field.

Krsto Pandza is a professor in strategy and innovation and ManETEI (Management of Emerging Technologies for Economic Impact) coordinator at Leeds University Business School, UK. In a broad sense, Krsto's research links the domains of strategic management, technology and innovation management, manufacturing strategy and research and development (R&D) policy. He strives to integrate the diversity of the above-mentioned research fields in order to achieve a more holistic understanding of the role of engineering and scientific knowledge, technology and operations in creating and sustaining a competitive advantage in technology-intensive and manufacturing organizations. He is especially interested in the role of managerial agency within dynamic and evolutionary organizational phenomena. His current research explores development dynamics in a nanotechnology innovation system with an aim to advance the debate on dynamic capabilities by addressing the relevance of managerial agency.

Andrew Parker is an associate professor at the Department of People, Organizations and Society at Grenoble Ecole de Management, France. He has conducted social network analysis research on a wide range of Fortune 500 organizations and government agencies. His research has covered top-level executive teams, functional departments, communities of practice and recently merged companies. His articles have appeared in *Social Networks*, *Sloan Management Review*, *Organizational Dynamics* and *California Management Review*. He completed his sociology PhD at Stanford University, USA, and also holds graduate degrees from the London School of Economics, UK and Northeastern University, USA. He joined the faculty of Grenoble Ecole de Management and LINC Lab in autumn 2010. His academic interests include social networks, organizational theory, careers and labour markets.

Mickael Pero, PhD, studied the economics of science and innovation at the Free University of Brussels (Belgium), Pierre Mendès-France University (Grenoble) and Bocconi University (Milan, Italy). In parallel, he conducted socio-economic evaluation of public research organizations in France (for example, CEA – Leti laboratory) and Italy (Elettra Sincrotrone Trieste) both from an academic and a policy angle (for example, the RIFI – Research Infrastructures: Foresight and Impact – European project). He is currently an experienced Marie Curie researcher (European Union ManETEI project – Management of Emerging Technologies for Economic Impact) at Fraunhofer ISI (Karlsruhe). His areas of expertise comprise the socio-economic impacts of public research and related indicators, incentives and constraints in the scientific and technological process, open innovation strategies, emerging technologies' scanning, and technological effects on complex systems.

Thomas Reiss is the head of the Competence Center 'Emerging Technologies' at the Fraunhofer Institute for Systems and Innovation

Research (Fraunhofer ISI) in Karlsruhe, Germany. He holds a PhD in molecular biology from Freiburg University, Germany (1983) and joined Fraunhofer ISI in 1987. His current research focuses on national and sectoral innovation systems; monitoring, foresight and impact assessment of new technologies; and innovation policies. He has coordinated numerous European and national projects focusing on innovation in biotechnology, biomedical engineering, health care systems, and nanotechnology. He is a member of the Management Committee of the European Techno-Economic Policy Support Network (ETEPS), of the Peer Review Panel of the European Science Foundation (ESF) and of the editorial boards of the *International Journal of Biotechnology* and the *International Journal of Learning and Change*. At the Karlsruhe Institute of Technology (KIT), Germany, he teaches a course on the management of new technologies.

Matti Rossi is an associate professor of information systems science at the Aalto University School of Business, Department of Information and Service Economy, in Finland.

Fabrizio Salvador is a professor of operations management at IE Business School in Madrid, Spain. He is also a founding member of the MIT Smart Customization Group, adjunct professor at the MIT–Zaragoza Logistics Program, and serves as senior scientific advisor for the IE Foundation. Dr Salvador has been visiting professor at the Massachusetts Institute of Technology (MIT), USA as well as adjunct faculty research associate at Arizona State University, USA. He received a PhD in operations management from the University of Padova, Italy, where he also graduated in industrial engineering.

Benjamin Schrempf is a PhD student at University College Dublin, Ireland, in complex systems and computational social science. In his PhD work he is investigating nanotechnology innovation networks with social network analysis methods. Based on the empirical data he is building an agent-based model to simulate the emergence and dissemination of general-purpose technologies. Benjamin graduated from the University of Hohenheim, Stuttgart, Germany in 2009 as Diplomoekonom. Prior to and during his studies, he worked for Landesbank Baden-Wuerttemberg, Volkswagen AG and Lux Research Inc. Benjamin is a member of the scientific staff at the EA European Academy of Technology Assessment GmbH, where he is involved in several European Union-funded projects.

Jatinder Sidhu is an associate professor at the Department of Strategic Management and Entrepreneurship, Rotterdam School of Management, Erasmus University, the Netherlands. After earning his first degree at the Delhi School of Economics, India, and an MSc in management sciences at the University of Hull, UK, Dr Sidhu studied for his PhD degree at the Tinbergen Institute, Erasmus School of Economics (ESE), the Netherlands. Former academic director (2007–2011) of the department's MSc programme in strategic management, Dr Sidhu currently teaches flagship courses in strategy at the graduate and undergraduate levels.

Henk W. Volberda is a professor of strategic management and business policy at the Department of Strategic Management and Entrepreneurship, Rotterdam School of Management (RSM), Erasmus University, the Netherlands. His work on strategic renewal, coevolution and new organizational forms has led to a large number of published articles in academic journals, including *Academy of Management Journal, Management Science, Decision Support Systems, European Business Forum, European Management Journal, European Management Review, Global Strategic Management, International Journal of Disclosure and Governance, International Studies of Management and Organization, Journal of Management Studies, Journal of Product Innovation Management, Long Range Planning, Omega, Organization Development Journal, Organization Studies, Organization Science, Strategic Management Journal* and *Technology Analysis and Strategic Management*.

Terence A. Wilkins is chief executive officer (CEO) of the NanoManufacturing Institute and Yorkshire Forward Professor of Nanomanufacturing Innovation at the University of Leeds, UK. He has 32 years' industrial and 10 years' academic experience as an innovator and director, engineering products and processes, exploiting emergent technologies, in the medical device, biotechnology, nanotechnology, chemicals, advanced materials and environmental engineering industries for substantial economic and societal impact. His first successful commercial product was a nanoparticle radioimmunoassay for metabolic diseases launched in 1982, yielding many €billions of sales. The invention lives on as a blood test for the early detection of prostate cancer. In September 2014, the UN Environmental Committee declared the ozone layer was now self-repairing as a result of HFC chemicals replacing CFCs equivalents in industrial applications. The HFC manufacturing business began in earnest in 1995 following innovations by his team in pilot plant developments. This sector now has global sales of €15 billion per year. This level of impact throughout his career has been recognised with eight major prizes, awards and medals for both new engineering science and commercial exploitation. He became Europe's first professor of Nanomanufacturing on joining Leeds University in 2005 where his personal research ranges across nanomedicine; industrial nanomaterials; nanoelectronics; EHS risks of engineered nanoparticles; and management and policy making for emergent technology manufacturing. Success as a serial industrial innovator has led to service *pro bono* as a high-level expert advisor (and committee chair) to the UK government since 1992 (BIS, DTI) including the Minister of Science & Universities, David Willetts (2011–14) and Greg Clarke (2015 to date); the European Commission EU (since 1994) and EU Parliament (since 2008), covering nanotechnology, advanced materials, production technologies (NMP) and the Marie Curie mobility programme. Uniquely, he has led large collaborative industrial engineering research and innovation projects in all seven EC framework programmes from 1982–2013 with a total grant income ~€125 million to date invested in the development of large numbers of able young engineers, scientists and new high-value product and manufacturing technologies for exploitation in his chosen fields of commercial application.

Miriam Wolf is a Marie Curie researcher at Leeds University Business School, UK. She holds an MA in social and cultural anthropology from the LMU Munich, Germany, and an MSc in sustainable resource management from the TU Munich and is currently finishing her PhD in management and organizational theory. Her research interests lie at the intersections between organizational theory and policy analysis. In particular she focuses on new forms of organizing, strategic agency and practice in European research and development (R&D) policy.

Preface

Smart, sustainable and inclusive growth is the pursuit of policy makers and practitioners in developed and developing economies alike. At the heart of it lie emerging technologies that may serve as game changers, market disruptors and catalysts of accelerated growth.

This book provides a number of views based on a European Union (EU)-funded project, Management of Emerging Technologies for Economic Impact (ManETEI), and derived from the work funded to explore issues related to emerging technologies and how best to manage them to unlock and leverage their potential and value added. Moreover, this book aims to present ongoing research on management of emerging technologies that impact upon society, economically and socially.

To this end, the volume addresses research questions, discusses empirical findings and puts forward recommendations that are applicable to the management of emerging technologies. Its chapters cover a variety of contemporary issues that lie at the heart of the emerging technologies management field, and address wide-ranging topics of interest to scholars, managers and policy makers. A diverse mix of methodologies and theoretical framings are applied to shed light on this complex socio-technical phenomenon.

In particular, the book aims to publish chapters that build up and/or test theory, explore common and unique phenomena relevant to management of emergent technologies, and craft out recommendations for practitioners. The broad set of applied methodological and theoretical frameworks range from statistical and network analysis, qualitative studies and theoretical pieces, to simulation models and best-practice analysis.

Elias G. Carayannis
Science, Innovation, Technology and Entrepreneurship (SITE) Series Editor

Acknowledgement

The book editors acknowledge financial support from the seventh framework – the People – programme of the European Commission – grant agreement number 238382 (February 2010–January 2014), for Marie Curie initial training networks on Management of Emergent Technologies for Economic Impact (ManETEI), www.manetei.eu.

1. Management of emerging technologies for economic and social impact: an introduction

Dimitris G. Assimakopoulos, Ilan Oshri and Krsto Pandza

PREAMBLE

From February 2010 to January 2014, the European Union (EU)-funded Management of Emerging Technologies for Economic Impact (ManETEI) – Marie Curie Initial Training Network, brought together a diverse group of leading European business schools, research institutes and industrial partners to investigate the challenges associated with the management of emerging technologies for economic and social impact.

Individual research projects progressed broadly in two directions. The first group of researchers embraced the notion that technological and organizational innovation unfolds in complex settings where a myriad of actors (companies, governments, universities and wider public) collaborate and interact in order to translate technological advances into solutions with impact of economic progress and social well-being. They followed the already recognized notion that innovation frequently depends on interplay and networking of many organizations and communities across different industries. Such ecosystems of interdependent organizations and scientific and technological communities are characterized by the wide dispersion of adequate knowledge that needs to be integrated and coordinated through collaboration and networking. This group could be further divided into researchers who investigated strategic challenges for policy makers in shaping technology development and those who centred their research on exploring collaborative and networking practices in technological and regional innovation networks.

The second group zoomed in and investigated how individual companies organize their innovation processes and identified what constitutes organizational capabilities that help organizations to capture value from emergent technologies. This group consists of researchers who investigated the foundations of organizational capabilities for technological innovation and those who explored how emerging technologies (mostly information and communication technology-enabled) help companies to strengthen their innovation capabilities.

The book is organized in four parts. Each part includes four chapters. Part I explores foundations of organizational capabilities for technological innovation. Part II explores collaboration and networking for shaping the emergence and progression of technologies. This is followed by Part III, addressing strategic challenges for policy makers that influence the sustainable and responsible development of technology. Part IV is less concerned with the technology dynamics; it investigates how emerging technology supports organizational capabilities and how novel technologies affect work and communication practices.

In the rest of this introductory chapter we discuss, in four sections: the changing conceptions of managing innovation and emerging technologies; the role of collaboration networks and communities in the process of emergence of emerging technologies; and then in the last two sections we discuss how emerging technologies relate to broader discussions on open innovation and changing work practices in organizations, with a particular reference to ICT innovations.

THE CHANGING NATURE OF MANAGING INNOVATION

Innovation is a word on everybody's lips. However, it often means different things to different people. Policy makers responsible for channelling governmental funds into research have recently rebranded science and technology policy into innovation policy. This should remind academic scientists and technologists that their inventions require transformations into something with value and impact. The wider public mostly experience innovation through improvements in products, new gadgets, and by consuming news about technology (mostly internet) entrepreneurs and their journeys from a garage to riches. For scientists and engineers, innovation is essentially a scientific discovery or engineering problem-solving. For managers in companies, innovation stands for new products and services that support the competitive advantage of their companies. They represent a group most likely sensitive to the fact that innovation requires strategic intent, appropriate structures, coordination and incentives, yet they still mostly believe that innovation is about a link between investment into research and development (R&D) and the number of new products and services introduced to market. For economists, innovation is about studying the relations between measurable inputs into a national innovation system and quantitative outputs captured in conventional national accounting. This suggests that different constituencies are united in seeing innovation as a set of inputs and outputs.

Innovation, however, is much better portrayed as a process: a journey (Van de Ven et al., 2008) characterized by temporal and relational complexity as well as extreme uncertainty. It encompasses the emergence of novelty and patterns among activities and events that enable the transformation of novel ideas into new products, services, organizational forms or industry that create value for customers and, more broadly, for society. This process resists stabilization and continuously develops in directions that challenge established theories and the very practice of innovators. In order to make sense of complex innovation processes Garud et al. (2013) divide the process into phases and levels. Innovation progresses from invention through development towards implementation and could be analysed on a level of a firm, multi-organizational networks and wider communities.

This multifaceted and multilayered nature of innovation is even more profound if the process that connects the emergence of technology and grand societal challenges is concerned. The twenty-first century poses major challenges to protect and enhance the quality of life, economic wealth and social stability (Pandza and Ellwood, 2013). It is deemed that advances in emergent technology fields offer opportunities to address the challenges in areas such as health, energy, environment, agriculture, transport, security and education, and that these in turn create business opportunities. Two influential reports were published at the beginning of the twenty-first century to guide policy makers and research funders when making research policy decisions.

A report sponsored by the USA National Science Foundation and edited by Roco and Bainbridge (2002) introduces nanotechnology, biotechnology, information technology and cognitive technologies (NBIC) as four general-purpose technologies that, through convergence, enable improvement in human potential. The Nordmann (2004) report that was supported by the European Commission (EC) focuses on nanotechnology, biotechnology and information communication technology (ICT). Nanotechnology is the understanding and control of matter at dimensions of roughly 1–100 nm, where unique phenomena potentially enable a huge variety of novel applications. Biotechnology explores and exploits chemical and physical processes and structures in living systems that are traced to their material basis in cellular and genetic organization; it has applications in health, medicine, agriculture, food and energy. ICT is the study, design, implementation, support or management of computer-based information systems, particularly software applications and computer hardware.

Schmidt (2007) highlights some major differences between both reports, especially in respect to understanding interdisciplinarity. The NBIC report argues for a holistic view of technology based on transformative tools, the mathematics of complex systems, and unified cause-and-effect understanding. By contrast, the European approach to emergent technologies assumes that nanotechnology, biotechnology and ICT are not the only enabling technologies[1] capable of supporting each other (Nordmann, 2004: 39). It recognizes and supports the contribution of the social sciences to emergent technologies. The report explicitly suggests integrating social science research into emergent technologies, which could then be promoted alongside science and technology research. By recognizing the social and dynamic nature of emergent technology-driven innovation, the report implicitly calls for management and business studies research that uncovers insights into the socially complex processes that lie between investment into emerging technologies and social impact.

The business and management studies literature has already contributed significantly to the understanding of technological and organizational innovation and its impact on firms' competitive positions. There is a general understanding that changing technological paradigms create environments of uncertainty and instability. It is also argued that the higher the complexity of innovation, the bigger the influence of non-technical factors on the adoption of technology (Anderson and Tushman, 1990). In the context of emerging technologies, this implies that non-technical, socio-economic factors warrant special consideration from the policy, practice and social science points of view. The knowledge about emerging technologies is also widely dispersed across disciplinary, sectoral, institutional and national borders. Knowledge about technology is therefore integrated, diffused and utilized in a complex technology innovation system (Carlsson et al., 2002), which can include individuals, firms, universities, research institutions, venture capitalists and public policy agencies (or parts or groups of each).

Building on the tenets of the sociology of technology (Knorr-Cetina, 1999), Garud and Karnøe (2003) emphasize the importance of distributed agency in the wider social construction of technology progression. From this perspective, it is impossible to distinguish between the social and technical spheres; technology itself, understood as a complex mix of social and technical elements, becomes a level of analysis. New technology is viewed as a driver for the dynamic emergence of new industrial architectures, disintegration of established industries, and changes in relations between co-specialized firms (Jacobides et al., 2006). Nanotechnology potentially enables applications that will spread through different industries (Rothaermel and Thursby, 2007) and may prove fruitful for studying the influence of emergent technology on emergent and established industrial architectures.

Literature on technology dynamics has traditionally been concerned with competitive dynamics between incumbent and new firms. The notion of disruptive technology (Christensen and Rosenbloom, 1995) emphasizes the difficulties incumbent firms face when emergent technology threatens to make their core capabilities obsolete. The emergence of ICT certainly had disruptive effects on areas such as music and media distribution and publishing. On the other hand, biotechnology shows that interfirm competitive dynamics could take the form of productive co-existence between different firms of the kind characterizing biotechnology firms and the pharmaceutical industry. Interfirm competitive dynamics introduce two additional perspectives on emergent technology innovation. Competitive interactions in the context of emergent technologies often switch into a collaborative dynamic, in which members of innovation networks, orchestrators of innovation ecosystems (Dhanaraj and Parkhe, 2006), strive to integrate and share the knowledge to foster innovation. The notion of open innovation (Chesbrough, 2003) highlights the relevance of externally available knowledge, but also creates challenges to establish appropriate business models to capture value from collaborative endeavours for large global firms.

The networked nature of technological and organizational innovation and dispersion of relevant knowledge challenges firms' dynamic capabilities (Eisenhardt and Martin, 2000). Established firms face the challenge of absorbing relevant external knowledge to accelerate their knowledge trajectories or create new knowledge paths in high-velocity industries. On the other hand, new technology-driven firms, which have established an identity based on their competency in a particular aspect of emergent technology, need transformative capacity (Garud and Nayyar, 1994) to transfer emergent knowledge into a potentially immense variety of applications.

Last but not least, individual agency is deemed to be instrumental for the process of technological innovation. Opportunity-seeking behaviour is a critical element of technology entrepreneurship (Shane and Vankataraman, 2000) and managerial agency is essential for competitive survival in the context of high uncertainty. This resonates with the recent call to understand better the role of managerial agency in creating new organizational capabilities (Barney and Felin, 2013; Gavetti, 2005). Moreover, entrepreneurial behaviour is not solely contained in entrepreneurial and established firms, but is also instrumental for initiating action and creating novelty in non-profit organizations such as universities, research policy institutions and community supported organizations.

COMMUNITIES AND NETWORKS IN EMERGING TECHNOLOGIES

Emerging technologies addressing grand societal challenges increasingly stem from neither any single company, nor a unique organizational setting. They stem from collaboration and networking within and across diverse communities, networks and regions bringing together and integrating knowledge and expertise in Europe and worldwide (Chesbrough et al., 2011; EC, 2012; Curley and Salmelin, 2013). Emerging technologies have progressively engaged a large number of organizations and people, which do not only compete, but also collaborate, forming diverse networks across all kinds of boundaries: geographical, institutional, demographic and disciplinary (Assimakopoulos, 2007; Schweer et al., 2012; Boudreau and Lakhani, 2013). In their process of emergence, emerging technologies also raise several challenges – technical, social and economic – that open up tremendous opportunities for solving the most complex and challenging problems of our era, while at the same time they create formidable barriers in understanding and managing precisely this process of emergence.

Students of emerging technologies need to acquire an in-depth understanding of an array of complex knowledge bases underlying the origins of emerging technologies vested in multiple scientific and technological communities (Constant, 2002). Such an understanding seems a vital requirement before any meaningful policy or strategy intervention is put in place for managing their emergence. As scientific and technological subject areas have become vastly complex, knowledge is accumulated in huge databases of publications and patents (Leydesdorff et al., 2013). The organization of collective knowledge in subject categories, patent classifications and the like seems a necessary evil for bounding disciplines and associated communities of experts in silos of vested interests. Experts and knowledge workers need first to absorb 'disciplinary' knowledge before they start transcending boundaries, collaborating for innovation and emergent technology across boundaries. Recent research has also shown that experts are rewarded if they broaden their expertise in several knowledge domains, and both publish and patent their knowledge in an intricate interplay of scientific communication and intellectual property protection (Murray, 2002; Meyer, 2006; Breschi and Catalini, 2010). As a result they have to make choices about whether they socialize and engage in multiple relevant communities involved in the social shaping of emerging technology, maintaining multiple memberships and appreciating different professional values and beliefs in the process of engaging with these communities.

Nowadays a cross- or multidisciplinary understanding of emerging technology is challenging to achieve on an individual basis. This is often grasped at the collective level, by both pre-existing and new emerging technological communities, professional associations, committees setting standards (Rosenkopf and Tushman, 1998), and securing both continuity and change in terms of pre-existing and emerging technologies (Assimakopoulos, 2000). Membership in multiple communities raises its own challenges as each subject area and scientific or technological community has accumulated a vast body of knowledge that unavoidably creates cognitive dissonance to outsiders. It also fosters patterns of collaborative behaviour that often embed and to some extent lock in experts to networks and communities where they talk the same language and share the same interests, values and beliefs (Bechky, 2003).

Nonetheless key actors and institutions in an array of communities try to shape the technological evolution of emerging technologies and socially construct their deployment and use. A myriad of actor networks (Latour, 2005) have tried for decades to create and take stock of necessary research infrastructures, testing scientific discoveries, and carrying out trials for new solutions in terms of diagnosis and treatment (Rosenberg, 2009). For example, in developing and diffusing new therapies for cancer there is a basic requirement for carrying out trials that they may last five to ten years minimum for establishing mortality rates or/and quality of life after treatment, before government administrations approve them for adoption and use with 'real' patients.

Key enabling technologies (KETs) also call for smart public policies and specialized clusters (EC, 2012; Wilkins et al., 2013), where orchestrators build interorganizational networks with small and large 'anchor' firms, leading universities and public research laboratories, as well as related supporting innovation services. The distributed nature for emerging technologies therefore requires the development of business models at the ecosystem rather than the organizational level of analysis, highlighting the interdependencies among emerging technologies, co-evolutionary ecosystem dynamics and the performance of key firms (Adner and Kapoor, 2010; Agouridas and Assimakopoulos, 2014). In a sense, key actors within such ecosystems need to develop dynamic capabilities and manage dispersed industrial architectures, global supply chains and distant markets, and co-create with lead customers.

Last but not least, as emerging technologies are diffusing across different geographies and cultures, they are reinvented from a broad range of actor networks who give situated meaning to such technologies based on different local interests, traditions and scientific and technological practices. Historians of science and technology as well as the field of social studies of science and engineering have highlighted the social construction of large technological systems for the past few decades (Hughes, 1983; Geels, 2002). We endeavour here to unravel the socio-economic impact of such emerging technologies for all kinds of stakeholders in order to manage them more effectively in their early critical days.

One of the main findings that cuts across several chapters of the book is the need to create, share and manage knowledge and research infrastructures at the network and community levels of analysis. This is characteristic of much European research and innovation since the early 1980s, when the first collaborative EU research and technological development framework programme was funded by the European Commission (EC) in the area of ICT (Georghiou, 1999; Assimakopoulos et al., 2004). No one organization or discipline on its own can solve these highly complex problems arising from practice (clinical or otherwise). 'Mode 1' knowledge production is rather limited scientific knowledge triggered and produced within disciplinary boundaries and scientific paradigms. 'Mode 2' knowledge production in the 'triple helix' of university–industry–government relationships has rather been the norm for more than 20 years (Gibbons et al., 1994). 'Mode 3' knowledge production in the 'quadruple helix' of innovation systems, including non-profit, non-governmental community-supported organizations, is the latest non-linear network innovation model, appearing in the early 2010s (Carayannis and Campbell, 2012).

EMERGING TECHNOLOGIES AND OPEN INNOVATION IN ORGANIZATIONS

The focus on innovation networks for managing emergent technology exposes intriguing questions of how organizations organize internally to search for adequate knowledge dispersed through wider networks. The openness (Chesbrough, 2003) suggests that organizations need to attach equal importance to knowledge generated outside the organization and knowledge produced within organizational borders. Managers within organizations are well advised to tap into the rich pool of externally available knowledge, at the same time paying serious attention to internally generated knowledge that resists commercialization through existing business models or current market channels. This focus on collaboration and the open nature of innovation has exposed questions of: (1) how organizations organize internally to search for adequate knowledge dispersed throughout an innovation network; (2) which strategies are required for capturing value in open innovation contexts; and (3) how new internal organizational structures emerge as a result of the need to participate in and benefit from innovation networks.

Many innovation networks are driven by grand societal challenges (for example, sustainability, environment, health and ageing), which increases their institutional diversity. Transnational institutions, national governments, local authorities and professional associations accompanied by organized members of the public play a powerful role in shaping the innovation efforts that aim to translate social challenges into growth opportunities. The changing nature of innovation in such issue-driven innovation networks requires the investigation of very complex interactions between agglomerations of organizations, groups and individuals. This further challenges the internal organizational design of innovating firms, their strategies and organizational capabilities. The societal issue-driven and open nature of innovation in emergent technology innovation networks radically shifts the logic and understanding of firms' innovation strategies and the organization of innovation processes.

Openness and the resulting knowledge complexity, interdependency and dispersion alter the logic of how innovation processes within a firm are strategically managed. These characteristics of emergent technology-driven innovation networks demand a simultaneous search for technological knowledge, development of organizational capabilities and creation of new business models. The openness of the innovation process suggests that knowledge and solution-providing competency is extremely widely distributed. Innovative organizations are, therefore, challenged to organize their search processes for the integration and recombination of distant (less familiar technologies and markets) and often very fragmented knowledge domains (Afuah and Tucci, 2012). New organizational structures are emerging that enable coordination across organizational borders, yet those that support open innovation processes are not sufficiently studied. The search for highly distributed knowledge is additionally complicated, because firms are not only searching for external knowledge that supports the development of their products or services. Many technology-enabled new products or services are themselves complex knowledge architectures (for example, cloud computing) with numerous firms contributing their competency only at a component level. If firms are facing architectural innovation (Henderson and Clarke, 1990) – components change only incrementally, but interdependencies among them alter radically – then search, no matter how distant, cannot only be focused on finding knowledge that supports improvement of a component. It also needs to provide adequate knowledge to understand the innovation on the level of the entire system. Architectural innovation changes competitive dynamics, division of capabilities and industrial structure, yet it is less studied how firms should organize their search activities in order to successfully shape emerging architectural innovation. In such circumstances it is unlikely that a network of contributing organizations will form a well-structured supply chain. Networks of organizations that attempt to shape complex architectures of new products or services are more akin to ecosystems that do not always have a clearly defined central actor. This suggests that firms need to organize their search activities not only to acquire adequate technological knowledge, but also to develop new organizational capabilities and business models in order to influence the direction of architectural innovation, differentiate themselves within the complex product or service, and capture adequate value from within the emerging industrial architecture.

It is intriguing that emerging research insights from the ManETEI project suggest that many technology-driven companies are reluctant to label collaboration with external partners as 'open' if new knowledge supports the existing core products. Regardless of whom they engage with or how uncertain or radically different the relevant technology is, collaboration is rarely described as open if it directly contributes to improvements of core products or technological competency. It is indicative, however, that companies use open innovation strategies when exploring integration of their products into radically new product or service architectures. Here, technological advances in products are of secondary importance to how a product fits into emerging product or service architectures. This is even more pronounced if the emerging industry architecture lacks a central firm to orchestrate integration efforts or when collaboration is needed between companies from different sectors to create a complex new product or service architecture. Companies have to develop new organizational capabilities to navigate emergent innovation networks that aspire to create complex architectural innovation from a variety of innovative products (for example, smart cities). In such an architectural innovation context it is unproductive to distingush between product, process, business model (Teece, 2010) and organizational innovation (Birkinshaw et al., 2008). Organizations often experiment with all these innovations simultaneously.

The institutional diversity of societal issue-driven innovation networks increases the number and variety of constituencies involved in innovation and changes the ways innovative firms organize their innovation processes to navigate multiple institutional logics (Thornton et al., 2012). The institutional diversity of the innovation process is recognized in the innovation management literature, inspired by insights from the sociology of technology (Garud and Gehman, 2012). Configurations of interdependent actors include not only firms and public research institutions, but also governments, regulators, professional associations, organized public and end-users. Such institutionally complex innovation networks are theoretically conceptualized as organizational fields that bring together various constituencies with often disparate purposes and institutional logics. It is often characteristic for such organizational fields to be driven by a central issue (for example, sustainability: smart cities) and not necessarily by a particular technology or market (Hoffman, 1999). Societal issues that drive institutionally diverse innovation ecosystems challenge firms' existing capabilities, yet it is rarely studied how developments at the level of an organizational field impact upon innovation strategies and capabilities at the level of an individual firm. It is even more scarcely investigated how an innovative firm organizes in order to shape and influence developments in an issue-driven innovation ecosystem, where the interactions are not only limited to interfirm or industry–university collaboration, but also include close interactions with governments, regulators and the wider public.

ManETEI researchers have identified the emergence of new organizational forms (for example, technology platforms supported by the EC) as typical meta-organizations (Gulati et al., 2012) that combine elements of interorganizational networks with structures and hierarchies as characteristics for single organizations. The ManETEI focus on exploring interactions between an individual firm and members of the wider organizational field is timely, because innovation managers with strategic responsibilities often participate in initiatives such as setting standards and collaborative technology development (for example, SEMATECH consortia), that bring together institutionally diverse constituencies. Moreover, they interact with governments, municipalities and the wider public in order to gain information, generate social and reputational resources and influence the direction of innovation. It is indicative that such interactions between institutionally diverse members often happen within new organizational forms such as European Technology Platforms (European Commission, 2004) and private–public partnerships that require adequate managerial skills and organizational capabilities in order to influence the agenda and support the strategic intent of a company that participates in such meta-organizations.

If innovation ecosystems are characterized by different degrees of institutional diversity and complexity, interdependency and dispersion of adequate knowledge, it is possible to argue that different capabilities and different internal organizational structures are needed within a single organization. This suggests that heterogeneity of external innovation ecosystems could be accompanied by internal heterogeneity of innovation structures. ManETEI researchers have identified that technology-driven companies are themselves becoming complex networks of internally distributed innovation initiatives. These decentralized groups are becoming increasingly autonomous in collaborating with external partners such as universities and technology-intensive small and medium-sized companies. ManETEI research has identified two intriguing and interrelated consequences. Firstly, complex organizations introduce new organizational units that specialize in supporting autonomous innovation initiatives undertaken in collaboration with external partners. Secondly, innovation management becomes a new profession distinct from R&D activities. This very intriguing trend enables the study of new organizational interdependencies (Siggelkow, 2011) and professional identities that influence the emergence of autonomous actions within complex organizations involved in innovating with emergent technologies.

WORK PRACTICES AND EMERGING ICT

For as long as technology has existed, it has affected the way people performed their work. In the beginning of the twentieth century, technology helped to automate manufacturing lines, resulting in the specialization of the workforce and efficiencies for car manufacturers. One hundred years later, ICT has emerged to enable collaboration and coordination between dispersed teams, redefining the grounds upon which distributed work can be carried out. However, such advances in ICT were not free of challenges, and indeed were at the centre of the ManETEI project. Various aspects such as geographical distance, time zone and cultural differences associated with global distribution have been examined as causing problems for globally distributed software teams in achieving successful collaboration. Indeed, a growing number of studies have reported problems regarding collaboration in distributed work, such as coordination breakdowns, lack of understanding of a counterpart's context (Cramton, 2001) and different language competencies across remote sites. Other studies have argued that globally distributed work may exacerbate the chance of misunderstandings, lack of trust (Jarvenpaa and Leidner, 1999), asymmetry in distribution of information among sites (Carmel, 1999), difficulty in collaborating due to different skills and training, and mismatches in information technology (IT) infrastructure.

By and large, the practices proposed in the literature have focused on two possible routes to ensure effective and efficient collaboration and coordination between remote sites. One stream of research has advocated for the division of work that minimizes the need for intersite coordination and, therefore, communication and synchronization (Herbsleb and Mockus, 2003). To achieve this, it was recommended that tightly coupled work items that require frequent coordination and synchronization should be performed within one site. Clearly, such a view highlights the importance of ICT as supporting the integration of components across sites.

Another stream of research has promoted the idea of rich and intensive interactions between various sites, supported by ICT and face-to-face meetings in which counterparts develop social and cultural contextual knowledge of their remote peers as part of enabling coordination and collaboration. This view, in contract to the former one, advocates the use of ICT as a communication tool that enriches coordinative activities via contextual knowledge, thus supporting distributed work.

The contrasting views presented here offer new avenues to understand the links between ICT and work practices. Arguably, firms cannot eliminate coordinative efforts through the practice of division of tasks between remote sites. While excessively investing in coordination activities may erode the advantages that distributed work may offer, such as low-cost locations and the utilization of time zones to speed up product development.

There are, therefore, two possible explanations through which the relationships between work practices and ICT can be further explored. First, understanding how socialization takes shape in such settings. Second, exploring the notion of 'expert systems' and how codification plays a role in such systems.

ICT and Socialization

Collaboration and team performance depends, to some extent, on the socialization of dispersed team members (Govindarajan and Gupta, 2001). Socialization refers to the process by which individuals acquire the behaviours, attitudes and knowledge necessary for participation in an organization (Ahuja and Galvin, 2003). Through socialization, the norms, identity and cohesion between team members develop, enabling team members to effectively communicate and perform (Hinds and Weisband, 2003). Most studies refer to organizational socialization as a process that is based on interactions between a newcomer and members of the organization (for example, colleagues, superiors or subordinates). Through such interactions an employee is taught and learns what behaviours and views are customary and desirable at their workplace, and becomes aware of those that are not, as well as acquiring the knowledge and skills needed to perform their job. While such a socialization practice is supported by human interactions when team members are co-located, distributed teams may face spatial, cultural and time zone challenges that may prohibit them from socializing.

Socialization in globally distributed teams, therefore, may take place through two key mechanisms. One is the application of ICT and the other is through face-to-face interactions. In terms of the application of ICT, various electronic media will be needed to support connectivity between remote sites and facilitate socialization. Additionally, generic collaborative technologies (for example, Groupware) will be needed to enable remote counterparts to connect and communicate. The most commonly suggested collaborative technologies are social media, e-mail, chat (for example, instant messaging), phone / teleconferencing, videoconferencing, intranet, group calendar, discussion lists and electronic meeting systems. More recent studies have focused on integrating collaborative technologies into an integrated development environment in order to offer solutions that deal with breakdowns in communication among developers in dispersed software development teams.

However, while ICT plays a role in socializing, face-to-face meetings are also important for the development of distributed teams through the establishment of interpersonal relationships. Furthermore, such meetings were found to positively affect team collaboration and team performance, mainly through the enhancement of interactions between team members. However, such face-to-face meetings are sporadic, short, selective and formal, therefore suggesting that their impact on socializing remote counterparts is likely to be minimal.

Recent studies have indeed acknowledged the difficulties in socializing through human interaction in distributed settings and therefore proposed the need to 'reacquire' norms, attitudes and contextual knowledge by combining social media with short-term visits, a term coined 're-socializing' (Oshri et al., 2007). Re-socializing in distributed contexts takes several steps in which both human interactions and the application of ICT play an equal role in moving the distributed team from introducing each other through social media means, meeting up for a preliminary meeting, but also renewing social ties via short visits and interactive social media platforms.

Transactive Memory and ICT

Socialization may increase remote counterparts' awareness of the islands of knowledge existing within the team; however, this may not suffice to effectively transfer knowledge between the sites. Indeed, studies have demonstrated repeatedly that, despite advances in technologies, ICTs do not prevent breakdowns in the transfer of knowledge across distributed sites (Cramton, 2001). While ICTs are critical for knowledge transfer processes in distributed teams, a neighbouring stream of studies has considered human-related factors, such as trust and interpersonal ties, which may act as facilitators for knowledge transfer between remote counterparts. But collaboration also requires remote counterparts to be aware of their peers' expertise and their mastery. In this regard, ICT are unlikely to support developing awareness for mastery, but rather assist in encoding, storing and retrieving information from and for experts when needed (Wegner, 1995).

Indeed, a transactive memory system (TMS) has been defined as the combination of individual memory systems and communications (also referred to as 'transactions') between individuals. The group-level TMS is constituted by individuals using each other as a memory source. Transactions between individuals link their memory systems; through a series of processes (that is, encoding, storing and retrieving), knowledge is exchanged. Individuals encode information for storing and retrieval, similar to a librarian entering details of a new book in the particular library system before putting it on the shelves. Through encoding, knowledge is categorized (that is, assigned labels that reflect the subjects of the knowledge) for systematically storing the location of the knowledge, but not the knowledge itself. Then, individuals store this information internally (building their own memory), or externally (storing it in ICT or indirectly in other people's memories). And lastly, information about the location of the knowledge or expertise is retrieved when someone else asks for it (Nevo and Wand, 2005).

Retrieval thus consists of two interconnected sub-processes: person A asks person B for information; person B retrieves the information. As Nevo and Wand (2005: 551) put it simply: 'knowledge is encoded, stored and retrieved through various transactions between individuals'. Wegner (1995) explains that for a TMS to work, three corresponding aspects to encoding, storing and retrieving should be considered. Creating a TMS that supports collaboration between team members is perceived to be less challenging than in distributed contexts (Oshri et al., 2008). Globally distributed teams often experience changes in membership that negatively affect the long-term development of a TMS. Furthermore, in many distributed settings, team members do not have any prior experience of working together. Their distributed mode of operation decreases communications and increases the possibilities for conflict, misunderstanding and breakdowns in communication. In teams that do not carry out joint training or arrange face-to-face meetings, the development of shared understanding is even more challenging because members of such teams do not stand on common ground. How then can a TMS benefit from ICT, thus enhancing collaborative practices in distributed modes?

Studies suggest that, similarly to socialization, there is interaction between human interaction and artefacts which creates the conditions for the development of a TMS in distributed contexts. Indeed, both personalized and codified directories are created during memory processes in a TMS which supports the encoding, storing and retrieving of information when remote counterparts exchange knowledge (Oshri et al., 2008). ICT and digital artefacts provide standard templates to capture information about individuals involved in knowledge exchanges, and central project repositories serve as storing artefacts to ensure the ability to retrieve information upon request.

Furthermore, we find that the three transactive memory processes – that is, encoding, storing and retrieving – play different roles in shaping a work practice such as knowledge transfer. Firstly, the development of collective expertise – that is, encoding – acts as a process for defining the procedure through which knowledge will be transferred. During the encoding, parties negotiate the meaning of knowledge (that is, the subject and location of the knowledge) following either a codified, standardized approach or by relying on an embedded routine developed within the organization. Secondly, the management of expertise – that is, storing – creates a pointer to the location where the knowledge is stored and from which it can later be transferred. In this regard, creating a pointer involves the actual storing activity during which remote counterparts attach particular labels to the knowledge stored within them or in a digital artefact. These labels, including for instance contextual information, make it possible to negotiate and clarify the meaning of this information, and its subsequent retrieval from its place of storage. And thirdly, the coordination of expertise – that is, retrieval – concerns the integration of knowledge by bringing together experts through search mechanisms and interpersonal contacts. For knowledge transfer to take place, teams rely, on the one hand, on the procedures and shared meanings established through encoding processes; and on the other hand, on interpretation and the use of labels attached to the transferred knowledge during the storing process. The coordination of expertise – and thus knowledge transfer – can be supported by relying on either the codified or the personalized directories, or both.

The question that we may pose at this juncture is: to what extent do personalized or codified directories matter to a certain practice such as knowledge transfer? Indeed, it would be wrong for either of these memory systems to be perceived as 'better' or 'worse'. In line with Cook and Brown's (1999) observation on epistemologies of knowledge, codified and personalized directories are best seen as two complementary, rather than competing, memory systems. Absence of the codified or digital directories would deprive the teams of shared methods for encoding, storing and retrieving information, which may strain the personalized directories beyond feasibility. Leaving out the personalized directories, on the other hand – for instance, due to high personnel turnover rates – would leave the project with independently working individuals who would find it difficult to agree on collaboration standards. In this regard, groups develop meta-routines that interlink the two types of directories. Further, there is a 'generative dance' (Cook and Brown, 1999) between these two memory systems that contributes to the transfer of knowledge between remote counterparts. The codified directories depend on interpersonal 'norming' processes for defining standards, templates and procedures. The personalized directories extend the codified system by offering additional avenues in cases when documents provide incomplete knowledge about a task. In these cases, individuals know who to contact and how to retrieve information. Evidently, the development and use of a TMS may change over time. During initial phases of the project, rudimentary parameters of transactive memory are defined (for example, which sites and individuals are responsible for which tasks and knowledge domains). These are extended and refined when people work together over prolonged periods of time, renegotiating meanings and regenerating learning around the knowledge transfer process.

The two explanations offered here about the relationships between work practices and emerging ICTs have been captured in the ManETEI network. Cloud services, as a recent example, pose serious challenges to firms' business models, in particular those that rely on a product business model, requiring such firms to reorganize their operations to consider a service-based business model. Similarly, the digital games generation, who bring gamification skills to their work, are shaping the way they interact with work artefacts and what would motivate them to perform. In this regard, the research carried out by ManETEI researchers is advancing our understanding of the complex relationships between technology and work practices, and how these two evolve over time and recursively shape each other.

NOTE

1. The report coined the term 'Nano-Bio-Info-Cogno-Socio-Anthro-Philo-Geo-Eco-Urbo-Orbo-Macro-Micro-Nano'.

REFERENCES

Adner, R. and Kapoor, R. (2010). Value creation in innovation ecosystems: how the structure of technological interdependence affects firm performance in new technology generations. *Strategic Management Journal*, 31 (3), 306–333.

Afuah, A.N. and Tucci, C. (2012). Crowdsourcing as a solution to distant search. *Academy of Management Review*, 37 (3), 355–375.

Agouridas, V. and Assimakopoulos, D. (2014). Collaborative business model innovation: genesis and prototyping in an aerospace setting. Presented at the R&D Management Conference, Fraunhofer IAO, Stuttgart, Germany, 3–6 June.

Ahuja, M.K. and Galvin, J.E. (2003). Socialization in virtual groups. *Journal of Management*, 29 (2), 161–185.

Anderson, P. and Tushman, M.L. (1990). Technological discontinuities and dominant designs: a cyclical model of technological change. *Administrative Science Quarterly*, 31, 439–465.

Assimakopoulos, D. (2000). Social network analysis as a tool for understanding the diffusion of GIS innovations. *Environment and Planning B*, 27 (4), 627–640.

Assimakopoulos, D. (2007). *Technological Communities and Networks*. Oxford: Routledge.

Assimakopoulos, D., Piekkari, R. and Macdonald, S. (2004). ESPRIT: Europe's response to US and Japanese dominance in IT. In Coopey, R. (ed.), *Information Technology Policy*. Oxford: Oxford University Press, pp. 247–263.

Barney, J. and Felin, T. (2013). What are micro-foundations? *Academy of Management Perspectives*, 27 (2), 138–155.

Bechky, B.A. (2003). Sharing meaning across occupational communities: the transformation of knowledge on a production floor. *Organization Science*, 14, 312–330.

Birkinshaw, J., Hamel, G. and Mol, M.J. (2008). Management innovation. *Academy of Management Review*, 33 (4), 825–845.

Boudreau, K.J. and Lakhani, K.R. (2013). Using the crowd as an innovation partner. *Harvard Business Review*, April, 3–13.

Breschi, S. and Catalini, C. (2010). Tracing the links between science and technology: an exploratorary analysis of scientists' and inventors' networks. *Research Policy*, 39, 14–26.

Carayannis, E.G. and Campbell, D.F.J. (2012). Mode 3 knowledge production in quadruple helix innovation systems. Springer Briefs in Business 7. Available at DOI: 10.1007/978-1-4614-2062-0_1.

Carlsson, B., Jacobsson, S., Holmén, M. and Rickne, A. (2002). Innovation systems: analytical and methodological issues. *Research Policy*, 31, 233–245.

Carmel, E. (1999). *Global Software Teams: Collaborating Across Borders and Time Zones*. Upper Saddle River, NJ: Prentice Hall.

Chesbrough, H. (2003). *Open Innovation: The New Imperative for Creating and Profiting from Technology*. Boston, MA: Harvard Business School Press.

Chesbrough, H., Vanhaverbeke, W., Bakici, T. and Lopez Vega, H. (2011). Open innovation and public policy in Europe. A research report commissioned by the ESADE Business School and the Science and Business Innovation Board AISBL, available at www.sciencebusiness.net.

Christensen, C.M. and Rosenbloom, R.S. (1995). Explain the Attacker's Advantage: technological paradigms, organisational dynamics, and the value network. *Research Policy*, 24, 233–257.
Constant, E.W. (2002). Why evolution is a theory about stability. *Research Policy*, 31 (2), 1241–1256.
Cook, S. and Brown, J. (1999). Bridging epistemologies: the generative dance between organizational knowledge and organization knowing. *Organization Science*, 10, 381–400.
Cramton, C.D. (2001). The mutual knowledge problem and its concequences for dispersed collaboration. *Organization Science*, 12 (3), 346–371.
Curley, M. and Salmelin, B. (2013). Open Innovation 2.0: a new paradigm. EC OISPG White Paper available at https://ec.europa.eu/digital-agenda/node/66731.
Dhanaraj, C. and Parkhe, A. (2006). Orchestrating innovation networks. *Academy of Management Review*, 31, 659–669.
Eisenhardt, K.M. and Martin, J.A. (2000). Dynamic capabilities: what are they? *Strategic Management Journal*, 21, 1105–1121.
European Commission (EC) (2004). *European Technology Platforms. From Definition to Implementation a Common Research Agenda*. Directorate-General for Research. Brussels: European Commission.
European Commission (EC) (2012). *NMP Expert Advisory Group Orientation Papers on Industrial Innovation*. Directorate-General for Research and Innovation. Brussels: European Commission.
Garud, R. and Gheman, J. (2012). Meta-theoretical perspectives on sustainability journeys: evolutionary, relational and directional. *Research Policy*, 41, 980–995.
Garud, R. and Karnøe, P. (2003). Bricolage versus breakthrough: distributed and embedded agency in technology entrepreneurship. *Research Policy*, 32, 277–300.
Garud, R. and Nayyar, P.R. (1994). Transformative capacity: continual structuring by intertemporal technology transfer. *Strategic Management Journal*, 27, 365–385.
Garud, R., Tuertscher, P. and Van de Ven, A. (2013). Perspectives on innovation processes. *Academy of Management Annals*, 7 (1), 775–819.
Gavetti, G. (2005). Cognition and hierarchy: rethinking the microfoundations of capabilities' development. *Organization Science*, 16, 599–617.
Geels, F.W. (2002). Technological transitions as evolutionary reconfiguration processes. *Research Policy*, 31 (2), 1257–1274.
Georghiou, L. (1999). Socio-economic effects of collaborative R&D: European experiences. *Journal of Technology Transfer*, 24, 69–79.
Gibbons, M., Limoges, C., Nowotny, H., Schwartzman, S., Scott, P. and Trow, M. (1994). *The New Production of Knowledge: The Dynamics of Science and Research in Contemporary Societies*. London: Sage.
Govindarajan, V. and Gupta, A.K. (2001). Building an EVective global business team. *MIT Sloan Management Review*, 42 (4), 63–71.
Gulati, R., Puranam, P. and Tushman, M. (2012). Meta-organization design: rethinking design in interorganizational and community contexts. *Strategic Management Journal*, 33 (6), 571–586.
Henderson, R.M. and Clark, K.B. (1990). Architectural innovation: the reconfiguration of existing product technologies and the failure of established firms. *Administrative Science Quarterly*, 35 (1), 9–30.
Herbsleb, J.D. and Mockus, A. (2003). An empirical study of speed and

communication in globally-distributed software development. *IEEE Transactions on Software Engineering*, 29 (6), 1–14.
Hinds, P. and Weisband, S. (2003). Knowledge sharing and shared understanding in virtual teams. In Gibson, C. and Cohen, S. (eds), *Virtual Teams that Work: Creating Conditions for Active Virtual Teams*. San Francisco, CA: Jossey-Bass, pp. 21–36.
Hoffman, A.J. (1999). Institutional evolution and change: environmentalism and the US chemical industry. *Academy of Management Journal*, 42, 351–371.
Hughes, T.P. (1983). *Networks of Power*. Baltimore, MD: Johns Hopkins University Press.
Jacobides, M.G., Knudsen, T. and Augier, M. (2006). Benefiting from innovation: value creation, value appropriation and the role of industry architectures. *Research Policy*, 35, 1200–1221.
Jarvenpaa, S.L. and Leidner, D.E. (1999). Communication and trust in global virtual teams. *Organization Science*, 10(6), 791–815.
Knorr-Cetina, K. (1999). *Epistemic Cultures: How the Sciences Make Knowledge*. Cambridge, MA: Harvard University Press.
Latour, B. (2005). *Re-Assembling the Social*. Oxford: Oxford University Press.
Leydesdorff, L., Carley, S. and Rafols, I. (2013). Global maps of science based on the new Web-of-Science categories. *Scientometrics*, 94 (2), 589–593.
Meyer, M. (2006). Are patenting scientists the better scholars? An exploratory comparison of inventor-authors with their non-inventing peers in nano-science and technology. *Research Policy*, 35 (10), 1646–1662.
Murray, F. (2002). Innovation as co-evolution of scientific and technological networks: exploring tissue engineering. *Research Policy*, 31(8–9), 1389–1403.
Nevo, D. and Wand, Y. (2005). Organizational memory information systems: a transactive memory approach. *Decision Support Systems*, 39, 549–562.
Nordmann, A. (2004). Converging technologies – sharpen the future of European societies. European Commission, DG Research, Report EUR 21357, available at http://ec.europa.eu/research/social-sciences/pdf/ntw-report-alfred-nordmann_en.pdf.
Oshri, I., van Fenema, P. and Kotlarksy, J. (2008). Knowledge transfer in globally distributed teams: the role of transactive memory. *Information Systems Journal*, 18 (6), 593–616.
Oshri, I., Kotlarsky, J. and Willcocks, L.P. (2007). Global software development: exploring socialization in distributed strategic projects. *Journal of Strategic Information Systems*, 16 (1), 25–49.
Pandza, K. and Ellwood, P. (2013). Strategic and ethical foundations for responsible innovation. *Research Policy*, 42 (5), 1112–1125.
Roco, M.C. and Bainbridge, W.S. (2002). Converging technologies for improving human performance. NSF/DOC report, Arlington, VA.
Rosenberg, N. (2009). Some critical episodes in the progress of medical innovation: an Anglo-American perspective. *Research Policy*, 38 (2), 234–242.
Rosenkopf, L. and Tushman, M. (1998). The coevolution of community networks and technology: lessons from the flight simulation industry. *Industrial and Corporate Change*, 7, 311–346.
Rothaermel, F.T. and Thursby, M. (2007). The nanotech versus the biotech revolution: sources of productivity in incumbent firm research. *Research Policy*, 36, 832–849.

Schmidt, J.C. (2007). Knowledge politics of interdisciplinarity – specifying the type of interdisciplinarity in the NSF's NBIC scenario. *Innovation*, 20, 313–328.
Schweer, M., Assimakopoulos, D., Cross, R. and Thomas, R.J. (2012). Building a well-networked organization. *MIT Sloan Management Review*, 53 (2), 35–42.
Shane, S. and Vankataraman, S. (2000). The promise of entrepreneurship as a field of research. *Academy of Management Review*, 25, 217–226.
Siggelkow, N. (2011). Firms as systems of interdependent choices. *Journal of Management Studies*, 48 (5), 1126–1140.
Teece, D.J. (2010). Business models, business strategy and innovation. *Long Range Planning*, 43 (2–3), 172–194.
Thornton, P.H., Ocasio, W. and Lounsbury, M. (2012). *The Institutional Logics Perspective: A New Approach to Culture, Structure, and Process*. Oxford: Oxford University Press.
Van de Ven, A., Polley, D., Garud, R. and Vankataraman, S. (2008). *The Innovation Journey*. Oxford: Oxford University Press.
Wegner, D.M. (1995). A computer network model of human transactive memory. *Social Cognition*, 13, 319–339.
Wilkins, T. Deliyaniakis, N., Maillaband, A. and van Neck, N. (2013). NMP EAG Orientation Paper on Best practice in innovation. Directorate-General for Research and Innovation. Brussels: European Commission.

PART I

Foundations of organizational capabilities for technological innovation

INTRODUCTION

It is generally accepted that in the twenty-first century, technological innovation will spell the difference between companies that succeed, and those that do not, in creating and holding on to a competitive-advantage position in their industry. There is, however, less certainty about the resources and tools that companies require to become masters of technological innovation. A central factor behind this uncertainty is the diversity of variables that can potentially influence the ability to innovate, which makes it difficult to ascertain the specific configuration of organizational capabilities most appropriate for a particular company.

Over the years, a number of potent ideas and frameworks – including the concepts of absorptive capacity (Cohen and Levinthal, 1990), dynamic capabilities (Teece, 2007) and ambidexterity (Raisch et al., 2009) – have been proposed which facilitate analysis of the challenge of technological innovation and how a company might become more innovative. One common message emanating from different studies on the topic is that the organizational capability to effectively integrate the knowledge which different divisions and specialized functions of a company possess plays a crucial role in technological innovation (Miller et al., 2007). To the extent that technological innovation represents the creation of something new by recombining and synthesizing discrete elements of knowledge in fresh ways (Carnabuci and Operti, 2013; Fleming and Sorenson, 2001), the importance of intra-organizational knowledge coordination is almost self-evident. Any new knowledge acquired in one part of the company can only possibly contribute to the creation of a novel way of satisfying market need if productively merged with complementary bits of knowledge in other parts of the company. Thus, new knowledge that research and development (R&D) specialists have of scientific and technical possibilities is likely to facilitate the development of commercially successful technological innovations when pooled with the knowledge of marketing specialists about trends in customer product-use patterns.

Although intra-organizational knowledge transfer and synthesis are vital (Argote et al., 2003; Zhou and Li, 2012), companies often face difficulties in relation to them. Cognitive as well as material factors might hamper knowledge assimilation. The former can encumber it in that knowledge exchange is made difficult by the different frames of reference and interpretations of different groups of specialists (Carlile, 2004). Material factors can inhibit knowledge assimilation in that existing structural arrangements and organizational practices and processes make it difficult to combine new knowledge with entrenched understanding and routines (Dougherty and Hardy, 1996; Nag et al., 2007). The chapters in this part of the book seek to expand the understanding of cognitive and material impediments to knowledge transfer and synthesis. In pursuing their specific objectives, the chapters complement one another in shedding light on distinct issues that need to be addressed when companies attempt to develop an organizational capability for technological innovation.

Using the notion of technology intelligence to showcase four modes of knowledge search, Maya (Chapter 5) presents a handy set of tools for promoting knowledge assessment and dissemination that could support a company's drive to become more innovative. In contrast, Chapter 4 by Cotta and Salvador suggests that the capability for technological innovation is more likely if a company has a more developed transactive memory system, because such a system arguably allows the management of a greater cognitive load through efficient storage, retrieval and sharing of knowledge. On a different but related note, Chapter 3 by Es-Sajjade and Pandza underscores the importance of recognizing that different groups in a company cooperate as well as compete with one another. As such, the building of a capability in technological innovation requires mechanisms which enable group-level competences to be shared and adopted across groups. That managers have an important role to play in initiating such mechanisms – through the establishment of appropriate routines – is a key theme in the discussion of Khanagha, Volberda and Oshri (Chapter 2).

Maya's analysis focuses on developing a toolbox for an integrated technology service provider (ITSP): a company founded for the specific purpose of providing technology intelligence to its holding company in the chemical industry. To set the stage for a discussion of the challenges facing ITSPs when implementing technology intelligence systems, Maya starts with a more general discussion of technology intelligence modes and processes. The author highlights that, in the case of ITSPs, the range of technological domains to be monitored is quite large, because the holding company typically has clients in a range of different industries. This makes the task of assessing and disseminating knowledge that can support innovation difficult. The tools useful for the purpose may vary from one context to another, depending on the amount of slack available, the culture of the organization and the level of commitment of the top management. To stay on top of emerging opportunities and threats, Maya suggests that companies should have an evolving technology intelligence approach. In particular, a mutable portfolio of both quantitative and qualitative tools – as might be appropriate under the changing circumstances – should be employed, such that the tools promote technological innovation by facilitating knowledge exchange among different units and groups.

The chapter by Cotta addresses the fundamental problem of balancing requisite flexibility driven by environmental dynamics with the everlasting goal of efficiency. The fundamental idea is that the concept of transactive memory system, originally developed as a team-level concept (Anand et al., 1998), can indeed be extended to the organizational level. According to this view, organizations can address environmental uncertainty and produce innovative solutions by activating multiple transactive memory systems, each one responsible for a specific task. Following this logic, organizations develop themelves in transactive memory systems in a response to environmental uncertainty, in order to achieve superior levels of efficiency. This is definitely a strong statement which begs for theoretical scrutiny, as it proposes that the challenge implied by an uncertain environment is actually an opportunity to thrive where most organizations fail. At the theoretical level, the main contribution of the chapter is that it posits that one organizational capability needed to react to environmental uncertainties, such as those associated with emergent technologies, consists of creating multiple interlocking transactive memory systems that synergistically connect different expertise focused on customers, development and execution activities.

Es-Sajjade and Pandza draw the reader's attention to the point that all products, services and innovations of a company are ultimately the outcome of the nature of interactions between different groups of specialists. Taking intergroup dynamics as their level of analysis when examining innovation at a medium-sized telecommunications company, the authors suggest that groups' social context is important in shaping intra-group cooperation and intergroup competition. These dynamics presumably arise because of how social context influences processes of social identification. Using data from archival sources, semi-structured interviews, and direct observations, Es-Sajjade and Pandza show that if there is balanced allocation of resources between intra-group cooperation and intergroup competition, the outcome is positive in terms of capability development. In light of the observation that capability emergence depends on sharing group-level competences, the authors suggest that management practitioners should consider both organizational design and strategy as levers to spur group cooperation and competition. In this regard, they particularly underscore the integration and institutionalization of relevant tools, routines and processes across different groups in the company.

Finally, the chapter by Khanagha, Volberda and Oshri offers an attempt to explain the link between high-order 'routine-changing' routines and lower-order routines meant as repetitive patterns of action. Examining this link is essential to understanding the mechanisms through which organizations adopt emerging technologies and embed their exploitation within the lower-order routines that make up product development and operational processes. They explore this problem by means of an in-depth case study of a telecom operator facing the opportunity but also the challenge of implementing cloud computing technology, moving from a manufacturing to a service-dominant logic. The detailed analysis of the adoption process shows how the role of the top management team is essential not only in setting up new high-order routines, but also in facilitating the derivation of lower-order routines that can exploit the potential of the new technology. The top management team ability to manage their attention, so as to link higher- and lower-order routines in the presence of technological turbulence, is another effective way to conceptualize the nature of the organizational capabilities needed to seize the opportunity of emerging technologies.

Whether it is the issue of partners and networks to organize for emergent technologies (see Part II of this book), or the challenges policy makers face in shaping technology development (see Part III of this book), or the utilization of new technologies for further innovation (see Part IV of this book), an informed analysis requires consideration of the underlying organizational capabilities that matter. In identifying and analysing a range of critical factors and processes having a bearing on organizational capabilities for technological innovation – from specific tools to intra- and intergroup processes and managerial agency – the chapters in this first part of the book set the stage nicely for a rich, fruitful discussion of how to manage technological innovation.

Jatinder Sidhu and Fabrizio Salvador

REFERENCES

Anand, V., Manz, C.C. and Glick, W.H. (1998). 'An organizational memory approach to information management'. *Academy of Management Review*, 23, 796–809.

Argote, L., McEvily, B. and Reagans, R. (2003). 'Managing knowledge in organizations: an integrative framework and review of emerging themes'. *Management Science*, 49, 571–582.

Carlile, P.R. (2004). 'Transferring, translating, and transforming: an integrative framework for managing knowledge across boundaries'. *Organization Science*, 15, 555–568.

Carnabuci, G. and Operti, E. (2013). 'Where do firms' recombinant capabilities come from? Intraorganizational networks, knowledge, and firms' ability to innovate through technological recombination'. *Strategic Management Journal*, 34, 1591–1613.

Cohen, W.M. and Levinthal, D.A. (1990). 'Absorptive capacity: a new perspective on learning and innovation'. *Administrative Science Quarterly*, 35, 128–152.

Dougherty, D. and Hardy, C. (1996). 'Sustained product innovation in large, mature organizations: overcoming innovation-to-organization problems'. *Academy of Management Journal*, 39, 1120–1153.

Fleming, L. and Sorenson, O. (2001). 'Technology as a complex adaptive system: evidence from patent data'. *Research Policy*, 30, 1019–1039.

Miller, D.J., Fern, M.J. and Cardinal, L.B. (2007). 'The use of knowledge for technological innovation within diversified firms'. *Academy of Management Journal*, 50, 308–326.

Nag, R., Corley, K.G. and Gioia, D.A. (2007). 'The intersection of organizational identity, knowledge, and practice: attempting strategic change via knowledge grafting'. *Academy of Management Journal*, 50, 821–847.

Raisch, S., Birkinshaw, J., Probst, G. and Tushman, M.L. (2009). 'Organizational ambidexterity: balancing exploitation and exploration for sustained performance'. *Organization Science*, 20, 685–695.

Teece, D.J. (2007). 'Explicating dynamic capabilities: the nature and microfoundations of (sustainable) enterprise performance'. *Strategic Management Journal* 28(13), 1219–1350.

Zhou, K.Z. and Li, C.B. (2012). 'How knowledge affects radical innovation: knowledge base, market knowledge acquisition, and internal knowledge sharing'. *Strategic Management Journal*, 33, 1090–1102.

2. Routines and technological discontinuity: an attention-based view on the organizational adaptation process

Saeed Khanagha, Henk W. Volberda and Ilan Oshri

INTRODUCTION

This chapter concerns the process of adopting emerging technologies within established organizations, focusing mainly on the role and implications of routines, as the repeated patterns of behaviour in firms (Feldman, 2000). Discussing the efficiency and replication of best practices as the ultimate outcome of organizational routines, many of the prior studies have studied routines as a source of rigidity in organizations (e.g. Gilbert, 2005). Conversely, another stream of research considers routines to be inherently dynamic and therefore a source of continuous change (Feldman, 2000). Despite a considerable amount of research work on the consequences of organizational routines as well as their emergence and co-evolution process, the question of in what circumstances and how routines act as enablers of adaptation and change is not fully explored. In particular, the role of the top management team (TMT) and their cognition, interpretation and sense-making in relation to an emerging technology seems to be an overlooked but important factor that puts organizational change routines in action. We focus on this aspect and discuss how managerial attention acts as a determinant of dynamicity and change in the organizational routines.

In the following, first we develop a theoretical framework by discussing ongoing debates concerning organizational routines as the sources of rigidity on the one hand, and dynamicity and change on the other. This framework incorporates discussions on the role of the top management team and managerial attention in connection with organizational adaptability. The chapter then presents the research context, the methodology and an analysis of the data. Finally, a discussion and recommendations for further research are presented.

THEORETICAL FRAMEWORK

Organizational Routines and Emerging Technologies

Emergence of a new technology is challenging for the organizations, in part because of the need for dealing with a mature technology that drives the revenues at the same time as exploring an emerging technology whose benefits and consequences are to a large extent uncertain. Despite frequent scholarly enquiry, we still do not know much about how organizations learn to manage the contradictory needs for stability and efficiency on the one hand, and flexibility and experimentation on the other (Farjoun, 2010). A promising avenue in researching this area seems to be to deviate from static perspectives towards organizations and delve deeper into the processes for the emergence and development of organizational routines.

The need for adaptations and change in the organizational routines – understood as the repeated patterns of behaviour that characterize much of an organization's ongoing activity (Nelson and Winter, 1982) – in response to environmental dynamism has been well documented across a range of industries (e.g. Barley, 1986; Tyre and Orlikowski, 1994; von Hippel, 1994; Szulanski, 2000). The basic features of routines are the possibility for repetition and the predictability of outcomes, both of which facilitate more effective decision-making through standard procedures, rules and patterns of behaviour, and by limiting the space of events that managers should scan (Baum and Singh, 1994; Parmigiani and Howard-Grenville, 2011). Nevertheless, as the environment changes, predictability and possibility for repetition become less relevant, and this issue is prominent in the adoption process of emerging technologies.

The design of a technology often leads to the establishment of habitual routine (Edmondson et al., 2001), defined as a similar pattern of behaviour in a given stimulus situation without explicitly selecting it over alternative ways of behaving (Gersick and Hackman, 1990). This results in situations where organizations' members try to act according to the familiar routines even when the external stimuli are considerably different (Edmondson et al., 2001). The development of new routines may prove to be very challenging for an established organization, because organizational and technological contexts support existing routines (Edmondson et al., 2001), and experimentation that is essential for the development of new routines remains under-resourced.

Woiceshyn (2000) compares the technology adoption process in two large oil companies and discusses the new routines as outcomes of a collective learning process which he defines as a feedback loop of observing, interpreting, integrating and acting. The author further explains that the variance in performance of established firms stems from the quality of the routines that are being developed through the above-mentioned learning loop. He further discusses how contextual factors such as superiority in capabilities, resources, incentives, motivation and effort together with favourable shared values result in better routines and, therefore, more success when adopting a new technology.

In other research, Edmondson et al. (2001) take this perspective that, when a new technology disrupts existing work routines, the adopting organization must go through a learning process for the development and establishment of new routines. According to their results, some organizational factors such as size, innovation history and management support have little impact on how organizations perform in this process; whereas the capabilities of the team leader and the team's psychological safety are suggested to be influential factors. While discussing the technology adoption process in general terms, this chapter is focused on a technology where the impact is limited to a number of small surgery teams in different hospitals.

A more recent study by Labatut et al. (2012) discusses the process of the co-emergence of new technologies and new organizational routines that results in the development and establishment of new practices in the context of the adoption of genetic selection technology in the French sheep industry. Viewing technologies as a combination of three dimensions – namely technical substrate, managerial philosophy and organizational model – the authors analyse the dynamics between technologies and ostensive and performative aspects of organizational routines that result in the creation of new practices.

Behavioural and evolutionary theories discuss the existence of meta-routines (Cyert and March, 1963; Nelson and Winter, 1982; Feldman and Pentland, 2003; Lewin et al., 2011) or higher-order routines in organizations that act as regulators of changes and adaptations in lower-order routines. However, similar to operating routines, it is possible that higher-order routines become anchored to the existing technological context. In other words, while meta-routines assist in the emergence of new operating routines, the persistence of a set of technology-specific routines for a long period of time results in the predisposition in the change routines (Ashmos et al., 1998). An upshot for the organization is to become more and more dependent on the existing technological path, which is caused by a gradual tapering process that narrows the scope of search and action of the incumbent (Schreyögg and Sydow, 2011). Subsequent path-dependency limits the performance of organizations' collective learning routines in the face of technological disruptions. So is it possible to consider routines as inherently dynamic, or there are external triggers for actuating the dynamicity and change in the routines?

Proponents of dynamicity in routines tend to relinquish consideration of routines as automatic, static objects and instead they see them as alive, with the potential for dynamicity and change (Pentland and Feldman, 2008). However, it seems that routine dynamicity is influenced by the cognition, interpretation and sense-making of an organization's members, and in particular the chief executive officer (CEO) and top management group. These influences result in path-breaking (or new path generation) decisions by managers through, for example, the introduction of new structures, processes and practices; which in turn activate the change process and related routines. For example, Schroeder et al. (2000) have discussed the importance of introducing new structures and procedures initiated by senior managers that facilitate change. Fleming (2002) relates Hewlett-Packard's (HP) success with inkjet technology to the introduction of practices such as social and technological mixing and juxtaposition, recycling of engineers, management by objective, strong socialization processes, rapid prototyping, and adaptations in organizational norms and processes that helped HP to deal with the increased risk of failure during exploration efforts. Such initiatives are to a large extent dependent on the top management team's involvement and initiatives, which will not occur unless they dedicate a sufficient amount of attention to the environmental change.

Managerial Attention and Dynamicity in Routines

The attention-based view of the firm discusses the relationship between a firm's behaviour and the ways in which attention is being channelled and distributed for decision-making in the firm (Ocasio, 1997). In this context, the CEO and top management team are known to be the most critical players in the regulation of attention within an organization (Hambrick and Mason, 1984). Similar to the studies by Li et al. (2013), we build our framework upon the complementary nature of upper-echelon and attention-based theories. The combination of the two theories is basically justified by the role that top executives play in processing the information from the external environment and then, through their cognitive lenses, influencing the behaviour of the organization. Characteristics of the senior executive team have been studied, therefore, as an important precursor of an organization's responses to environmental triggers. However, the relationship between dynamicity in the routines and the TMT is not fully understood.

Managerial attention, a core concept in the attention-based view of the firm, is likely to play an important role in determining the organization's response to an emerging technology. The decision to implement the organizational changes that are necessary for the adoption of a new technology is to a high degree dependent on the matters that key decision makers perceive to be the most important, and focus their attention on (Ocasio and Joseph, 2005).

Emerging technologies pose additional complexity dimensions for managers who are already exposed to more information than they can absorb. Dealing with ongoing businesses of the organization requires allocation of attention-based resources to the activities that directly contribute to the survival of the organization (Gilbert, 2005). Although the allocation of attention and other resources to exploratory innovation is somewhat routinized in today's organizations, technological disruption often requires going beyond the regular balance between exploration and exploitation, and taking resources out of existing technologies and allocating them to the alternative, emerging one. Deciding upon the timing and scope of such shifts in the allocation of resources allocation is, however, very challenging, mainly because of the uncertainties that are associated with emerging technologies.

Developed in response to questions on why firms behave in certain ways, and when, why and how they respond to environmental change (Ocasio, 1997), the attention-based view seems to be very relevant in studying the antecedents of variance in the performance of organizations for managing an emerging technology. We utilize the attention-based theory in connection with our discussion on organizational routines to portray the process through which new routines emerge and co-evolve.

TMT Attention and Dynamicity in Routines: A Framework

The discussions on higher-order and lower-order organizational routines highlighted the possibility for inherent dynamicity in the routines. However, as Schreyögg and Kliesch-eberl (2007) have discussed, without the consideration of an exogenous player it is not possible to consider routines as a source of both stability and dynamicity. Such an external role, for example in terms of boundary-spanning or gatekeeping functions (Cohen and Levinthal, 1990), is a solution for constant monitoring of environmental change and evaluating the implications for the adopting organization. One may wonder whether such activities may also be routinized in the organization. Indeed, it is reasonable to believe

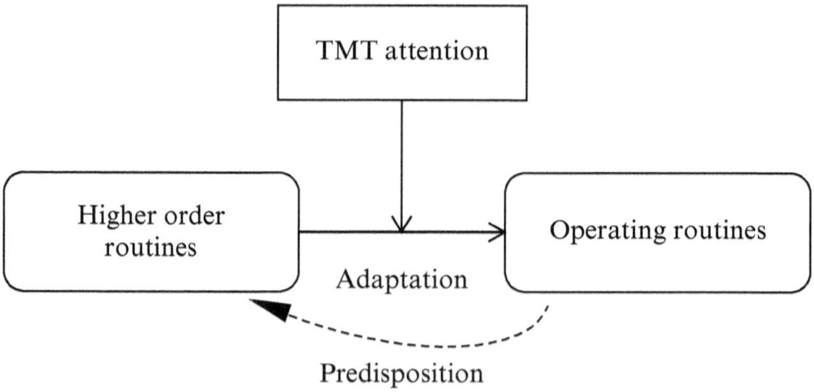

Figure 2.1 In uncertain and complex situations, TMT attention is a key driver of adaptation and change in organizational routines

that in many cases it is possible to predict the intensity and likelihood of changes in the environment. But when the degree of uncertainty and complexity increases, especially in the rather rare cases of technological discontinuity, routines are less likely to act effectively. This is mainly because of the fact that the routines are, in principle, the organization's memory of the best prior practices and are not developed for dealing with the unprecedented. Therefore, the top management team and their attention to a new technology play a key role in guiding the behaviour of the organization towards adaptation for embracing the new technological paradigm. Figure 2.1 illustrates the key assumptions of this research.

RESEARCH CONTEXT

Focused on technology leadership and having a strong presence in the market, Telco (a pseudonym) has been a key player in the telecom infrastructure sector for more than a century. It has survived so many crises during its lifetime that, when asking a senior manager about the consequences of technology or market shifts, he replied, 'we have successfully passed many similar situations in the past and this is not an exception'. The emergence of cloud computing technology (CCT), however, seems to be perceived rather differently. The findings of this study are best understood by elaborating the extraordinary challenges that are associated with the adoption process of CCT.

CCT and the Telecom Sector

CCT and related concepts are based on 'service logic' that is the prevalent playground of IT giants such as Google and Microsoft. This logic is clearly far from the comfort zone of Telco and other telecom players who are basically manufacturers of complicated and expensive telecommunication infrastructure equipment. For many years, Telco and a few other very large firms have continuously developed very complicated and reliable mobile equipment according to globally accepted standards. Owing to the technological and financial entry barriers, telecom vendors are able to produce and sell their complex products with high profit margins to customers (mainly mobile network operators). This equipment must be designed to be durable enough to work for many years in order to justify investment. During their life cycle, the infrastructure vendors busily continue with further research and development (R&D) to make the next generation of hardware and software ready for market, while ensuring value propositions such as extra efficiency, capacity, features and reliability. CCT totally disrupts this model of revenue generation.

In a service-dominant technology such as CCT, customers become so empowered that, in theory, they should be able to pay exactly and only for what they really need and can use instantly. An example can better illustrate the nature of this shift: imagine a scenario in which someone decides to establish an e-business website. In the past, this person had to purchase, install and maintain servers and operating software before being able to launch the website. Nowadays, the required capacity can be purchased from cloud providers and instantly accessed and launched via the internet. Many of the cloud providers offer a large extent of customizability to their clients in such a way that they can have access to the services that ideally fulfil their interests. When this model finds its way into the telecommunication sector, similar flexibility and speed would be expected by the customers, and the scope of implications on the infrastructure providers would be remarkable.

This makes it quite alarming for Telco and its peers because they clearly realize that many of their core competencies as hardware manufacturers will sooner or later be obsolete. The issue that makes the situation even more complicated concerns the fact that there has been a large amount of uncertainty associated with the architecture of the technology as well as the timing and scope of impacts on the markets. Given the number of areas that are affected by the new technology, such uncertainties make it very difficult for managers to justify the challenging steps to prepare the organization for the new context. In this chapter we cover the life cycle of CCT between 2008 and 2012 and investigate the organizational challenges of the adoption of a new core technology through the lens of organizational routines.

DATA AND METHODS

Because of the key interest of this study, which provides a deeper view on dynamicity in routines, the grounded theory development approach based on a longitudinal in-depth case study was found to be appropriate (Glaser and Strauss, 1967; Martin and Turner, 1986). This method provided us with an opportunity to delve deeper into a longitudinal case study on the adaptation process in the case study. The concurrency of data gathering and final phases of transition process in this case study was a unique opportunity for observation of events, and considerably increased the reliability of collected data. The study covers the time period of 2007 to 2012, starting from the rise of cloud computing technology in 2007. While the case study approach is limited in terms of generalizability, the possibility for scrutiny of the structures, processes and practices of the organization perfectly matched our aim to provide a better understanding of the links between management innovation and technology adoption in established firms.

Five sources of data were primarily used in this research: (1) two rounds of semi-structured interviews, one starting from September 2010 and ending in June 2011 and the other starting from August 2011 and ending in November 2011; (2) three focus group sessions with key stakeholders; (3) field study and observation by one of the researchers in the period of September 2010 to November 2011; (4) searching internal archival data for the period of 2007–2011, including all internal intelligence reports, knowledge-sharing tools, minutes of meetings, and so forth; and (5) searching external intelligence reports available to Telco, as well as academic scientific databases. Therefore a combination of external and internal perspectives on Telco's experience with cloud computing accommodated an in-depth study on organizational routines and management innovations that support the technology adoption process.

A total number of 23 interviews and two focus group sessions took place in this phase, mostly in person, except for two interviews which took place over the phone. The duration of interviews ranged from 45 to 90 minutes. Transcripts of the interviews were communicated with the interviewees via e-mail and following comments modifications were applied. Very early in this phase of the data collection process, the researchers' attention shifted towards the need for adaptations in the product development and innovation processes according to the specific needs of the emerging business

Table 2.1 Overview of interview participants in 'Telco'

Role	No. of interviews
Senior VP strategy	1
VP strategy and business development	1
Director of Telco's cloud programme (from 2009 to mid-2011)	1
Director of Telco's cloud programme from mid-2011	1
Product managers	6
Product management staff	18
R&D staff	11
Technology intelligence experts	7

model. This resulted in the initiation of a second round of data gathering after completion of the original research in June 2011. A total of 18 managers, all involved with CCT from a technology or business perspective, participated in one focus group and subsequent semi-structured interviews (see Table 2.1 for a list of interviewees).

Technology Adoption Process at Telco

From the technological architecture perspective, CCT concerns the centralization of processing power and providing customers with access to a pool of processing power through the internet. The new architecture of technology eliminates the possibility of selling large amounts of small but highly reliable equipment to the customers, and replaces it with selling access to the pool of hardware resources. In the new architecture, it is relatively easier to secure the required level of reliability and precision. Moreover, the installed base and footprint that are key determinants of success for Telco and other incumbents and act as entry barriers for newcomers will become less relevant. These factors inevitably shift the focus of competition toward providing customers with better, more flexible and more efficient services. That is why selling 'offerings as services' is a concept that is being used almost interchangeably with CCT among telecom infrastructure vendors. The scope of the impact on the systems, structures, processes and practices of Telco is subtle. This can be best understood by looking at some of the routines and areas that are affected by the new technology. The intention here is not to be exhaustive, but we would like to present some examples of operating routines that need to be adapted in the new technological context.

Firstly, research and development. Securing the required level of performance, reliability and capacity according to the existing technology architecture has always been accomplished via heavy investment in research and then development of products according to the research findings. This entails an idea generation process by star scientists which results in inventions and patents. The next step is, typically, embedding the research findings in the next generation of products, and in Telco this used to be done through the development of prototypes and a long process of testing laboratories and selected customer sites. A stage gate process model for product development was a natural choice for Telco in the past in order to increase the efficiency of the product development activities. With the progress of CCT and service logic, the process will become somewhat upside down. When there is no hardware purchase in between, the customer will seek to define the services that they require, and this means that a large portion of the ideation process happens within the customer premises. The development process will also become more similar to IT practices such as agile development, which implies an iterative development and feedback process. Therefore, there would not be much room for lengthy processes of prototype development and testing. Clearly, there is a need for revising some of the key organizational routines in order to design the research and development function in the new context.

Secondly, sales and marketing. With the increased power of customers and their role in the initial stages of the innovation and product development process, the role of sales and marketing people changes as they are no longer responsible for selling already developed products. Instead, they have to start the discussion with customers in order to understand what they need, and provide the initial feedback for the development process. A totally different set of competences and practices would be required in the CCT context, which would give customer accounts a central role which spans from initial steps of ideation to the final delivery stage.

Thirdly, pricing and contract management. The CCT model entails a totally different pricing model in contrast with what is prevalent now in the telecom sector. Currently, the revenue margin on selling telecommunication equipment is a percentage of the revenue that the customer is likely to make during the life cycle of the purchased equipment. The customer may not use the whole capacity or all features of the equipment, or may decide to swap the hardware earlier than its regular end of life, but they still pay the indicated price according to the contractual amounts. On top of this, maintenance and support services tend to be sold on top of each hardware sales contract. The consequences of CCT on such a model are remarkable: the customer would no longer pay for the capacities and features that they don't need; and, moreover, they would not pay for the whole life cycle of the physical resources. They would basically pay for the capacity, functionalities and duration that they actually need. In addition, there would no longer be much in terms of maintenance and customer support sales on top of the original contract. CCT-based contracts are usually based on service-level agreements (SLAs) which indicate a certain level of service quality, and falling below that threshold usually leads to penalties for the vendor. It is easy to imagine how different the pricing and contract management practices will look in the CCT context.

Fourthly, supplier and network relationships. In the existing model of the telecommunication sector, firms do not cooperate intensively with other firms and they seldom cooperate with their competitors. The adoption of CCT in the telecom sector brings about a great deal of similarity to the IT sector, in which quite a number of competitor firms need to cooperate in order to launch a project. This mandates a totally different set of cultures, practices and processes to those currently in use among telecom players. Table 2.2 provides some examples of quotations with reference to changes in the operating routines.

These examples highlight some areas in which Telco needed to develop substantially different new routines. There are more instances of impact areas, and almost certainly some which are yet to be identified. CCT has not yet been fully explored by Telco, and details of new or adapted routines are not yet known by Telco or any other firm in the industry. Therefore, managers soon realized the need for experimentation in new ways of working, and then revising and adapting according to the outcomes. Such an experimentation process entails the generation or adoption of new ideas for CCT-based offerings, and applying those ideas for product development and commercialization purposes. The required adjustments in the routines, or the need for the development of totally new operating routines, are best identified during a process of learning by doing. Such a learning process proved to be quite challenging for Telco in practice. Obviously, an organization that is running at full speed and efficiency in a different direction is not designed to experiment with routines that are fundamentally different to the existing ones. In fact, the organization was not prepared to act like this before 2011, during which a series of TMT-level decisions changed the situation in favour of the new technology.

According to many of the interviewees, there was a large gap between the time when Telco understood and planned for what was needed to be done on CCT, and the time when it actually operationalized those plans. As early as 2008, at the onset of the technology, the technology intelligence functions in different levels of the organization spotted CCT as an important prospect for Telco, and the company immediately started to study the technological and organizational impacts of the new technology. A bit later, in 2009, these speculations found their way into the highest levels of Telco's strategic plans, and intense discussions and presentations

Table 2.2 Some examples of the impacts of CCT on Telco's organization

Quotation	Impact area
'[in our normal business] we have almost no partners, but in the context of CCT it is totally opposite where you cannot almost implement a big project without having 20–40 different partners . . . you need to partner with your enemies almost all the time and that is something that we are not so used to'	Product development and delivery routines
'We have the budget processes and other business processes that are optimized to carry out the current business. And the evolution is in quite a straight line. When it comes to treating completely new technology or business phenomena that happens, our preparedness for accepting that and acting on that is a bit low.'	Finance and support processes
'Support systems and financial support for this kind of sales [CCT] is different. Contracting, roles and responsibilities when product, SW [software], and service are sold as one.'	
'New sales mentality and approach is required to sell hardware and software as services'	Sales and delivery routines and approaches
'It [CCT] requires another way of working or acting . . . It is not the product itself that is important, it's actually the knowledge that you have and how you can turn it into discussions with customers. So it's not that you can respond to a product requirement, so saying that you need this interface, this capacity and so on. That's not what it is. It is much more about knowing added value that we are bringing from that.'	
'Existing structure of business units are not helpful for Cloud, because most of the time we need involvement of all three business units [for the context of CCT]'	Organizational structure
'looking into cloud, you can see a fundamental difference in required mindset and culture and ways to make it happen, and different types of people you need. You need different engineers; telecom engineers and IT engineers are different and this might be a huge challenge for us.'	Human resources

at different layers of management were ongoing from that time forward. These plans looked very accurate and realistic when compared to the new plans in 2012, and there were people in the organization who were persistently pursuing the plans, although without much success. In addition to the frustration that resulted from the fact that actions were not executed from the plans, the opportunity to gain a competitive advantage from early adoption and involvement of the technology was missed, and being behind competitors in sending a market message and influencing customer plans was quite dissatisfying for Telco's managers, who did not conceal their dissatisfaction.

Top Management Attention and Organizational Adaptation in Telco

A number of factors in Telco were acting as hindrances in experimentation and learning in the CCT context. But on a high level, the tension and mismatch between the requirements of existing businesses and the requirements of CCT manifested itself as the main cause of these difficulties. Telco has evolved during its lifetime into a decentralized organization where the resources and authority have been distributed between the business and product units, each of which are responsible for the realization of their own share of total revenues. Responsibility for the commercialization of the outcomes of research activities is also with the business units and product areas within each unit. In most situations, the units are able to find a good way to adopt and embed a given new innovation in the roadmap of products. In cases where there is no existing unit with a perfect fit for the technology, it is possible to form a cross-unit team that drives the new initiative. Such initiatives may be driven by some enthusiastic researchers, or an organizational unit that has sufficient interest to support such activities. In the case of CCT, however, none of these scenarios took place, for two main reasons.

The first reason was the scope of impacts on Telco's products and organization. CCT is an alternative to the existing architecture of Telco's products, and the ways by which the products are modularized and divided between the business units and product areas are not valid in the new context. Not only is there a need for participation from many different areas across the organization, but also the impacts are usually acting against existing short-term and long-term goals of the organization. In other words, in the long run, CCT is not going to exist in parallel to the existing businesses of Telco; it will replace some important parts of their offerings. In this situation, allocating any resources to CCT not only equates to taking resources out of the existing businesses, but it also implies making the lifespan of profitability from existing products shorter by promoting the alternative. Moreover, the requirements for the development of new operating routines which are totally different to the existing ones are obviously challenging within the existing business. Therefore, no matter how quickly the trends on CCT were identified by Telco's technology gatekeepers, the existing product areas did not have the motivation and resources to realize the plans. Incentive systems and goal structures were acting as filters in the way of developing knowledge on CCT and commercializing it.

The second reason was technology and market uncertainty. When the existing business units are unable to secure required resources and motivation for a strategic issue, naturally the issue will be escalated to the senior management level, and the same thing happened for CCT. Presentations to the management team and frequent reports served to communicate the importance of the issue, and its consequences, to the executive management. At that time, however, there was no clear forecast on the market of potential revenues from investing in CCT. Although management was convinced about the importance of the matter, absence of evidence for demand combined with the scope of required actions made for a tough decision-making situation for the top management team. Before 2011, Telco's TMT made no decision in favour or against CCT.

It can be argued that if the scope of impacts was not so large (and therefore required actions were not so costly), or there were stronger signals from the market, it was more likely that Telco's TMT would have acted differently. However, TMT's time and attention were naturally being allocated to other matters that were more certain and had a direct effect on existing goals based on which shareholders evaluated the performance of the management team. This situation persisted until early 2011, when a number of events initiated a new chapter in the story of Telco and CCT.

In the earlier months of 2011, a couple of major customers approached Telco's TMT members with questions about their strategies and plans for CCT. This type of signal (together with some rumours about competitors' movements) is always taken seriously, and in this case triggered a totally new way of dealing with technology in the following months. Indeed, the biggest trigger to move forward with CCT from 2011 was the full attention of the TMT that persisted during the whole process of the technology adoption. Knowing the requirements for involvement from different organizational areas and the possibility for the filtration of CCT-related opportunities in existing organizational units based on existing goals and priorities, there was a need to secure the required resources and to eliminate the organizational hindrances so as to create a fertile environment for innovation in the cloud context. In the following paragraphs we discuss the actions that were taken by Telco during 2011–2012 to fulfil this goal.

Because of the high levels of resource requirements and the absence of any revenue expectation in the near future, creating an independent organizational unit with all the required resources and authorities was not a viable option. The alternative way was setting up parallel organizations in the form of cross-unit projects and programmes. Yet finding the composition and structure of such a programme proved to be very complicated and challenging in practice. The previous experiences of Telco with similar types of cross-unit initiatives were related to considerably smaller projects with limited impact and time frames for execution. The CCT initiative, as we have discussed, was perceived to cause alterations in many of the core organizational routines, but the nature, scope and time frame of the impacts were unknown. Therefore, without the involvement of the TMT it was difficult to secure the required level of motivation and resources in the programme. For the same reason, the TMT representation was a basic presumption in the formation of a parallel organization.

At the very first step, Telco's management nested the programme into one of the business units that was perceived to have a higher level of interest and resources to drive the programme. Resources from other organizational units were included in this initiative and a number of projects started on the development of a pilot offering based on CCT. Going through the development process, the organizational difficulties of bringing a large number of resources from other profit centres to contribute to the goals of the host business unit caused complications in securing the required resources and motivation. This triggered a need for restructuring the programme in order to resolve the participation issues. The TMT decided to take the cloud programme out of the business units and shape an independent programme with the participation of members from all affected business units and product areas. Therefore, in the second step, a large and independent parallel organization with the participation of around 70 engineers and managers was formed. This clearly secured a decent amount of resource availability in order to speed up the innovation process in the CCT context. Nevertheless, many of the interviewees believed that this setting was ineffective in terms of coordination and collaboration. Some people believed that this setting deprived the participants of the freedom that they had otherwise had when attempting to work within the boundaries of their business units.

In the third step, the TMT decided to dissolve the large programme into a few smaller subprojects assigned to different business units, and defined a small steering group that was responsible for driving and coordinating the activities. This new setting was perceived positively by the respondents, because they believed it provided the best possibilities to progress with the identified areas in the previous formats while minimizing the organizational frictions and challenges getting in the way of experimentation with the new technology. The steering group served as a facilitator for the allocation of attention-based and financial resources, and the smaller groups had the possibility to experiment with the new technology. In this period, a special organization reporting to the management team was responsible for systematically responding to the needs for new operating routines. Eventually the company found it more suitable to separate the two tasks and to keep the strategic steering team closer to the TMT and to leave much of the technical coordination to the business units which were, at that time, more familiar with the technology and its significance in comparison to the initial steps.

In sum, during the period of 2011 and 2012, Telco managed to go through a number of adaptations in the organization which were mainly aimed at the facilitation of experimentation with the new technology and the development of new operating routines. Through these activities, Telco gained a large amount of control and progress in dealing with CCT and, given the fact that it may take years before the actual impacts of CCT overturn the telecom sector, the company feels confident about its position in the new era. As we have discussed, two distinct but closely interrelated sets of routines were the subject of innovation. A summary of the distinct characteristics of these routines is presented in Table 2.3.

DISCUSSION AND CONCLUSION

An organization's ability to adapt to the requirements of managing an emerging technology is dependent upon the ability of managers to shift the focus of their attention from ongoing activities towards the long-term competitiveness of the organization. The threat posed by technological discontinuity cannot be neglected if executives want their organization to learn and become prepared for realizing the potential benefits of the new technology. Organizations often dedicate the task of technology monitoring to the other members of the organization who serve as boundary spanners. In the cases of technological disruption, these groups are able to spot, evaluate and communicate the potential of the new technology to senior managers. However, much of the information about the nature of implications can be acquired only by experimentation and learning by the organization's members, and boundary spanners have limited influence over the rigidities in routines and resources. Hence, unless and until senior managers understand the gravity of the situation and focus their attention on the sources of rigidity, the incumbent remains handicapped when attempting to deal with disruption.

Table 2.3 Distinctive characteristics of operating and higher-order routines undergoing innovation in Telco

	Operating (technology use) routines	Higher-order organizations' collective learning routines
Representations (affected systems, structures, processes and practices)	Product development, sales and marketing, delivery, accounting and pricing, partnership and alliancing, etc.	Resource allocation, goal-setting, profitability requirements for assessments, strategic priority-setting, communication channels with top management team, corporate-wide directives, etc.
Trigger(s)	Technological gatekeeping activities	Market and competitor pressure
Area(s) of influence	A single technology (or a family of the technologies)	Telco's portfolio of technologies
Key role(s)	Middle managers and technicians, with support of TMT	TMT
Enabler(s)	Experimentation with the emerging technology, organizational learning routines	TMT awareness and attention
Consequence(s)	Knowledge accumulation on the emerging technology	Possibility for experimentation with the emerging technology

Because of limitations in resources, senior executives are unable to attend equally to all of the threats that are being communicated by the gatekeepers. Therefore, managers need to allocate their attention to those threats that seem to be more dangerous if not tackled quickly. The time lag between the emergence of a new technology and its commercial impact, which could even be as lengthy as a decade, causes a tendency among managers to prioritize ongoing businesses. An ongoing financial crisis and competitive pressure resonates with this tendency.

In the case study that we have presented, management attention did not shift towards the emerging technology until strong indications from markets and competitors were received by the top management team. The perception of threat motivated changes in the structure and processes, that in turn facilitated the process for the allocation of resources required for experimentation with and learning about the new technology. The crucial role of managers in initiating dynamicity and change in organizational routines undermines assumptions about the inherent adaptability of routines.

While it is possible to routinize the regular technology monitoring, evaluation and strategy development, the case study revealed that in cases of discontinuity in the core technologies of an organization it becomes a managerial problem-solving task to revise and adapt the search routine. This can be done by the introduction of new systems, structures, processes and practices that allow for experimentation and knowledge accumulation in parallel with driving the ongoing businesses of the organization.

REFERENCES

Ashmos, D.P., Duchon, D. and McDaniel, R.R., Jr (1998). Participation in strategic decision making: the role of organizational predisposition and issue selling. *Decision Sciences*, 29 (1), 25–51.
Barley, S.R. (1986). Technology as an occasion for structuring: evidence from observation of CT scanners and the social order of radiology departments. *Administrative Science Quarterly*, 31, 78–108.
Baum, J.A.C. and Singh, J.V. (1994). Organization–environment coevolution. In Baum, J.A.C. and Singh, J.V. (eds), *Evolutionary Dynamics of Organizations*, New York: Oxford University Press, pp. 379–402.
Cohen, W. and Levinthal, D. (1990). Absorptive capacity: a new perspective on learning and innovation. *Administrative Science Quarterly*, 35, 128–152.
Cyert, R. and March, J. (1963). *A Behavioral Theory of the Firm*. Englewood Cliffs, NJ: Prentice- Hall.
Edmondson, A.C., Bohmer, R.M. and Pisano, G.P. (2001). Disrupted routines: team learning and new technology implementation in hospitals. *Administrative Science Quarterly*, 46 (4), 685–716.
Farjoun, M. (2010). Beyond dualism: stability and change as a duality. *Academy of Management Review*, 35, 202–225.
Feldman, M.S. (2000). Organizational routines as a source of continuous change. *Organization Science*, 11 (6), 611–629.
Feldman, M. and Pentland, B. (2003). Reconceptualizing organizational routines as a source of flexibility and change. *Administrative Science Quarterly*, 48 (March), 94–118.
Fleming, L. (2002). Finding the organizational sources of technological breakthroughs: the story of Hewlett-Packard's thermal inkjet. *Industrial and Corporate Change*, 11 (5), 1059–1084.
Gersick, C. and Hackman, R. (1990). Habitual routines in task-performing groups. *Organizational Behavior and Human Decision Processes*, 47, 65–97.
Gilbert, C. (2005). Unbundling the structure of inertia: resource versus routine rigidity. *Academy of Management Journal*, 48 (5), 741–763.
Glaser, B. and Strauss, A. (1967). *The Discovery of Grounded Theory: Strategies for Qualitative Research*. London: Wiedenfeld & Nicholson.
Hambrick, D.C. and Mason, P.A. (1984). Upper echelons: the organization as a reflection of its top managers. *Academy of Management Review*, 9, 193–206.
Labatut, J., Aggeri, F. and Girard, N. (2012). Discipline and change: how technologies and organizational routines interact in new practice creation. *Organization Studies*, 33 (1), 39–69.

Lewin, A.Y., Massini, S. and Peeters, C. (2011). Microfoundations of internal and external absorptive capacity routines. *Organization Science*, 22 (1), 81–98.
Li, Q., Maggitti, P. and Smith, K. (2013). Top management attention to innovation: the role of search selection and intensity in new product introductions. *Academy of Management Journal*, 56 (3), 893–916.
Martin, P.Y. and Turner, B.A. (1986). Grounded theory and organizational research. *Journal of Applied Behavioral Science*, 22 (2), 141–157.
Nelson, R.R. and Winter, S.G. (1982). *An Evolutionary Theory of Economic Change*. Cambridge, MA: Belknap Press/Harvard University Press.
Ocasio, W. (1997). Towards an attention-based view of the firm. *Strategic Management Journal*, 18, 187–206.
Ocasio, W. and Joseph, J. (2005). An attention based view of strategy formulation: linking micro- and macro- perspectives in strategy processes. In Szulanski, G., Porac, J. and Doz, Y. (eds), *Strategy Process*, Advances in Strategic Management Vol. 22, Oxford: Elsevier, pp. 39–61.
Parmigiani, A. and Howard-Grenville, J. (2011). Routines revisited: exploring the capabilities and practice perspectives. *Academy of Management Annals* (September), 37–41.
Pentland, B.T. and M.S. Feldman. (2008). Designing routines: on the folly of designing artifacts, while hoping for patterns of action. *Information and Organization*, 18(4), 235–250.
Schreyögg, G. and Kliesch-eberl, M. (2007). How dynamic can organizational capabilities be? Towards a dual process. *Strategic Management Journal*, 28, 913–933.
Schreyögg, G. and Sydow, J. (2011). Organizational path dependence: a process view. *Organization Studies*, 32 (3), 321–335.
Schroeder, Roger G., Andrew, H. and Van de Ven, G.D. (2000). The development of innovation ideas. In Van de Ven, G.D. and Angle, H.L. (eds), *Research on Management of Innovation: The Minnesota Studies*, New York: Harper & Row, pp. 107–134.
Szulanski, G. (2000). The process of knowledge transfer: a diachronic analysis of stickiness. *Organizational Behavior and Human Decision Processes*, 82, 9–27.
Tyre, M. and Orlikowski, W. (1994). Windows of opportunity: temporal patterns of technological adaptation in organizations. *Organization Science*, 5, 98–118.
von Hippel, E. (1994). Sticky information and the locus of problem solving: implications for innovation. *Management Science*, 40 (4), 429–439.
Woiceshyn, J. (2000). Technology adoption: organizational learning in oil firms. *Organization Studies*, 21 (6), 1095–1118.

3. Will the real innovator please stand up? Claiming ownership of an organizational capability

Abdelghani Es-Sajjade and Krsto Pandza

INTRODUCTION

The concept of organizational capability has been central for explaining interfirm heterogeneity, processes of growth, and reconfiguration of resources (Eisenhardt and Martin, 2000; Grant, 1996; Penrose, 1959; Teece et al., 1997). Scholars have largely embraced the collective nature of organizational capabilities (Dosi et al., 2000) and remained intrigued with structures of their constitutive elements and underlying developmental mechanisms. This search for constitutional elements of organizational capability has seen researchers identifying experiential learning patterns (Winter, 2000, 2003; Zollo and Winter, 2002) as the foundational mechanism for the development of organizational knowledge of how to get things done (Collis, 1994). More recently the focus has shifted to unveiling individual micro-foundations (Felin and Foss, 2005; Felin and Hesterly, 2007) with special attention being paid to the role of managerial cognition (Gavetti, 2005; Hodgkinson and Healey, 2011) as a foundational element of organizational capability.

Undoubtedly, the strategy process tradition has long recognized the role of individual actions in developing organizational capabilities. In this school of thought, resource allocation decisions of middle managers (Burgelman, 1994, 1996) and autonomous actions of groups (Burgelman, 1983; Kanter, 1985; Stopford and Baden-Fuller, 1994) are all deemed responsible for developing new capabilities. The complexity of social interactions has been acknowledged but its underlying behavioural foundations remain less well understood.

Additionally, scholars highlight the importance of proximity to a organizational capability, suggesting that individuals or groups that work closely with capabilities are more likely to initiate actions aimed at increasing capabilities' performance levels (Gavetti, 2005; Noda and Bower,

1996). Some recent work conducted in the strategy process tradition has made a more explicit effort to understand the role of group-level social interaction in resolving contests between competing strategic initiatives (Kaplan, 2008), the implementation of strategic initiatives (Huy, 2011) and the development of organizational capabilities (Pandza, 2011).

We believe that a focus on groups in examining capabilities is valuable because we take seriously the assertions from social psychology that people derive part of their identity and sense of self from the work groups to which they belong (Hogg and Vaughan, 1995; Hogg and Terry, 2000). Moreover, it has been argued that the development of capabilities involves a coordinated effort by teams to align and combine individual tasks (Helfat and Peteraf, 2003; Teece et al., 1997); *ergo* organizational groups are pivotal in capability development. Thus, we assert that the focus on individual attributes as prevalent in the emerging micro-foundations research stream can be extended by examining the potential role of organizational groups as constituents and drivers of capabilities. In this chapter we strive to contribute a number of factors that explain group-level action and interaction in respect of the development of organizational capabilities. We are particularly concerned with understanding how perceptions of a capability's value might drive the actions of groups aiming to influence capability development. Additionally, we are interested in understanding the social interaction within groups in parallel with the interaction between groups. Finally, we focus on innovation capability – that is, the development and commercialization of new technologies and products – as a specific organizational capability (Salvato, 2009) to impose clear empirical focus.

The aim of this research is to build and extend theory on the processes through which groups contribute to the development of capabilities and for this purpose we have used a single case study of a medium-sized firm in the telecommunications industry. The remainder of this chapter is organized as follows: in the next section we review literature on capabilities as processes and the role of groups in organizations. Subsequently, we introduce our methodology. We report the insights gained from field research, and then we discuss our chapter's theoretical and practical implications. The final section draws our conclusions.

THEORETICAL BACKGROUND

Social (Inter)action and Foundations of Organizational Capabilities

The strategy process literature has emphasized the effects of individual action on strategic change by demonstrating how local decisions made by lower-level managers can be of paramount importance to the course of development of organizational capabilities (Burgelman, 1994, 1996; Floyd, 1992; Floyd et al., 2011; Galunic and Eisenhardt, 2001; Pettigrew, 1987; Wooldridge and Floyd, 1990). This school of thought acknowledges the role of patterns of social interaction between individuals and groups in the development of organizational capabilities. One of the main contributions of this work is the focus on suborganizational groups in lieu of a firm-level perspective on capabilities. More specifically, researchers from the strategy process stream elaborate upon and emphasize the role of proactive courses of action and interaction driven by intra-group and intergroup dynamics in the development of capabilities. As these are central themes to our study, we shall hence adhere to the process perspective on organizational capabilities to better understand their foundational elements.

Burgelman (1994, 1996) argues that middle managers have the authority to make decisions that deviate from an old pattern of capability development, after which top managers assess these decisions and their consequences. The author refers to social complexity, but does not further elaborate on the elements that drive action, assuming that the authority given to middle managers, their proximity to capability development and their unique interests sufficiently explain proactive actions. We believe however that further elaboration of the drivers of action is appropriate to better understand the intricacies of social interaction.

A more recent strategy process study on capabilities by Salvato (2009) identifies how individuals contribute to the creation and development of capabilities. The author emphasizes the crucial role of ordinary activities in a product development process and how local deviations from this process are mindfully selected and recombined by top management to pursue alternative or similar paths of capability development. However, although this researcher neatly analyses how deviations and mindful selection occur, the 'Why?' question remains unanswered. Moreover, product development in larger firms is a cross-functional discipline requiring the management and contribution of distinct groups, whilst in this study a seemingly homogeneous workforce is presented. Kaplan (2008) investigates intergroup conflicts and builds on social movement theory to understand the importance of framing for resolving contests between competing strategic initiatives in periods of uncertainty. She argues that strategic response is constructed through conflicts in frames and the deliberate action of managers seeking to make their own frames triumph over others. A useful complement to this study would be an examination of the origins of intraorganizational heterogeneity in frames to probe the underlying motivations of actors or groups who purposefully act and interact to have their views prevail over those of others. In sum, albeit the strategy process tradition acknowledges social interaction and complexity in capability development, it does not however deeply expand upon the foundations of social interaction.

The micro-foundations literature stream has created a research agenda that focuses on the micro-level building blocks of organizational capabilities (e.g. Augier and Teece, 2009; Gavetti et al., 2007). Studies within this strand of thought have investigated different constituents of capabilities, *inter alia* individual cognition (e.g. Gavetti and Levinthal, 2000), routines (e.g. Salvato and Rerup, 2011) and emotions (Hodgkinson and Healey, 2011). We assert that a logical analytical step in microfoundations research would be to explicate the role of groups – in addition to individuals – as a unit of analysis in capability development. Indeed, organizations are known to accommodate a highly heterogeneous internal community represented by several groups connected by salient identities (Huy, 2011). Individuals who cooperate on a daily basis, interact frequently and share practices are likely to develop some form of collective thinking (Weick and Roberts, 1993). Furthermore, it has been argued that group cognition varies across business units and functions, which is an indicator for heterogeneity on a more collective level (Tyler and Gnyawali, 2009).

The role of organizational groups in developing organizational capabilities is explicitly examined in a paper by Pandza (2011). This researcher argues that autonomous action is likely to occur when a group has a highly distinctive cognitive frame for the strategic merit of a capability and the motivation to enhance the group's identity. The author further asserts that when this group does not possess the authority to directly shape the course of capability development, a socially complex process will take place to legitimize the group's action in light of a given capability. Social identity and cognitive frames are the two main building blocks of group-level action in this paper and the author acknowledges that in other organizations under different circumstances alternative forms of self-interest could add to identity as a driver of autonomous action. Our interest in the antecedents of social identity in capability development would build on the findings of this work. Moreover, whilst Pandza's paper focuses on intergroup interaction, we believe we could extend his work by investigating the role of social (inter)action within groups.

Another central study integrating the dynamics between identity and interaction is a paper by Nag et al. (2007). The authors present an intriguing case study of a high-technology, research and development (R&D)-driven organization aiming to develop its innovation capability by incorporating knowledge pertaining to the commercialization of innovation. The authors find that these two subprocesses of innovation actually accommodate a distinctive organizational identity which is enduring. The R&D identity and commercial identity drive the individuals belonging to each of the two to resist any change in knowledge exchange, work practices or norms. The central role of identity in shaping intergroup interaction toward competition, and ultimately inhibiting a strategic change trajectory, is revealed by the authors. *Ergo*, this study highlights the importance of incorporating identity in research on capability development and the knowledge exchange processes underlying it. In essence, this study demonstrates the resilience of identity and how organizational members can pursue (or resist) interaction based on dynamic interpretations of changes in their social environment.

Scholars have observed how members develop new interpretations about their identity, which changes their beliefs about distinctive qualities of their group in the face of internal stimuli (Dutton et al., 1994; Fiol and Connor, 2005; Gioia and Thomas, 1996; Gioia et al., 2000; Ravasi and Schultz, 2006). Huy (2011) shows how internal stimuli can enkindle middle managers' group social identities which influence outcomes on the organizational level. The author succeeds in explaining the social matters involved in organizational processes through a detailed analysis of the relationships between symbols as antecedents of group social identity. Our research potentially broadens these insights by focusing on the consequences of social identity process; that is, the social interaction between and within groups.

Social Identity in Organizations

Individuals derive a considerable part of their identity and sense of self from the organizational units to which they belong (Hogg and Terry, 2000), which are considered as 'groups' in social psychology. In this chapter we will also refer to organizational units as 'groups'.

Tajfel (1972: 292) defines social identity as: 'The individual's knowledge that he belongs to certain social groups together with some emotional and value significance to him of this group membership'. In organizational research, social identity is usually explained as being shaped by two psychological processes: self-categorization and self-enhancement (Huy, 2011). The basis for self-categorization is the accentuation of perceived intra-category similarities and intercategory differences which increases in strength when the categorization has benefit to a group member (Turner et al., 1987). Categorization causes group members' cognition and actions to be in accordance with the relevant in-group prototype; a subjective representation of salient characteristics (for example, beliefs, attitudes and behaviour) of a social (group) category. This representation is derived from social information resting on intergroup social comparisons that seek to confirm or establish in-group-favouring evaluative distinctiveness. Thus, social identity is a dynamic phenomenon (Gioia et al., 2000; Ravasi and Schultz, 2006) contingent upon context as it interacts with immediate out-group comparisons that become relevant in particular situations; for example, when several groups contest a valued organizational resource (Hogg and Terry, 2000). Furthermore, categorization and social identity produce stereotyping. Stereotypes are shared generalizations about members of a social group and are usually uncomplimentary simplified images when applied to an out-group (Hogg and Vaughan, 1995).

Hogg and Terry (2000) argue for additional antecedents of social identity which are not commonly included by organization researchers using social identity, namely, uncertainty and social context. They argue that social identity processes are motivated by a need to minimize uncertainty about an individual's perceptions, behaviour, feelings, and one's self-concept and place within the social world. In addition, they emphasize the responsiveness of social identity to social contexts as a central feature of social identity processes. This is attributable to the cognitive system which matches social categories to properties of the social context and activates that category, making the social context and one's place within it subjectively most meaningful. Group members may therefore draw on accessible categories and assess how well they fit their social environment. This implies that within organizations, groups may use a number of categories over time depending on the development of their organizational field and its members' perceptions, beliefs, attitudes and behaviour. Consequently, this seems to cause competition between and attraction within groups, leading to the motivated influencing of symbols (for example, through communication, appearance and action) by groups to manipulate the frame of reference within which accessibility (of category) and fit (with social context) of social identity are justified (Hogg and Mullin, 1999).

In this chapter, we intend to explore additional antecedents of social identity processes specific to organizations to elucidate how organization-related contextual features may influence the behaviour of groups and how this impacts upon the development of organizational capabilities. Also, while considerable attention has been given to the cognitive aspects of social identity, less is known however about how social identity translates into intergroup interaction in organizations, and hence this research is expected to shed light on how groups (inter)act respective to each other on the one hand, and how their members (inter)act within their groups on the other.

In sum, we synthesize between the strategy process tradition, literature on organizational groups (in particular the role of group social identity) and the microfoundations school of thought to better understand how organizational capabilities develop. In the following section we present our research design and the method we mobilized to answer our research question.

METHODS

We used an inductive theory-building approach (Eisenhardt, 1989) to explore how two groups act and interact in order to influence the development of innovation capability. By adopting a single case study strategy (Siggelkow, 2007) we were able to produce insights into complex social processes that are not easily captured through the use of quantitative data (Eisenhardt and Graebner, 2007).

Research Setting

For the purpose of building and extending theory we looked for an organization that would fall under the category of 'extreme case' (Yin, 2003). Extreme cases are particularly useful because the investigated phenomena are likely to be more prominent in both frequency and intensity than they might be in other settings. Having this criterion in mind, we came across a candidate – an international telecommunications firm which we shall refer to as 'WIRELESS' to guarantee company anonymity – which drew our attention because of two initial qualities relevant to the subject of interest: (1) a strategic emphasis on the role of innovation capability which had been central to this firm's growth; and (2) its innovation processes ran through multiple locations involving different units, which suggested the possibility of an interesting interplay between social and strategic processes shaped by the action and interaction of different groups After a series of interactions, access was negotiated in late 2010 with data collection commencing early in 2011.

WIRELESS was founded in 2002 in the UK by a group of entrepreneurs and specialists in radio telecommunications technology. It operates in the telecommunications sector where technological innovation is key to survival and success (Porter and Stern, 2001). 'To say that an organization has a capability means only that it has reached some minimum level of functionality that permits repeated, reliable performance of a capability' (Helfat and Peteraf, 2003: 999), and hence we observe this to be the case for this firm since it has been winning competitive bids over time against large corporations such as Nokia-Siemens and Cisco, based on its ability to innovate. In the early stages after the company's founding, significant patent-awarded technological innovations were developed and sold to gain entrance to the transportation sector where the company could apply its technology. Thenceforth the company has been successful in winning numerous competitive bids around the world through product and technology innovation, becoming the exclusive on-board wireless communication services supplier for numerous train operators in particular. After the first three years WIRELESS started developing its technology into additional fields within the transportation sector: media services, technological infrastructure, on-board equipment and data management. Despite a pivotal role given to innovation, the company did not assign ownership of innovation to a formal innovation unit. Instead, several teams located in different offices in the UK are involved in innovation processes. These teams aggregated into two main groups, one group focusing on the technical aspects (R&D, intellectual property development) and another group focusing on the business and marketing aspects of innovation (technology scouting and sourcing; product development and commercialization). We shall refer to these two groups as 'POLIS' and 'METRO' to ensure anonymity. Competition between these groups was acknowledged by top management but no decisions were made to allocate formal authority to either of the groups in leading innovation processes, which allowed autonomous action of groups and their individual members. More specifically, the lack of coordination between the groups due to the lack of involvement of top management in operations allowed space for each of the groups to act autonomously and proactively. This context provided a unique opportunity to study the role of groups and social interaction pertaining to innovation capability.

Data Collection

The data sources we used to empirically investigate the role of individuals and groups in innovation capability can be divided into three categories.

Semi-structured interviews

The primary method of data collection involved semi-structured interviews (n = 27). Table 3.1 provides a summary of informants' roles, location and number of interviews per informant. We identified numerous knowledgeable informants who addressed and experienced the focal phenomena from diverse perspectives (Eisenhardt and Graebner, 2007).

We planned interviews in a series of waves. The first wave of interviews with members from the executive team was based on purposive sampling (Patton, 1990) to provide a rich account of WIRELESS's innovation capability, how it had developed over time and what its constituent processes

Table 3.1 Informant's position, location and number of interviews

Position	Location	No. of interviews
Founder/chairman	France	3
CEO	METRO	1
Co-founder/CTO	Americas	1
COO	POLIS	2
CFO	METRO	1
Product manager	METRO	2
Business development manager	METRO	1
Product testing engineer	METRO	1
Pre-sales engineer	METRO	1
Sales manager	METRO	1
Applications developer	POLIS	1
Team leader systems admin.	POLIS	1
Software engineer	POLIS	1
Onboard systems manager	Americas	1
Head of software development	POLIS	1
Manager aftersales	METRO	1
Head of programme office	POLIS	1
Programme director	Europe	1
Team leader embedded development	POLIS	1
Regional sales director	Europe	1
Project engineer	POLIS	1
Former alliance partner	Europe	2
	Total	27

Note: CFO = Chief Financial Officer; CTO = chief technology officer ; COO = chief operating officer

were. Subsequently we changed to theoretical sampling (Eisenhardt, 1991) when we developed more specific theoretical research interests. We then realized that we needed to interview members from the METRO group, because the first two waves indicated: (1) METRO's direct involvement in the innovation process; and (2) contrasting accounts of the importance of METRO with respect to innovation. Furthermore, to mitigate the risk of potentially biased accounts we interviewed individuals from different hierarchical layers and organizational groups, external informants and former employees. Moreover, we conducted the interviews in two-person teams with one researcher dealing with the interview questions while the other made notes. This allowed data to be viewed from both contrasting and converging perspectives of two observers. Interviews lasted between 45 minutes and two and a half hours, and were all tape-recorded and transcribed.

Archival data: internal sources
In order to enhance validity we used archival data which we corroborated with interview data throughout the collection and analysis stages. These data comprised project documentation including templates and process flows (164 pages), PowerPoint presentations (34 slides), and product and technology design documentation (16 pages). Furthermore, we analysed press releases retrieved from the company website, which were available from 2007 and included 37 articles.

Archival data: external sources
External sources included available online material about WIRELESS identified using ABI/Inform Complete. Our search resulted in 57 records using the firm name as a keyword. Using these extensive archival data we derived a case history for WIRELESS to complement our primary data.

Observations
We also made use of observations in different settings to further increase validity and reduce potential bias. Furthermore, we were invited to workshops organized for the introduction of a new product which were attended by staff from both groups and one or two executives (May 2011). Moreover, we spent a week working from two of the company's offices, which allowed us to study day-to-day interaction and practices. Last, we attended product demonstrations (September 2011) in which pre-sales engineers explained the specifications of a new design, how it was utilized to win a competitive bid and how it was implemented in a current project.

Data Analysis

Data analysis was handled by interweaving collection and analysis of data in three high-level steps during which we continuously worked from an evolving juxtaposition of new information and the emerging structure of our theoretical arguments (Miles and Huberman, 1994). Figure 3.1 depicts the process of open coding, axial coding into theoretical categories, and the abstraction of categories and their relationships into aggregate theoretical dimensions:

- Step 1: Constructing initial categories and first-order codes. We initiated data collection and data analysis following prescriptions for inductive theory building (Locke, 2001). Shortly after the start of our fieldwork we identified provisional categories and started the open coding process by gathering statements that represented how our informants perceived the importance of innovation capability

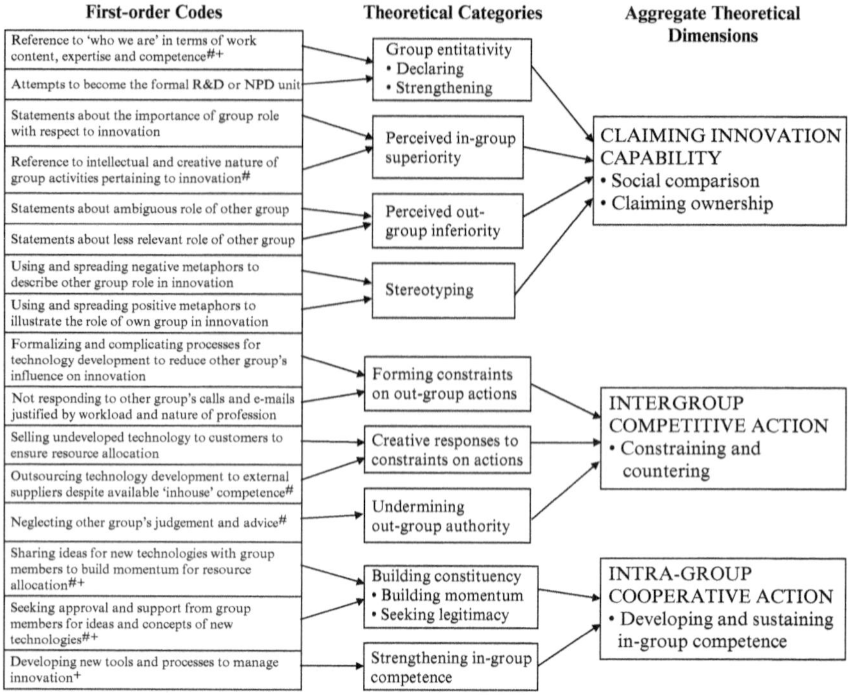

Note: #Augmented by archival data ; +Augumented by observations; NPD = New Product Development

Figure 3.1 Overview of data structure

and how they perceived different groups' roles with respect to innovation capability. In order to gradually capture provisional categories we completed a contact summary form after each interview which we adapted from Miles and Huberman (1994). The categorization processes in which we edited, elaborated or removed categories guided subsequent interviews within and between waves.

- Step 2: Linking first-order codes and developing theoretical categories. After each wave of interviews we grouped first-order codes per initial category and repeated this process for each wave after the third wave. The contact summary forms guided this process during which we analysed the within-group and intergroup dynamics around WIRELESS innovation capability. As we progressed, categories become more theoretical through a process of axial coding as we now were building on discovered relationships between first-order codes.

- Step 3: Determining theoretical structure through aggregate theoretical dimensions. Subsequent to the articulation of theoretical categories we reviewed data in light of these categories to motivate and initiate further abstraction and structuring. An example is the relationship between 'perceived in-group superiority' versus category 'perceived out-group inferiority'. After a series of discussions regarding how these (and other) categories related to each other, we defined the 'claiming innovation capability' theoretical dimension that adequately captured data fragments around both in-group superiority and out-group inferiority. In parallel we integrated the ongoing literature review to examine how our findings extended or built theory.

In the next section we present our findings, in which we highlight processes through which two groups in WIRELESS engaged in claiming ownership of innovation capability based on their group social identity. Furthermore, we elaborate on how social identity produced 'intergroup competitive action' and 'intra-group cooperative action'. Last, we shed light on how these forms of group-level action influenced capability development.

FINDINGS

Our data suggest that group social identity processes influence capability development through social action and interaction between and within groups. In Table 3.2 we provide empirical evidence to illustrate group identity processes around innovation capability. We identified the following organizational, contextual antecedents of group social identity: (1) an organizational capability which was perceived as strategically important across groups; and (2) organizational ambiguity around formal ownership of the valued capability. These antecedents triggered groups to engage in claiming innovation capability and mutual comparison in light of their contribution and importance to this capability, because both wanted to be associated with this central capability, and both saw the opportunity to autonomously make an impact because of a lack of formal coordination by top management. Furthermore we discovered two types of group-level action as a result of social identity processes: (1) intergroup competitive action; and (2) intra-group cooperative action. These actions we found to impact upon capability development in several ways, which will be explained in the next section.

We commence our findings with how organizational antecedents

Table 3.2 Illustrations of group identity process around innovation capability

Quote	Tone and origin	Reference to group	Theoretical dimension and category
'There's a disconnect within WIRELESS between our own and others' understanding of what software development actually is. It's a very creative process, the idea of translating the concept into something that actually does something. We were talking about it just today but you can consider akin to painting, you're an artist, or a poet writing a poem.'	Negative. This quote from a team leader of software developers at POLIS believes the complexity and distinctiveness of their activities are not well understood within the company.	POLIS	Claiming innovation capability – group entitativity.
'The METRO team are not developers down there and I'm quite happy if they're all about hardware procurement rather than anything else. That's a different level of activity to what we are doing, which is really about the generation of intellectual capital.'	Negative. The head of software development feels the activities at METRO are more about purchasing hardware than about the development of core (software) technologies.	METRO	Claiming innovation capability – perceived in-group superiority.

'When people don't understand all of the obligations of what POLIS has to deliver, they see them as resistant. Same is when you don't see all the sales guys are facing you don't understand the need for continuous innovation.'	Neutral. The COO explains the misunderstanding between the two groups because of the distinctive nature of their respective contribution to innovation.	METRO and POLIS	Claiming innovation capability – perceived out-group inferiority.
'We have rather a lot of teenagers in this company. You can't make things happen the official way therefore you go the executive way. That really is the spoilt teenager's approach, isn't it?'	Negative. A software developer at POLIS compares the METRO group to teenager behaviour because they do not always follow the processes underlying innovation capability.	METRO	Claiming innovation capability – stereotyping.
'POLIS is currently busy building a wall, that wall they refer to as technical release schedule in order for them to feel they've fulfilled their task. This way they're isolating themselves.'	Negative. The programme director refers to a local tool/process created by POLIS to instil more structure around innovation activities.	POLIS	Intergroup competitive action-forming constraints on out-group actions.
'If people ask POLIS and the door is closed then they go and get it somewhere else.'	Neutral. A regional sales director explains that METRO addresses external technology suppliers if POLIS does not provide resources for innovation projects.	METRO	Intergroup competitive action – creative responses to constraints on action.

Table 3.2 (continued)

Quote	Tone and origin	Reference to group	Theoretical dimension and category
'They don't like taking no for an answer and if they don't get what they want then they just ring the CEO. It's very hard. Then the CEO will ring me to say, "What can we do to sort this out?" They just ignore the impact and push their thing forward.'	Negative. The head of programme office complains about how METRO does not accept and respect POLIS's recommendations, by evading these if they are not favourable.	METRO	Intergroup competitive action – undermining out-group authority.
'You quite often see e-mails from one person who saw this neat thing, describes it and sends out an e-mail to basically the technical people here. People look at it and that'll lead to conversations amongst people. If it builds up enough momentum and it looks like something that could be really useful to us, then we'll use it.'	Positive. The CTO describes how knowledge and ideas are further developed to tangible outputs as a result of a process of social interaction within the POLIS group.	POLIS	Intra-group cooperative action – building constituency.
'The thing that technology roadmap is missing is the collaboration with everybody in the organization, the buy-in of everybody. It's very much internally generated.'	Negative. A business development manager at METRO refers to an innovation tool created by the POLIS group which he feels is not sufficiently shared outside POLIS.	POLIS	Intra-group cooperative action – strengthening ingroup competence.

influenced group identity processes of two groups involved in innovation at the focal firm. We then describe how social identity caused groups to compete with each other and how this affected social interaction within groups. Finally, we present the process through which group-level action influenced innovation capability development.

Perception of Capability Value and Organizational Ambiguity

WIRELESS started in 2002 as a hi-tech start-up promoting its emergent technology in the public transportation industry. The chief executive officer (CEO) believed technological innovation to be the key to success in this industry. One of the firm's alliance partners, an international mobile telecommunications corporation, confirmed that its choice to work with WIRELESS on a country-scale project in Europe was based on its innovative technology:

> I had the freedom to choose any supplier for our project. Because WIRELESS could do something that we couldn't, we both got the deal with this government train operator. But they told us it would go under our responsibility as a large corporate brand; 'If things go wrong, your brand will be held liable'. We took the risk because we believed in the technology. (Marketing executive at alliance partner)

Eventually, the company's breakthrough project with a UK rail operator in 2004 proved WIRELESS's capability to provide innovative solutions to the transportation industry. The perception of innovation capability being paramount to success was strongly advocated by the executive team. This made innovation capability a useful and logical context to be associated with for individuals within the firm (Hogg and Terry, 2000). In fact, employees were strongly aware of innovation capability as a defining element on an organizational level, as a business development manager from the METRO group observed: 'Innovation is so important that you want to belong to it. I think it's a positive thing for the organization that people view innovation as flag to gather around. "We're the most innovative!" "No, we're the most innovative!" That's great. (Emphasis on "we".)'

Start-ups have been found to entirely base their identity on the new technology they offer and attempt to either create a new industry or become established in an existing industry (Pandza and Thorpe, 2009); the latter being the case for this firm. As is often the case, start-ups create uncertainty and ambiguity not only for their environment but also for themselves. Whilst emphasizing the importance of innovation capability as a strategic driver, the executive team neither implemented a formal organizational design nor assigned formal ownership to any group for managing innovation. They simply stated a set of technology-related problems that needed to be addressed over the next year(s) and left it up to middle managers to autonomously solve these, which caused uncertainty. At the same time innovation processes ran through several units. The delivery of new technologies to customers required contributions from two main groups: (1) a group specialized in software development, applications development and technical project management (POLIS); and (2) a group handling product and business development, commercialization of new technologies, and technology scouting and sourcing (METRO). Ultimately, the absence of clear ownership of innovation and a formal structure or process provided these groups with considerable autonomy in managing their activities within the sphere of their group, and in relation to coordinating their respective efforts toward innovative solutions for the organization.

Social Identity and Group-Level Action around Innovation Capability

The two groups directly involved in WIRELESS's innovation capability were created in 2004 and continued to grow in size, scope and complexity since the company's first large-scale project in the same year. Figure 3.2 depicts each group's high-level contribution in different stages of the innovation life cycle. As ownership of innovation capability and its underlying processes was unclear, both groups attempted to emphasize their group-level contribution to innovative performance through processes which can be explained by organization-related social identity (Huy, 2011). These processes are reported next.

Claiming and comparing

Whereas METRO differentiated itself in terms of the business and commercial aspects of innovation, POLIS did so in terms of technology development and the generation of intellectual capital. Each group tried to align its distinctiveness (Hogg and Vaughan, 2011) with WIRELESS's valued innovation capability. Members from POLIS kept referring to their role in creating the company's intellectual property (IP), and how crucial technology development is to innovative performance. They used words such as 'poetry' and 'artist' when describing their technology development activities with respect to innovation. Furthermore, they felt that the creative nature of software development activities was not well understood by others in the organization, which caused uncertainty as to their social identity. Moreover, POLIS's members perceived the contribution of METRO to innovation as relatively marginal because of their reliance on external technology suppliers and their inability to understand software

	Ideation	Design	Development	Productization	Commercialization
METRO	• Market-driven ideation • Technology scouting	• Internally created proof of concept • Also third-party design	• Dependent on POLIS • Also third-party technology sourcing	• Product development • Business development	• Responsible for all sales activities
POLIS	• IP-driven ideation • Refinement of current technology	• Internally created proof of concept	• Software development • Application development	• Dependent on METRO	• Not involved in sales

Note: Commercialization by METRO often occurred directly after ideation to ensure the cooperation of POLIS in terms of providing resources for technology development.

Figure 3.2 Group activities throughout the innovation life cycle

development; the alleged core competency of the company according to POLIS. They used stereotypes in referring to METRO: for example, commercializing products and technology was a simple task, involving merely chatting, wining and dining with customers. As a result of social comparison and the perceived superiority, POLIS group members tried to claim ownership of innovation capability to enhance their group's position with respect to innovation in attempts to establish a formal innovation unit: 'We've been talking about actually having a separate R&D activity to the main thrust of development. And I think that's necessary' (head of software development).

Similarly, members of METRO explained how their activities were pivotal to innovation capability, which they argued from their social identity related to the business and commercial aspects of innovation. They mentioned how innovative performance could not be sustained without exposure to technological developments in the firm's environment and how they enabled capitalizing on this knowledge. Furthermore, they made POLIS's contribution to innovation seem unimportant by expressing their activities in stereotypes such as 'factory', and perceived POLIS as only being able to execute ideas and concepts prepared by METRO. Hence, METRO also engaged in claiming ownership of WIRELESS's innovation capability in ways similar to POLIS. In addition, they also pleaded for a dedicated innovation unit to be built in and by their group. As a business development manager argued, this innovation unit would be responsible for developing new products and technologies, and use POLIS for technical support. The ownership claiming and social comparison initiated by both groups affected their actions, which we bifurcated into 'intergroup competitive action' and 'intra-group cooperative action'.

Intergroup competitive action Given their motivation to claim ownership of innovation capability, both groups competed to control resources and processes to minimize the influence of the other group by constraining their actions. This intergroup competitive action emerged in a number of ways. First, POLIS members built local processes and structures in order to remain as autonomous as possible and to make innovation more efficient, structured and planned in order to reduce impact on their resource planning. An example is the Program Office which was set up to reduce ad hoc development – as often triggered by METRO – and to allocate resources to innovation projects based on a software release schedule.

Second, METRO observed the gradual increase of formal resource allocation around innovation projects. They discovered creative ways to deal with this bureaucracy and hence they did not voluntarily comply with the constraints on their actions. One of the few simple strategic rules dictated by the executive team was to win every competitive bid, and customer-driven innovation was central to achieve this. METRO members therefore regularly sold concepts of new technologies and products to customers *ex ante* development to gain access to POLIS's resources: 'When you want to do some proactive development and you don't have the resource to do it, then you make that proactive development mandatory by selling it. Then by its nature, you need to deliver it and today that's what's happening' (product manager from METRO).

In addition to constraints and responses to those constraints, both groups also undermined each other's authority, which is related to mutual perceptions of superiority and inferiority with respect to contributing to innovation. The chief financial officer (CFO) disclosed how he disliked the way METRO frequently ignored the impact of their requests on POLIS's resource planning. An example is a new technology which would reduce the overhead investment per project through virtualization of a central component of the company's hardware solution installed on trains. METRO pushed this idea to development – in particular the business development manager – despite the impact on other concurrent innovation projects.

Intra-group cooperative action In contrast with competitive action between groups, we found that individuals within groups cooperated strongly by sharing knowledge and ideas leading to the development and strengthening of group-level competences. This type of group-level action was based on similarities among group members regarding organization-related group social identity. Intragroup cooperative action was triggered by organizational ambiguity and high group entitativity (that is, group distinctiveness) which inspired group members to resort to each other to autonomously enhance competences related to managing innovation. Moreover, intra-group cooperative action was also affected by intergroup competitive action because each group would try to control innovation capability by increasing their own influence and decreasing the other group's influence on innovation processes. As a consequence, groups would locally develop and routinize new practices, tools and processes driving innovation capability. See Table 3.3 for a list of artefacts produced through intra-group cooperative action.

Furthermore, we observed that in both groups, members would build constituency and seek legitimacy after coming up with new concepts and ideas for the development of new products and technologies, before going to the rest of the organization. This often occurred in informal ways and outside office hours.

Based on our evidence we found intra-group cooperative action and

Table 3.3 Examples of intra-group cooperative action and capability development

Related quote	Developed artefact	Developed by group	Impact on capability development
'With the technology roadmap, POLIS can now design products that will serve several solutions. Then we can resell it.' (Head of software development)	A strategic planning tool for the development of the company's core technology by aligning capabilities and market development.	POLIS	This tool was positively appraised by the CEO and COO and therefore integrated as a central element of innovation capability.
'It's a reference point for these discussions, on some items I will choose to take them further, either to research them, to say what has been developed, what is in the pipeline, and how we go to market and sell whatever new functionality or new product we have built.' (Product manager)	A detailed formal product development and supplier management process.	METRO	This process added more marketing and business elements to the innovation capability and created technology sourcing as a valid avenue. It became common practice after the COO championed it.
'We get it down to two, maybe three, suppliers. Then the evaluation forms are completed, we evaluate them in the table format; on technical, commercial, support. We evaluate them on how easy it is to get along with that supplier and then we either have only one left, or we use two suppliers to qualify them off each other.' (Product manager)	Supplier evaluation tool to assess external technology suppliers.	METRO	This tool further formalized the technology sourcing element of the firm's innovation capability by building on elements of the product development and supplier management process.
'This QMS process ensures from the technical point of view that we're responsible for all of the software development life cycle. There are certain documents that everybody has to produce at a certain time. We're responsible for making sure that they produce them in the right format.' (Head of programme office)	A quality management system (QMS) to ensure developed products meet specified criteria and project milestones requirements are met.	POLIS	The embedding of this QMS process in the firm's resource planning system by the COO enhanced the firm-level dissemination of gained efficiency and quality monitoring of new products and technologies.

Note: Augmented by archival data; developed artefacts were provided in codified form as a company document.

intergroup competitive action to influence innovation capability development. In the next section we elaborate on the process that links group-level action to capability development.

An Impetus to Capability Development: The Role of Organizational Groups

Prima facie, the group social identity processes at the focal firm seemed to invoke a negative atmosphere when groups engaged in social comparison and competitive action in light of which group was most influential for innovation capability. However, the impact of social identity and the consequent competitive and cooperative action on innovation capability proved beneficial for two main reasons. First, intergroup competitive action challenged people's social identity, or 'professional pride' as the CFO described it. This caused groups to perform to their best ability to position themselves as (more) important to innovation capability. For instance, when the company was experiencing pressure from its most prominent competitor in 2006, the executive team believed it needed a strategic leap driven by innovation and pushed POLIS to develop the next-generation software for communication networks. This technology would aggregate a set of available mobile networks and connect to the most optimal network during the journey of a train. The firm succeeded in delivering this solution in a relatively short time frame because it challenged POLIS by offering the task to both POLIS and an external technology supplier in the United States; an idea originating from a METRO manager.

Second, albeit there was considerable competition between the two groups, it did not overshadow all activity, because within groups a positive atmosphere existed with strong cohesion among group members (Hogg, 1993) which allowed the sharing of knowledge and ideas for the benefit of innovation capability. Group members sought legitimacy for ideas concerning new technologies or approaches to innovation management within their group to develop the constituency needed for resource allocation. In essence, the balanced allocation of time and effort between competitive action and cooperative action by both groups allowed the creation and development of competences. Because these competences were subsequently shared across groups they positively impacted upon innovation capability development on the organizational level. Generally, the POLIS group added more efficiency, structure and quality monitoring to innovation capability; whereas METRO aligned innovation with external developments and customer preferences. An example is the technology roadmap which was introduced by POLIS early in 2010 and adopted by

METRO in September 2011. Interestingly, though, METRO attempted to claim ownership of this tool because they did not deem POLIS sufficiently equipped for managing it properly because of their lack of exposure to the company's technological environment. This implies that intra-group cooperative action leading to capability development also fed back into social identity processes. See Table 3.3 for more examples of developed competences and how they influenced capability development.

Executives played an important role in integrating group competences into innovation capability by championing specific artefacts and by making them common practice within the firm. Column 4 of Table 3.3 provides illustrations of how this took place for several artefacts. It seemed that the executive team believed that the deliberate creation of organizational ambiguity in managing innovation resulted in productive outcomes for WIRELESS's innovation capability: 'Any tension where you're challenging people's professional pride is not a bad thing. But you need to be careful with how to manage it so it doesn't become vindictive' (CFO).

DISCUSSION

We began this chapter by arguing for the importance of a focus on groups in studies on the development of organizational capabilities. Our research reveals that groups do indeed matter in the development of capabilities, and we contribute to theory on organizational capabilities in several ways. First, our findings reveal how a capability that is perceived as strategically important can be the social context that instigates group social identity processes. We found these social identity processes to lead to intergroup competitive action and intra-group cooperative action. Furthermore, we show that intergroup competitive action and intra-group cooperative action can have a positive impact on capability development if there is a balanced allocation of resources between intergroup competitive and intra-group cooperative action. Finally, we suggest that group competences developed through intra-group cooperative action, which are shared with other groups, drive capability development. To help make sense of the concepts and their relationships in our data we constructed Figure 3.3 which represents our main findings.

Theoretical Implications

Organizational capability and group identity processes

We find that the perception of capability value can be an antecedent of social identity processes. The responsiveness of social identity

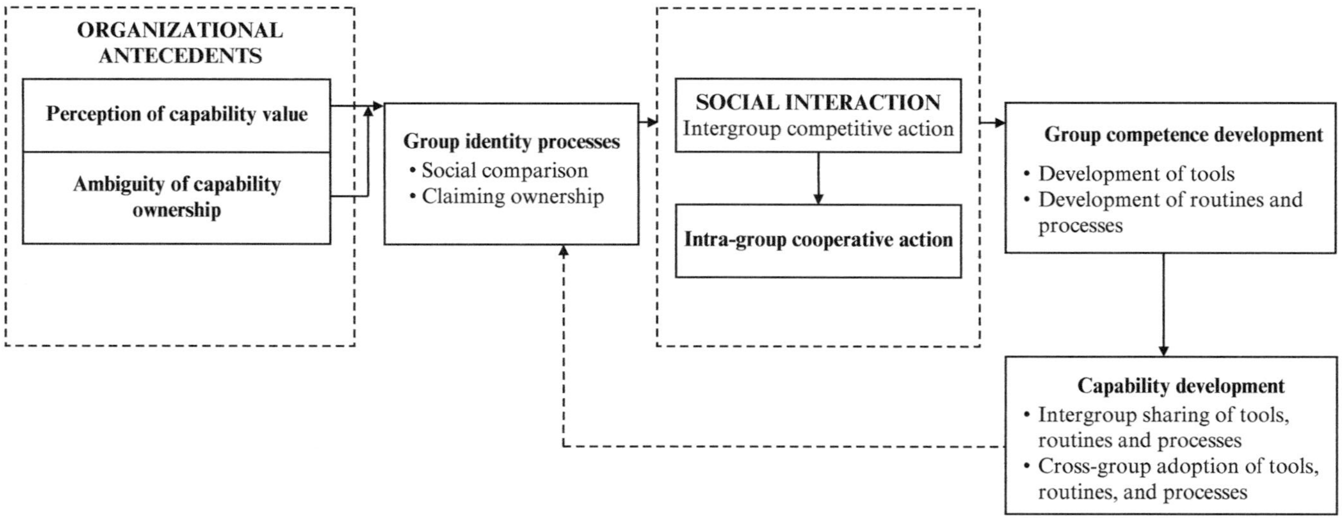

Note: The dotted arrow indicates a relationship for which we could collect only limited evidence.

Figure 3.3 A process model of the links between organizational antecedents, group identity, group (inter)action, and capability development

to immediate social contexts is an important feature of social identity theory (Fiske and Taylor, 1991) that has remained underinvestigated in management research. We propose responsiveness to organizational context to also play a role in shaping group social identity. Researchers have explained how organizational identity can offer a coherent reference of how organizational actors should behave (Ravasi and Schultz, 2006; Whetten and Mackey, 2002). These studies report that through formal identity claims, organizational leaders attempt to influence how internal audiences define and interpret the organization by positioning it within a set of legitimate social categories. Similarly, our research shows how the sense-giving function of organizational identity (that is, a company with innovation capability) was amplified by top management's claims about the importance of that identity. We add to theory by explaining how, due to intraorganizational heterogeneity, claims about organizational identity motivated group-level social identity which caused members of different groups to claim ownership of the organizational capability upon which organizational identity was built.

Furthermore, we contribute to theory on group social identity in organizations by introducing the role of ambiguity and uncertainty. In addition to the perception of capability value, group social identity was also shaped by organizational ambiguity in terms of formal ownership of innovation capability. This finding provides an opportunity for extending social psychology's 'uncertainty-reduction hypothesis' (Hogg and Mullin, 1999) to organizational contexts. This theory suggests that in addition to the notion of self-enhancement, social identity processes are also motivated by a need to reduce uncertainty about behaviours, perceptions and the position of an individual and their group in the social world, including organizations. In sum, we argue that if a capability is perceived as valuable on an organizational level, then it is likely that – acting from their social identity – different organizational groups will attempt to claim ownership of this capability. This relationship will be stronger in the presence of organizational ambiguity around formal ownership of a highly valued capability.

Competitive and cooperative action

This study's outcomes indicate a juxtaposition of two types of group-level action as a result of social identity processes: intergroup competitive action and intra-group cooperative action. Intergroup behavior is partially motivated by the need to maintain a positive social identity. This is an important underlying determinant of patterns of interaction in intergroup contexts (James et al., 1999; Ouwerkerk et al., 1999; Witt et al., 2002).

More specifically, based on our observations we contribute the distinguishing of particular forms of intergroup competitive action. The first form pertains to activities aimed at protecting groups from the influence of each other's power (that is, constraining). Recent work in entrepreneurship (Rindova et al., 2009; Goss et al., 2011) has discussed constraining as the creation of impediments to action to constrain the individual agency of others. Our research indicates that constraining and agentic responses also exist on a group level as a form of intergroup competitive action. The second form of competitive action we contribute relates to creative responses to those constraints where groups resort to utilizing hierarchical relations or dominant logic to work around constraints.

In addition to competitive action between groups we also discovered intra-group behaviour which was much more cooperative. We reject references to this form of action as 'routines' because we did not observe 'highly automatic behaviors of groups of individuals' (Salvato and Rerup, 2011: 472). Instead, we observed mindful, social interaction between group members revolving around the creation and sharing of ideas which gave birth to the development of group competences. Considering the above we suggest that a balanced allocation of resources between competitive and cooperative action can lead to productive outcomes within groups.

Group-level action and capability development

A key contribution of our work is on the role of group-level action in capability development. Where recent studies propose that groups influence strategy-making (Kaplan, 2008) and strategic change trajectories (Huy, 2011), our findings highlight how the actions of groups can have an impact on capability development. This study potentially adds to further elaboration of the micro-foundations school by highlighting the importance of groups as a separate unit of analysis in research on capabilities.

Teams in organizations tend to search viable alternatives for capability development based on their 'initial configurations of human capital, social capital, and cognition' (Helfat and Peteraf, 2003: 1002). We also observed how groups contributed to innovation capability in distinctive ways. Whereas one group added more competences in terms of efficiency, structure and the generation of intellectual property, the other group contributed competences concerning absorptive capacity (due to external exposure) and marketing (due to customer proximity). Whilst capability development has been attributed to organizational learning (Huber, 1991; Nelson and Winter, 1982) or learning-by-doing (for an overview, see Helfat and Peteraf, 2003), based on our evidence we contribute to the capability development literature by suggesting the incorporation of group social interaction as a driver of capability development.

Additionally, we suggest that social identity influences social interaction among members of the same group. An early work by Munsterberg (1913 [1999]) elucidates how group membership brings about more effective decision making. Indeed, it is 'the ability to think in terms of "we" and "us", not just "I" and "me", that enables people to engage in meaningful, integrated and collaborative organisational behaviour' (Haslam, 2001: 16). The type of collaborative interaction we observed in this case is explained by 'depersonalized social attraction' (Hogg, 1993) which connects group members based on social identity in lieu of individual characteristics. The social attraction hypothesis in social psychology posits that when uncertainty or group distinctiveness are high, or when the group is confronted by intergroup competition, then depersonalized social attraction within groups increases (Hogg and Terry, 2000). Although social psychology researchers have explained the psychological processes underpinning depersonalized social attraction in groups, we build theory by specifying how this translates into specific forms of (inter)action. In particular, we contribute to the process through which social interaction was used by group members as a platform for developing group competences. When these competences became established on the group level, top management interfered by selecting and disseminating group competences across different groups which led to the development of organizational capability.

Thus, based on our analysis we propose that if both intergroup competition and intra-group cooperation are high, this will likely lead to the development of group-level competences. Additionally, we suggest that the more group-level competences are shared and adopted across groups, then this is likely to be associated with capability development.

Practical Implications

We suggest that executives and middle managers can influence and enhance group dynamics to render them productive to capability development by configuring a number of factors related to strategy and organizational design. First, by creating ambiguity in terms of ownership of innovation, managers can instigate group social identity processes. Gioia and Chittipeddi (1991) introduced ambiguity-by-design as a powerful strategic and tactical tool to influence, facilitate and symbolize change. Similarly, we argue that a deliberate infusion of organizational ambiguity can create the uncertainty required to stimulate competition between groups and cooperation within groups for the sake of capability development. Managers should however carefully observe the allocation of resources to both competitive and cooperative action to maintain a balance that remains favourable on the organizational level.

Furthermore, we emphasize that managers need to encourage and facilitate the sharing of developed group competences with relevant actors and units within the firm in order to translate new innovation-related knowledge into meaningful action, beneficial to organization-level capability development. This is an essential step through which knowledge transits from the group level to the organizational level, and managers can achieve this through the integration, championing and institutionalization of tools, routines and processes across groups.

Limitations

Our study is prone to the usual limitations of case study research, which trades scope and generalizability for depth, detail, richness, accuracy and insight into social processes that are otherwise difficult to capture (Pratt, 2009). This implies that generalizing findings from this single case study should be handled with caution. Our process model however emerged from different data sources and a highly differentiated interviewee sample, which consolidate our confidence in the relevance of our interpretations beyond the specific boundaries of our study. In addition our research setting, which revolves around a medium-sized firm, may have influenced the emergence of investigated group-level phenomena.

These limitations however also present possibilities for future research. Future researchers could further examine the contextual antecedents – ambiguity and capability perception – to explore their relative strength and whether their impact is of a mediating or a moderating nature. In addition, future research could explore the presence and role of deviant behaviour in groups and how this affects intra-group cooperative action and capability development. Social psychologists have theorized the beneficial and detrimental effects (e.g. Wann et al., 1995) of deviants as subgroups and individuals within groups, but these have not yet been empirically investigated in organizational contexts.

CONCLUSION

The objective of our research was to build and enrich theory in the area of organizational capabilities and we believe we contribute to this literature in several important ways. First, we show how a capability that is perceived as valuable on the organizational level can motivate different organizational groups to attempt to claim ownership of this capability. We also demonstrate how this is amplified in the presence of organizational ambiguity around formal ownership of a highly valued capability. Second, we suggest that group social identity can lead to two forms of group-level action: (1) intergroup competitive action; and (2) intra-group cooperative action. Consequently, based on our findings it seems that if intergroup competition and intra-group cooperation are high, this will likely result in the development of group-level competences. Last, we highlight the importance of intra-group cooperative action as an important outcome of group social identity processes that has remained overlooked by organizational researchers. Intra-group cooperative action can lead to the development of group-level competences, and we suggest that the more group-level competences are shared and adopted across groups, the more likely it is that this will influence capability development.

REFERENCES

Augier, M. and Teece, D.J. 2009. Dynamic capabilities and the role of managers in business strategy and economic performance. *Organization Science*, 20(2): 410–421.
Burgelman, R. 1983. A model of the interaction of strategic behavior, corporate context, and the concept of strategy. *Academy of Management Review*, 8(1): 61–70.
Burgelman, R. 1994. Fading memories: a process theory of strategic business exit in dynamic environments. *Administrative Science Quarterly*, 39(1): 24–56.
Burgelman, R. 1996. A process model of strategic business exit: implications for an evolutionary perspective on strategy. *Strategic Management Journal*, 17: 193–214.
Collis, D.J. 1994. Research note: how valuable are organizational capabilities? *Strategic Management Journal*, 15(S1): 143–152.
Dosi, G., Nelson, R.R. and Winter, S.G. 2000. *The Nature and Dynamics of Organizational Capabilities*. Oxford: Oxford University Press.
Dutton, J.E., Dukerich, J.M. and Harquail, C.V. 1994. Organizational images and member identification. *Administrative Science Quarterly*, 39(2): 239–263.
Eisenhardt, K.M. 1989. Building theories from case study research. *Academy of Management Review*, 14(4): 532–550.
Eisenhardt, K.M. 1991. Better stories and better constructs: the case for rigor and comparative logic. *Academy of Management Review*, 16(3): 620.
Eisenhardt, K.M and Graebner, M.E. 2007. Theory building from cases: opportunities and challenges. *Academy of Management Journal*, 50(1): 25–32.
Eisenhardt, K.M and Martin, J.A. 2000. Dynamic capabilities: what are they? *Strategic Management Journal*, 1121: 1105–1121.
Felin, T. and Foss, N.J. 2005. Strategic organization: a field in search of microfoundations. *Strategic Organization*, 3(4): 441–455.
Felin, T. and Hesterly, W.S. 2007. The knowledge-based view, nested heterogeneity, and new value creation: philosophical considerations on the locus of knowledge. *Academy of Management Review*, 32(1): 195–218.
Fiol, C.M. and Connor, E. 2005. Identification in face-to-face, hybrid, and pure virtual teams: untangling the contradictions. *Organization Science*, 16(1): 19–32.
Fiske, S.T. and Taylor, S.E. 1991. *Social Cognition*. New York: McGraw-Hill.

Floyd, S.W. 1992. Middle management involvement in strategy and its association with strategic type: a research note. *Strategic Management Journal*, Special Issue: Strategy Process: Managing Corporate Self-Renewal, 13(Summer): 153–167.
Floyd, S.W., Cornelissen, J.P., Wright, M. and Delios, A. 2011. Processes and practices of strategising and organising: review, development and the role of bridging and umbrella constructs. *Journal of Management Studies*, 48(5): 933–952.
Galunic, D.C. and Eisenhardt, K.M. 2001. Architectural innovation and modular corporate forms. *Academy of Management Journal*, 44(6): 1229–1249.
Gavetti, G. 2005. Cognition and hierarchy: rethinking the microfoundations of capabilities' development. *Organization Science*, 16(6): 599–617.
Gavetti, G. and Levinthal, D.A. 2000. Looking forward and looking backward: cognitive and experiential search. *Administrative Science Quarterly*, 45(1): 113–137.
Gavetti, G., Levinthal, D. and Ocasio, W. 2007. Perspective – neo-Carnegie: the Carnegie School's past, present, and reconstructing for the future. *Organization Science*, 18(3): 523–536.
Gioia, D.A. and Chittipeddi, K. 1991. Sensemaking and sensegiving in strategic change initiation. *Strategic Management Journal*, 12(6): 433–448.
Gioia, D.A., Schultz, M. and Corley, K.G. 2000. Organizational identity, image, and adaptive instability. *Academy of Management Review*, 25(1): 63–81.
Gioia, D.A. and Thomas, J.B. 1996. Identity, image, and issue interpretation: sensemaking during strategic change in academia. *Administrative Science Quarterly*, 41(3): 370–403.
Goss, D., Jones, R., Betta, M. and Latham, J. (2011). Power as practice: a micro-sociological analysis of the dynamics of emancipatory entrepreneurship. *Organization Studies*, 32(2): 211–229.
Grant, R.M. 1996. Toward a knowledge-based theory of the firm. *Strategic Management Journal*, 17: 109–122.
Haslam, S.A. 2001. *Psychology in Organizations: The Social Identity Approach*. Thousand Oaks, CA: Sage Publications.
Helfat, C.E. and Peteraf, M.A. 2003. The dynamic resource-based view: capability lifecycles. *Strategic Management Journal*, 24(10): 997–1010.
Hodgkinson, G.P. and Healey, M.P. 2011. Psychological foundations of dynamic capabilities: reflexion and reflection in strategic management. *Strategic Management Journal*, 32: 1500–1516.
Hogg, M. 1993. Group cohesiveness: a critical review and some new directions. *European Review of Social Psychology*, (4): 85–111.
Hogg, M.A. and Mullin, B.A. 1999. Joining groups to reduce uncertainty: subjective uncertainty reduction and group identification. *Social Identity and Social Cognition*. Oxford: Blackwell, pp. 249–279.
Hogg, M.A. and Terry, D.J. 2000. Social identity and self-categorization processes in organizational contexts. *Academy of Management Review*, 25(1): 121–140.
Hogg, M.A. and Vaughan, G.M. 1995. *Social Psychology*. Harlow: Pearson.
Hogg, M.A. and Vaughan, G.M. 2011. *Leadership and Decision Making*. Harlow: Pearson.
Huber, G.P. (1991). Organizational learning: the contributing processes and the literatures. *Organization Science*, 2(1): 88–115.
Huy, Q.N. 2011. How middle managers' group-focus emotions and social identities influence strategy implementation. *Strategic Management Journal*, 32(13): 1387–1410.

James, K., Clark, K. and Cropanzano, R. 1999. Positive and negative creativity in groups, institutions, and organizations: a model and theoretical extension. *Creativity Research Journal*, 12(3): 211–226
Kanter, R. 1985. Supporting innovation and venture development in established companies. *Journal of Business Venturing*, 1(1): 47–60.
Kaplan, S. 2008. Framing contests: strategy making under uncertainty. *Organization Science*, 19(5): 729–752.
Locke, K. 2001. *Grounded Theory in Management Research*. London: Sage Publications.
Miles, M.B. and Huberman, A.M. 1994. *Qualitative Data Analysis*. Beverly Hills, CA: Sage Publications.
Münsterberg, H. 1913 [1999]. *Psychology and Industrial Efficiency*. Boston, MA and New York: Houghton Mifflin. Reprinted 1999, Bristol: Thoemmes Press.
Nag, R., Hambrick, D.C. and Chen, M.J. 2007. What is strategic management, really? Inductive derivation of a consensus definition of the field. *Strategic Management Journal*, 28(9): 935–955.
Nelson, R.R. and Winter, S.G. 1982. *An Evolutionary Theory of Economic Change*. Cambridge: Belknap.
Noda, T. and Bower, J.L. 1996. Strategy making as iterated processes of resource allocation. *Strategic Management Journal*, 17(S1): 159–192.
Ouwerkerk, J.W., Ellemers, N. and de Gilder, D. 1999. Group commitment and individual effort in experimental and organizational contexts. In Ellemers, N., Spears, R. and Doosje, B. (eds), *Social Identity: Context, Commitment, Content*. Oxford: Blackwell Science, pp. 184–204.
Pandza, K. 2011. Why and how will a group act autonomously to make an impact on the development of organizational capabilities? *Journal of Management Studies*, 48(5): 1015–1043.
Pandza, K. and Thorpe, R. 2009. Creative search and strategic sense-making: missing dimensions in the concept of dynamic capabilities. *British Journal of Management*, 20: S118–S131.
Patton, M.Q. 1990. *Qualitative Evaluation and Research Methods*, 2nd edn. Newbury Park, CA: Sage Publications.
Penrose, E. 1959. *Theory of the Growth of the Firm*. Oxford: Oxford University Press.
Pettigrew, A.M. 1987. Context and action in the transformation of the firm. *Journal of Management Studies*, 24(6): 649–670.
Porter, M.E. and Stern, S. 2001. Innovation: location matters. *Sloan Management Review*, 42(4): 28–36.
Pratt, M.G. 2009. From the editors: For the lack of a boilerplate: tips on writing up (and reviewing) qualitative research. *Academy of Management Journal*, 52(5): 856–862.
Ravasi, D. and Schultz, M. 2006. Responding to organizational identity threats: exploring the role of organizational culture. *Academy of Management Journal*, 49(3): 433–458.
Rindova, V., Barry, D. and Ketchen, D.J. 2009. Entrepreneuring as emancipation. *Academy of Management Review*, 34(3), 477–491.
Salvato, C. 2009. Capabilities unveiled: the role of ordinary activities in the evolution of product development processes. *Organization Science*, 20(2): 384–409.
Salvato, C. and Rerup, C. 2011. Beyond collective entities: multilevel research

on organizational routines and capabilities. *Journal of Management*, 37(2): 468–490.
Siggelkow, N. 2007. Persuasion with case studies. *Academy of Management Journal*, 50(1): 20–24.
Stopford, J.M. and Baden-Fuller, C.W.F. 1994. Creating corporate entrepreneurship. *Strategic Management Journal*, 15(7): 521–536.
Tajfel, H. 1972. Social categorization. In Moscovici, S. (ed.), *Introduction a la psychologie sociale*, Paris: Larousse, pp. 272–302.
Teece, D.J., Pisano, G. and Shuen, A. 1997. Dynamic capabilities and strategic management. *Strategic Management Journal*, 18(7): 509–533.
Turner, J.C., Hogg, M.A., Oakes, P.J., Reicher, S.D. and Wetherell, M.S. 1987. *Rediscovering the Social Group: A Self-Categorization Theory*. Oxford: Blackwell.
Tyler, B.B. and Gnyawali, D.R. 2009. Managerial collective cognitions: an examination of similarities and differences of cultural orientations. *Journal of Management Studies*, 46(1): 93–126.
Wann, D.L., Hamlet, M.A., Wilson, T.M. and Hodges, J.A. 1995. Basking in reflected glory, cutting off reflected failure, and cutting off future failure: the importance of group identification. *Social Behavior and Personality: An International Journal*, 23(4): 377–388.
Weick, K. and Roberts, K. 1993. Collective mind in organizations: heedful interrelating on flight decks. *Administrative Science Quarterly*, 38(3): 357–381.
Whetten, D.A. and Mackey, A. 2002. A social actor conception of organizational identity and its implications for the study of organizational reputation. *Business and Society*, 41(4): 393–414.
Winter, S.G. 2000. The satisficing principle in capability learning. *Strategic Management Journal*, 21(10–11): 981–996.
Winter, S. 2003. Understanding dynamic capabilities. *Strategic Management Journal*, 24(10): 991–995.
Witt, L.A., Patti, A.L. and Farmer, W.L. 2002. Organizational politics and work identity as predictors of organizational commitment. *Journal of Applied Social Psychology*, 32(3), 486–499.
Wooldridge, B. 1990. The strategy process, middle management involvement, and organizational performance. *Strategic Management Journal*, 11(3): 231–241.
Wooldridge, B. and Floyd, S.W. 1990. The strategy process, middle management involvement, and organizational performance. *Strategic Management Journal*, 11(3): 231–241.
Yin, R.K. 2003. *Case Study Research, Design and Methods*, 3rd edn. Newbury Park, CA: Sage Publications.
Zollo, M. and Winter, S. 2002. Deliberate learning and the evolution of dynamic capabilities. *Organization Science*, 13(3): 339–351.

4. Emergence, transactive memory systems and efficiency: a contingency approach

Diogo Cotta and Fabrizio Salvador

INTRODUCTION

Organizing for emergent technologies poses significant conceptual and practical challenges. Conceptually, the processes that most essentially constitute characterizations of organization imply non-trivial degrees of predictability, control and replication; not exactly properties one would associate with entities that are not yet fully discernible. This inherent tension is entirely acknowledged by several brands of organizational research, from early contingency theorists recognizing the particularities of dealing with the unpredictable (Burns and Stalker, 1961; Lawrence and Lorsch, 1967; Galbraith, 1973), to more contemporary frameworks documenting the severe difficulties of exploring the potential while simultaneously exploiting the actual (March, 1991; Levinthal and March, 1991; Benner and Tushman, 2002). What we know about emerging technologies seems to be exactly at odds with what we know about effective organizing. Unsurprisingly, organizational practice has persistently struggled with the handling of the new, the unexpected, the uncertain; all facets that fundamentally typify emergent technologies.

In order to overcome this twofold deficiency, observation has to be directed at organizational phenomena that are manifested as both process and structure. Accordingly, our conceptual efforts will be forced to adopt frameworks able to capture the dual nature of certain organizational entities that evenly reveal static and dynamic features. Emergence entails novelty, in particular new sets of relationships whose governing principles and degree of stability are still only vaguely perceptible (Hodgson, 2000). Thus, organizational structures and processes devised to deal with emergent technologies need to embed in themselves the capacity to flexibly confront the unforeseen. In that spirit, the present chapter seeks a more fine-grained understanding of how organizing might simultaneously achieve efficiency and flexibility objectives, by proposing a theoretical lens that conceives of organizations as potential transactive memory systems (Wegner, 1986). Succinctly, a transactive memory system is a set of inter-individual processes for encoding, storing and retrieving knowledge that creates specialized yet integrated individual directories of expertise (Lewis and Herndon, 2011). It is a concept that entails both processual and structural dimensions, and whose significance for group performance is empirically well established (e.g. Lewis, 2003; Ren et al., 2006; Peltokorpi, 2008). In the context of this chapter we contend, however, that its importance for organizational performance has thus far been largely neglected (notable exceptions are Jackson and Klobas, 2008; Argote and Ren, 2012). Hence, the central purpose of this chapter is to conjecture how organizational-level transactive memory systems may arise and, in particular, how they allow organizations to react efficiently to novelty. The key point of departure is that, contrary to what is implied by much research on organizational theory, an environment that is complex and unstable does not necessarily correlate with poorer organizational efficiency outcomes.

LITERATURE REVIEW

Organizations and their Context

Structural contingency theory generically claims that organizational effectiveness is determined by how congruently internal structures and processes fit with the external context (Drazin and Van de Ven, 1985).[1] In this sense, it postulates that there is no universal best way of designing an organization to successfully perform the work tasks required by particular environmental conditions. In complex and uncertain environments, organizations would do well to adopt flexible and adaptable configurations; whereas in stable and homogeneous circumstances they would be better served by employing structured and uniform arrangements (Burns and Stalker, 1961). This normative stance evidently presumes that the context provides the organization with consistent, straightforward requirements.

On the other hand, much of the literature drawing on the tensions between exploration and exploitation is based on the assumption that at least some organizations may face contradictory demands from their task environment (March, 1991; Raisch et al., 2009). Thus, research conducted on this topic supposes that the external context does not always possess a unidimensional nature that allows the organization to develop optimal structures. Instead, organizations are said to commonly encounter paradoxical contingencies that compel them into establishing operational structures and processes that strive for high levels of efficiency and, concurrently, provide them with the ability to remain flexible and adjustable (Adler et al., 2009). Unfortunately, these sorts of structures and processes are not easy to come by and the solutions offered by the literature have, we argue, remained at too abstract a level to induce a satisfactory awareness of the challenges imposed on organizations facing multidimensional environments on a routine basis. By and large, the literature agrees that, at least in the short term, flexibility-enhancing structures are negatively correlated with efficient performance; while measures to improve efficiency tend to result in less adaptable organizational processes. For this reason, many analyses end up recommending either a structural separation of activities (Tushman and O'Reilly, 1996; O'Reilly and Tushman, 2004) or a sequential partitioning of their performance (Duncan, 1976; Adler et al., 1999; Gupta et al., 2006). This chapter seeks to address organizational contexts that are more nuanced in order, hopefully, to offer a richer view of this relationship.

In our view, the structuring processes that are quintessentially part of organizing are juxtaposed to the uncertainty intrinsic in emergent technologies, and the conclusion is that efficiency and flexibility are not necessarily mutually exclusive. The chapter is fundamentally grounded in the tenets of contingency theory in that it surmises that the level of stability inherent in the environment effects particular internal structures and processes. By 'internal structures' is meant the set of social relationships existing between members of the organization (Luthans and Stewart, 1977), which would include authority and reporting relationships, behaviours mandated by rules, patterns in decision-making and knowledge-sharing practices (Donaldson, 1996). By 'internal processes' is meant the set of formalized and actual activities performed by the organization in order to achieve its goals.

Specifically, the chapter argues that an organization's transactive memory system is capable of mediating the tensions between environmental dimensions that require flexibility and swift adaptability, and those that demand reliability and efficiency. In this vein, the chapter rejects the assumption that these inevitably represent incompatible organizational objectives. It suggests, instead, that to the extent that environmental contingencies mandating flexible dispositions for effective performance induce the surfacing and development of transactive memory systems, instability may inadvertently generate internal structures and processes able to give rise to optimal efficiency outcomes. In so doing, this conjecture tries to go beyond traditionally proposed frameworks, such as those provided by structural (Puranam et al., 2006) and cyclical (Venkatraman et al., 2007 [2005]) framings of ambidexterity issues. These streams do their utmost to circumvent the apparent incongruity between certain task requirements inherent in the environment by recommending their spatial or temporal separation within an organization's operational structuring. The ingenuity of these solutions notwithstanding, they fail to take into account endeavours for which both emergence and efficiency goals represent equally important and synchronous strains on the workflow.

The occurrence of these situations is obviously not solely restricted to organizations that experiment with emergent technologies, in the traditional sense of the term. Organizations that, for instance, devote a non-trivial part of their production efforts to tailor-made customer requirements are likewise unable to break up their operational functioning into distinctly structured chunks, each designed to address specific environmental characteristics. A customized software design firm cannot temporally sequence its new product development endeavours and its exploitation activities, because each customer request may, to a non-trivial extent, entail the development of a new product or, at the very least, dealing with something novel for the organization. A construction firm is not able to spatially disconnect the structures directed at dealing with repetitive and homogeneous tasks from those devised to handle uniqueness. The workflow interdependencies which their environment subjects them to are so intricate that their structures need to be able to time and again generate new products and production processes. This chapter proposes that, in order to fully understand how effective performance can result from environmental heterogeneity, one needs to deploy concepts that accommodate both structural and processual ontological features. Transactive memory systems represent one such concept.

Transactive Memory Systems: From Team to the Organization

The notion of transactive memory was first introduced by Wegner (1986, 1995) and Wegner et al. (1991) in a series of studies about the division of cognitive labour that tends to arise among individuals engaged in intimate contact. It was noticed that when people with intersecting concerns were somehow bound to each other, they increasingly leant on one another to access and store information from different domains. This process of codification, preservation and retrieval progressively created a system of differentiated expertise that was integrated by the fact that each member was aware of the others' knowledge. Further studies provided empirical support for the thesis that the operation of these systems had positive impacts for collective performance, particularly within organizational settings (Faraj and Sproull, 2000; Hollingshead, 2000; Lewis, 2004; Peltokorpi, 2008).

Transactive memory systems are composed of task–expertise–person units (Brandon and Hollingshead, 2004). These are said to structure the group processes that make the need to perform given tasks correspond with the recognition of specific knowledge requirements and the concomitant identification of respective individual holders. The natural development of the system creates increasingly refined directories of expertise, together with more responsive processes for effectively allocating and communicating relevant knowledge. Thus, a transactive memory system entails both a structure of defined, differentiated responsibilities for knowledge work existent at a given point in time, and a set of processes that stimulate its continuous updating and employment in multiple settings. The proper operation of transactive memory systems allows individual components to focus in circumscribed information and knowledge domains, thereby eliciting permanent specialization. It also implies that individuals are aware and make use of each other as cognitive aids, thus increasing the informational content available to a single person. Processes of encoding, storing and retrieving information ensure that the system is adjustable to the specificities of each task, in that it swiftly facilitates coordinated action to guarantee that the knowledge brought to bear on it is the most relevant and accurate possible. Research indicates that familiarity (Agkun et al., 2005; Lewis, 2004) and frequency of communication (Kanawattanachai and Yoo, 2007; Peltokorpi and Manka, 2008; Jackson and Moreland, 2009) are observable antecedents of the development of transactive memory systems. Still, the ultimate antecedent for the emergence of these systems of shared knowledge within groups is interdependence (Levine and Moreland, 1999; Moreland, 1999; Hollingshead, 2001; Zhang et al., 2007). It is not group membership per se that inspires the development of transactive memory systems, rather it is the interdependence implied by group membership that originates the mechanisms by which it thrives. This interdependence may be determined by formal reward structures or, as often occurs in complex and non-routine endeavours, by convoluted task interdependencies that conspire to merge one member's output with another's input in not always predictable ways.

Although the concept began to be deployed in the context of small team research, and even though most of the empirical research has been conducted at that level of analysis, the potential usefulness of transactive memory systems for organizational theory has been widely recognized (Jackson and Klobas, 2008; Lewis and Herndon, 2011; Ren and Argote, 2011), even if still conceptually underspecified. This chapter explicitly generalizes the concept to the organizational level of analysis and suggests that an environment that generates non-repetitive and uncertain requirements might stimulate the emergence and development of organization-wide transactive memory systems. Through this effect, this type of environment may actually positively impact upon an organization's efficiency outcomes, to the extent that the former induces quick, coordinated expertise transactions within the production workflow.

STATEMENT OF PROPOSITIONS

When the environment determines task requirements that are heterogeneous, not fully predictable and complex, contingency theory contends that effective performance is achieved by means of organic-type structures (Burns and Stalker, 1961). Underlying this conclusion lies the reasoning that if what is asked of the organization is largely non-routine, mechanistic designs are bound to prove too rigid to deal with frequent novelty. In a consultancy firm, for instance, whenever a new customer project is submitted it has firstly to be evaluated. Then, if accepted by the firm, a formal definition of the project's aims and resources has to be established. A detailed plan for implementation has to be devised and, finally, it has to be executed. To a non-trivial extent, every single materialization of these otherwise typical processes is unique. In a manufacturing enterprise in which a significant proportion of orders require design and engineer-to-order activities, fully formalized workflow processes become obsolete as soon as unexpected events begin to occur. Sales cannot agree a product definition with a client without being aware of the technical and manufacturing capabilities of the firm. The engineering function is not able to develop product and process design plans without being fully informed about what exactly was agreed upon with the client; and manufacturing is not able to thwart unexpected production contingencies without the continuous support of sales and engineering. This prevents these sorts of organizations from fully departmentalizing their workflow, because interdependencies are just too large and unpredictable. Since relevant expertise is bound to be organizationally distributed, it is not possible to fully predetermine what sort of expert knowledge will be useful in a specific project. People in sales have to continually understand what is technically feasible and profitably deliverable; and people in manufacturing have to be enduringly aware of what product definition was discussed with the client. No department can possess all relevant knowledge and information in order to deliver the one-of-a-kind product or service. Moreover, a lot of expertise may remain tacit, either because it is not used frequently enough to become externalized in a formal process, or because it is tacit by nature. In any case, departments have to be able to leverage expertise located in other departments or, to put it more generally, other formal elements in the workflow.

These contingencies create a huge degree of interdependence among organizational members. In a case study of legendary product design firm IDEO, Hargadon and Sutton (1997) describe how the delivery of each challenging customer request had to leverage expertise and casual collaboration that stretched way beyond the team formally responsible for the project. Complexity tends to drive greater need for information processing (Galbraith, 1974; Kmetz, 1984). The relevance of particular pieces of information is transient and contextual and, therefore, no component of the order-performing process can afford to act independently. To the extent that every component recognizes this mutual epistemic dependency, and insofar as intra-organizational communication avenues are not irredeemably stifled, this requirement has the potential to create internal dynamics that stimulate the exchange of knowledge and expertise across the workflow of task performance.

In this sense, this chapter proposes that high degrees of cognitive interdependency generated by a complex and unpredictable environment may give rise to the operation of organization-wide transactive memory systems, which at once structure knowledge-sharing activities and envelop the processes by which these adjust to novelty. In contexts in which complexity, non-routineness and need to coordinate effort coalesce, distinct organizational functions have to mutually rely on one another to encode, store and retrieve the appropriate expert knowledge and information (Brandon and Hollingshead, 2004). In this sense, the chapter draws on contingency theory's assertion that organizations develop structures and processes that, in order to achieve effective performance, evolve towards fit with the characteristics of their environment (Siggelkow, 2002). Furthermore, the chapter accepts Weick's (1988) conception that internal perceptions of environmental contingencies are the chief drivers of enacted organizational structures and processes. From this perspective, the operation of convoluted knowledge-exchanging practices is, thus, seen to reflect the lack of task formalizability implied by emergence. The fact that complex transactive memory systems do in fact seem to disguisedly arise in contexts of high uncertainty (Hargadon and Sutton, 1997) is evidence that particularly acute environmental pressures may counteract natural internal tendencies towards simplicity and routinization of intra-organizational expertise management practices (Aldrich and Pfeffer, 1976). Hence, this chapter contends that:

Proposition 1: The more complex and unpredictable the environment facing an organization, the greater the potential for the development of its transactive memory system.

In addition to representing flexible, dynamic structures of organizational knowledge distribution, transactive memory systems also allow organizational members to specialize in their respective expertise domains. This contributes to more efficient and reliable processes. However, in the context of these systems specialization is not necessarily related to a breakdown in integration levels (Lawrence and Lorsch, 1967), because a well-operating transactive memory system makes sure that relevant expertise can be efficiently located and retrieved. Interfunctional cognitive division of labour is accompanied by the organization's ability to swiftly leverage knowledge stored in every relevant element of the workflow. By reaping the benefits of functional specialization and enjoying the advantages from an informal, flexible and effective coordination mechanism (Faraj and Xiao, 2006), the organization necessitates fewer resources than it otherwise would. In the example of the engineering firm, if the function involved in the process of deciding to accept or decline the customer's request neglects the knowledge embedded in the other two functions, the organization runs the risk of wasting valuable time and money. Conversely, if the technical or manufacturing functions go about their business without losing a continuous awareness of the customer knowledge embedded in the sales function, the organization can save considerable resources in start-overs and maintenance.

Hence, this chapter also proposes that transactive memory systems are associated with better organizational efficiency outcomes. Empirical studies at the group level of analysis have long established the links between team transactive memory systems and efficient performance (Faraj and Sproull, 2000; Rau, 2005; Ren et al., 2006). A recently performed computer simulation replicated the results at the organization level of analysis (Miller et al., 2012). This chapter formally advances that flexible structures may equally be associated with efficient outcomes to the extent that they stimulate the development of organization-wide transactive memory systems:

Proposition 2: The greater the level of development of an organization's transactive memory system the more efficient its productive outcomes will be.

DISCUSSION

This chapter drew on a theoretical concept that simultaneously entailed structural and processual components. An organizational transactive memory system implies both a differentiated structure of knowledge specialization and the dynamic processes that ensure its repeated updating and usage. In this sense, the chapter hopes to refresh the application of contingency theory to organizational research, by inviting the use of concepts that concurrently denote structure and agency. The deployment and study of such concepts allows for a more adequate consideration of multidimensional environments that seemingly carry paradoxical contingencies. Given that there exist several available instruments to measure the existence and development level of transactive memory systems in teams (Moreland and Myaskovsky, 2000; Austin, 2003; Lewis, 2003), which could easily be translated into the organization level of analysis, the chapter offers two straightforward propositions for empirical consideration.

By conjecturing how, through its effects on the development of a system of knowledge specialization and sharing practices, an uncertain environment may actually generate efficiency, the chapter also rejects the assumption that flexibility and optimal resource utilization are necessarily negatively correlated. Furthermore, the chapter offers some hope in improving our understanding of organizations that are not able to separate their exploitation from their exploration activities, their mechanistic from their organic structures and processes. In contexts in which past experience cannot serve as sole input for organizational formalization, as organizations are not able to entirely foresee the exact contours of the products or services they will be asked to deliver, the concept of transactive memory systems helps us to make sense of how effective and efficient outcomes are actually achieved.

If uncertainty is the name of the game, then any managerial action directed at eliciting and stimulating the development of systems of coordinated expertise transaction will foster the organization's ability to deal with the unforeseen. Even though the chapter has opted to focus on environmental drivers, there are evident organizational actions capable of nurturing the sustained development of transactive memory systems. Familiarity between members is a necessary condition for an integrated directory of expertise to emerge. In this sense, organizations are incentivized to maintain a stable set of employees in order to provide the conditions for each member's knowledge about one another's expertise to emerge and gain in accuracy. Similarly, employees who are intrinsically open to cooperation with others and have the requisite intellectual ability to be aware of different informational domains are more likely to prompt the emergence and operation of transactive memory systems. Hence, organizations would do well to integrate these psychological characteristics in their recruitment and training policies. More fundamentally, the chapter aimed to propose a way out of the conceptual difficulties that plague the issue of organizing for emergent technologies. The first step is to recognize the potential of these systems as theoretical lenses through which to conceive of organizational knowledge exchange activities. This chapter constitutes a contribution towards that objective.

NOTE

1. Albeit from a different perspective, fit is also a key concept in evolutionary theories of organization (Hannan and Freeman, 1984; Aldrich and Pfeffer, 1976). However, within this strand managerial agency in adaptation is neglected in favour of environmental selection as the principal fit mechanism.

REFERENCES

Adler, P., Goldfotas, B. and Levine, D. (1999), 'Flexibility versus Efficiency? A Case Study of Model Changeovers in the Toyota Production System', *Organization Science*, 10(1): 43–68.

Adler, P., Benner, M., Brunner, D.J., MacDuffie, J.P., Osono, E. Staats, B., Takeuchi, H.,Tushman, M, and Winter, S. (2009), 'Perspective on the Productivity Dilemma', *Journal of Operations Management*, 27: 99–113.

Agkun, Byrne, J., Keskin, K., Lynn, G. and Imamoglu, S.Z. (2005), 'Knowledge Networks in New Product Development Projects: A Transactive Memory Systems Perspective', *Information Management*, 42(8): 1105–1120.

Aldrich, H.E. and Pfeffer, J. (1976), 'Environments of Organizations', *Annual Review of Sociology*, 2: 79–105.

Argote, L. and Ren, Y. (2012), 'Transactive Memory Systems: A Microfoundation of Dynamic Capabilities', *Journal of Management Studies*, 49(8): 1375–1382.

Austin, J. (2003), 'Transactive Memory in Organizational Groups: The Effects of Content, Consensus, Specialization and Accuracy in Group Performance', *Journal of Applied Psychology*, 88(5): 866–878.

Benner, M.J. and Tushman, M.L. (2002), 'Process Management and Technological Innovation: A Longitudinal Study of the Photography and Paint Industries', *Administrative Science Quarterly*, 47: 676–706.

Brandon, D. and Hollingshead, A. (2004), 'Transactive Memory Systems in Organizations', *Organization Science*, 15(6): 633–644.

Burns, T.M. and Stalker, G.M. (1961), *The Management of Innovation*, London: Oxford University Press.

Donaldson, L. (1996), 'The Normal Science of Structural Contingency Theory', in Clegg, S. and Hardy, C. (eds), *Handbook of Organization Studies*, Thousand Oaks, CA: Sage Publications.

Drazin, R. and Van de Ven, A. (1985), 'Alternative Forms of Fit in Contingency Theory', *Administrative Science Quarterly*, 30: 514–539.

Duncan, D.B. (1976), 'The Ambidexterous Organization: Designing Dual Structures for Innovation', *Management of Organization*, 1: 167–188.

Faraj, S. and Sproull, L. (2000), 'Coordinating Expertise in Software Development Teams', *Management Science*, 46(12): 1554–1568.
Faraj, S. and Xiao, Y. (2006), 'Coordination in Fast-Response Organizations', *Management Science*, 52(8): 1155–1169.
Galbraith, J. (1973), *Designing Complex Organizations*, Boston, MA: Addison-Wesley.
Galbraith, J. (1974), 'Organization Design: An Information Processing View', *Interfaces*, 4(3): 28–36.
Gupta, A.K., Smith, K.G. and Shalley, C.E. (2006), 'The Interplay between Exploration and Exploitation', *Academy of Management Journal*, 4: 693–706.
Hannan, M.T. and Freeman, J. (1984), 'Structural Inertia and Organizational Change', *American Sociological Review*, 49: 149–164.
Hargadon, A. and Sutton, R.I. (1997), 'Technology Brokering and Innovation in a Product Design Firm', *Administrative Science Quarterly*, 42: 716–749.
Hodgson, G.M. (2000), 'The Concept of Emergence in the Social Sciences: Its History and Importance', *Emergence*, 2(4): 65–77.
Hollingshead, A. (2000), 'Perceptions of Expertise and Transactive Memory in Work Relationships', *Group Processes and Intergroup Relations*, 3: 257–267.
Hollingshead, A. (2001), 'Cognitive Interdependence and Convergent Expectations in Transactive Memory', *Journal of Personality and Social Psychology*, 81(6): 1080–1089.
Jackson, M. and Moreland, R.L. (2009), 'Transactive Memory in the Class Room', *Small Group Research*, 40(5): 508–534.
Jackson, P. and Klobas, J. (2008), 'Transactive Memory Systems in Organizations: Implications for Knowledge Directories', *Decision Support Systems*, 44: 409–424.
Kanawattanachai, P. and Yoo, Y. (2007), 'The Impact of Knowledge Coordination on Virtual Team Performance Over Time', *MIS Quarterly*, 31(4): 783–808.
Kmetz, J.L. (1984), 'An Information Processing Study of a Complex Workflow in Aircraft Electronics Repair', *Administrative Science Quarterly*, 29(2): 255–280.
Lawrence, P.R. and Lorsch, J.W. (1967), 'Differentiation and Integration in Complex Organizations', *Administrative Science Quarterly*, 12(1): 1–29.
Levine, J.M. and Moreland, R.L. (1999), 'Knowledge Transmission in Work Groups: Helping Newcomers to Succeed', in Thomson, L.L., Levine, J.M and Messick, D.M. (eds), *Shared Cognition in Organizations: The Management of Knowledge*, London: Lawrence Erlbaum.
Levinthal, D. and March, J.G. (1991), 'The Myopia of Learning', *Strategic Management Journal*, 14: 95–112.
Lewis, K. (2003), 'Measuring Transactive Memory Systems in the Field: Scale Development and Validation', *Journal of Applied Psychology*, 88(4): 587–604.
Lewis, K. (2004), 'Knowledge and Performance in Knowledge-Worker Teams – A Longitudinal Study of Transactive Memory Systems', *Management Science*, 50(11): 1519–1533.
Lewis, K. and Herndon, B. (2011), 'Transactive Memory Systems: Current Issues and Future Research Directions', *Organization Science*, 22(5): 1254–1265.
Luthans, F. and Stewart, T. (1977), 'A General Contingency Theory of Management', *Academy of Management Review*, 2(2): 181–195.
March, J.G. (1991), 'Exploration and Exploitation in Organizational Learning', *Organization Science*, 2(1): 71–87.
Miller, K., Pentland, B. and Choi, S. (2012), 'Dynamics of Performing and

Remembering Organizational Routines', *Journal of Management Studies*, 49(8): 1536–1558.

Moreland, R.L. (1999), 'Transactive Memory: Learning Who Knows What in Work Groups and Organizations', in Thompson, L.L., Messick, D. and Levine, J. (eds), *Sharing Knowledge in Organizations*, Hillsdale, NJ: Lawrence Erlbaum.

Moreland, R.L. and Myaskovsky, L. (2000), 'Exploring the Group Benefits of Group Training: Transactive Memory or Improved Communication', *Organizational Behavior and Human Decision Processes*, 82: 117–133.

O'Reilly, C. and Tushman, M. (2004), 'The Ambidextrous Organization', *Harvard Business Review*, 82(4): 74–81.

Peltokorpi, V. (2008), 'Transactive Memory Systems', *Review of General Psychology*, 12(4): 378–394.

Peltokorpi, V. and Manka, M.L. (2008), 'Antecedents and the Performance Outcomes of Transactive Memory in Daycare Work Groups', *European Psychologist*, 13(2): 103–113.

Puranam, P., Singh, M., Zollo, M. (2006), 'Organizing for Innovation: Managing the Coordination–Autonomy Dilemma in Technology Acquisition', *Academy of Management Journal*, 49(2): 263–280.

Raisch, S., Birkinshaw, J., Probst, G. and Tushman, M. (2009), 'Organizational Ambidexterity: Balancing Exploitation and Exploration for Sustained Performance', *Organization Science*, 20(4): 685–695.

Rau, D. (2005), 'The Influence of Relationship Conflict and Trust on the Transactive Memory: Performance Relation in Top Management Teams', *Small Group Research*, 36(6): 746–771.

Ren, A. and Argote, L. (2011), 'Transactive Memory Systems 1985–2010: An Integrative Framework of Key Dimensions, Antecedents and Consequences', *Academy of Management Annals*, 5(1): 189–229.

Ren, D., Carley, K.M. and Argote, L. (2006), 'The Contingent Effects of Transactive Memory: When is it More Beneficial to Know What Others Know', *Management Science*, 52(5): 671–682.

Siggelkow, N. (2002), 'Evolution Towards Fit', *Administrative Science Quarterly*, 47: 125–159.

Tushman, M. and O'Reilly, C. (1996), 'Ambidextrous Organizations: Managing Evolutionary and Revolutionary Change', *California Management Review*, 38(4): 8–30.

Venkatraman, N., Lee, C. and Lyer, B. (2007 [2005]), 'Strategic Ambidexterity and Sales Growth: A Longitudinal Test in the Software Sector', unpublished manuscript (earlier version presented at the Academy of Management Meetings, 2005).

Wegner, D. (1986), 'Transactive Memory: A Contemporary Analysis of the Group Mind', in Mullen, B. and Goethals, G. (eds), *Theories of Group Behavior*, New York: Springer.

Wegner, D. (1995), 'A Computer Network Model of Human Transactive Memory', *Social Cognition*, 13(3): 319–339.

Wegner, D., Erber, R. and Raymond, P. (1991), 'Transactive Memory in Close Relationships', *Journal of Personality and Social Psychology*, 61(6): 923–929.

Weick, K.E. (1988), 'Enacted Sensemaking in Crisis Situations', *Journal of Management Studies*, 24(4): 305–317.

Zhang, Z., Hempel, P.S., Yu-Lan, H. and Tjosvold, D. (2007), 'Transactive Memory System Links, Work Team Characteristics and Performance', *Journal of Applied Psychology*, 92(6): 1722–1730.

5. Tools to support technology intelligence processes in integrated technology service providers

Natalia Maya

INTRODUCTION

This chapter contributes to the literature on technology intelligence (TI) by describing the TI selected tools for an integrated technology service provider (ITSP) in the chemical and pharmaceutical area. ITSPs are defined here as companies created and owned by a mother company, and their main mission is to offer technological services to create value for their holding company. The goal of TI processes is to identify potential business opportunities and threats arising from technology development (Rohrbeck, 2006). The practical application of TI in industry lags, though, despite the increasing interest in TI from both scholars and industry practitioners, and not too many companies seem to be applying intelligence techniques systematically or running a regular intelligence system (Arman and Foden, 2010).

As TI is a complex process, it needs to be streamlined and supported by tools. There is a big variety of them, both qualitative and quantitative, and companies usually struggle in choosing the tools to satisfy their needs. The main goal here is therefore to create a tool box for the analysed ITSP to identify emerging technologies (that is, opportunities and threats) outside the company boundaries. The first step is to identify the type of tools needed, by understanding the particular characteristics of the ITSP operating method and defining the main challenges faced when implementing TI processes. Then potential appropriate tools are analysed and selected, and finally a pilot test is carried out with the tools chosen.

This chapter is organized in the following way. It first introduces the main concepts, then gives a short overview of TI and ITSP, describing the TI process and how companies usually approach it. The chapter then focuses on the challenges that ISTPs usually face when implementing TI systems. The tools analysed are summarized, and conclusions drawn.

TECHNOLOGY INTELLIGENCE AND ITSP

TI is a core process of technology management and can be defined as the acquisition, assessment and communication of information on the development of new technologies in both the internal and external environments, so that an organization becomes aware of technology threats and opportunities. This is in turn a key input for strategy planning and innovation processes (Lichtenthaler, 2006; Kerr et al., 2006). In other words, TI provides organizations with a means to monitor developments, look for new technologies, and pass this information to the decision makers and planners in the organization (Kerr et al., 2006).

It is important to acknowledge that TI is complemented by other company functions related to business intelligence such as market intelligence, competitor analysis, regulations, and so on (Kerr et al., 2006). In fact, the significance of business intelligence increases at the expense of technology intelligence as the innovation process develops (Runge, 2006).

Due to the fast pace of technological, social and economic change that generates uncertainty and complexity, TI is a challenging task and companies are finding it increasingly difficult to keep abreast of the latest technology development and trends (Runge, 2006). Nowadays, the same strategic planning period encompasses more generations of new innovations (Caldwell et al., 2005). For this reason, having the appropriate tools to collect and analyse large amounts of data, as well as to communicate and act on the intelligence (obtained from those data) is of critical importance (Behkami and Daim, 2012).

Usually companies deploy a combination of in-house and external tools to streamline TI processes. According to a study carried out by the European Union in 2002 the use of simple tools that do not require too much effort to implement seems to predominate. It shows that qualitative tools such as brainstorming, intuitive thinking, or expert consultations are widely used; likewise quantitative tools such as patent and publication analyses, benchmarking exercises or market forecasts are among the most popular ones. Scenarios, simulation-based methods and mini-Delphi are some of the most complex tools used.

Selecting the right tools is a challenging process not only because there are many variables to consider but also because there is a wide range of options to choose from. It is difficult to judge which are more appropriate, which better fit companies' needs, or how they can be linked together, as this depends on the maturity level of the technologies to watch, the uncertainty level, the time frame, the personnel and budget available, the organizational structure, previous experience, and so on. For this reason it is necessary to look at TI systems in the context of specific groups of companies to be aware of their conditions and challenges, and to identify what supporting tools may be adopted. As mentioned earlier the focus here will be on ITSPs.

When implementing a TI system to serve the ITSP itself and its clients, ITSPs face particular challenges when compared to independent technology service providers (IDTSPs). Unlike IDTSPs, where the TI services are part of the core business, TI services within ITSPs are in most cases created to support the core businesses of the mother company, therefore their focus cannot be only on growth and profits. Thus, while an IDTSP deals with each customer independently and can customize its service according to the needs of its client, and get specialists and proprietary tools for this purpose, the TI system of an ITSP has to integrate the intelligence needs of different customers, always taking into account what is more likely to strengthen the mother company's core competences and create value for it.

TECHNOLOGY INTELLIGENCE PROCESS

The concept of TI is better understood by taking a look at how TI is actually done. In this case, a conceptual model developed by Kerr et al. (2006) is used. The model consists of three tiers: framework, system and process. The framework level maps the information requirements and knowledge gaps of the decision makers to the intelligence activities of an organization. This allows determining who needs information and what they want to know. The system level refers to the system architecture and its operational modes: mine, trawl, target and scan (Table 5.1). The process level is how a technology intelligence system actually runs.

A TI system uses all four modes concurrently, but each organization will have a different mix of them as TI activities must be adapted to the needs and strategic orientation of a company. According to the European Commission, there is a notable lack of future intelligence activities within the European private sector, despite companies' efforts to improve their ability to detect weak signals and anticipate future trends. In the analysed ISTP there was also a special interest in the external TI activities, and for this reason attention was focused on the scan and target modes, as highlighted in Table 5.1. Besides, they were considered to have the greatest potential for identifying threats and opportunities for the business.

There are several descriptions of the TI processes in the literature, which differ mainly in the number of steps. The steps proposed by Kerr et al. (2006) are: (1) coordinate; (2) search for information; (3) filter; (4) analyse; (5) document; and (6) disseminate. According to Lichtenthaler (2006), the basic steps all process descriptions have in common are: (1) determine

Table 5.1 Intelligence search modes: different ways of looking for information

	Internal	External
Open-ended	Trawl Information may be contained within the company but searchers do not know where it is kept	Scan The company is looking for relevant information that has not been identified
Directed	Mine Information is contained within the company and searchers know where to find it	Target The company is looking for specific information outside its boundaries

Source: Mortara et al. (2007).

Table 5.2 Steps usually followed in an intelligence process

Clarification of needs	Data collection	Analysis	Communication

information needs; (2) acquire information; (3) analyse information; and (4) communicate information (Table 5.2). These steps will be taken as the reference for this chapter.

Companies do TI for a variety of reasons. Usually, as mentioned before, it is seen as a means to detect and warn companies about potential technology-related business threats and opportunities, but companies also do TI to better understand the social and cultural context where the technology is meant to be applied, to generate new ideas, to open the company to the outside world and find strategic partners, and so on.

Generally speaking, companies carry out TI activities in an ad hoc or continuous basis. The ad hoc approach is an on-demand approach. For example, when a company carries out TI because a business or service is losing its share in the market, more information is required to make a decision as to what new R&D projects to undertake, or the business or technology strategy is being redefined. When necessary, temporary groups are formed to carry out these activities. The alternative approach occurs when TI activities are done regularly, and there is a variety of approaches, from a completely centralized to a decentralized approach. In the centralized approach, companies have a unit focused on this activity, which coordinates the whole intelligence process. These units are usually long-term-oriented and are supported by internal and external networks. Generally this is done at corporate level by specialists (Ahlgrimm, 2008). An alternative and decentralized approach is, for example, when intelligence activities are embedded in the research and development (R&D) activities. In this case TI is mainly concerned with providing background information (such as competitor analyses) for the decision-making processes and is carried out by the groups (and subgroups) and the business units themselves. There are usually few people who perform TI tasks full-time (if any), but there are many people who have intelligence roles as part of their job.

People who perform TI activities with the goal of discovering new technologies outside the company are usually referred to as 'scouts'. Some companies have created a scout network worldwide to keep abreast of what competitors are doing. Some examples are Deutsche Telekom and the Kodak European Research (closed in 2009). The former established a network of technology scouts worldwide to create monitoring and scanning capabilities which consists of a balanced distribution of internal and external, part-time or full-time scouts around the world. Having a scouts-based TI strategy offers many advantages: scouts are usually located where technology development is happening, and this may reduce the time lag between development and detection of technologies; scouts are more likely to have personal contact with the inventions and inventors, and this may facilitate the sourcing of technology (Rohrbeck, 2006). However, many companies cannot afford to have scouts spread worldwide, as is the case for the analysed ITSP, and the use of TI tools to support the already available staff to strengthen their monitoring and scanning capabilities becomes even more critical.

CHALLENGES FACING ITSPs WHEN IMPLEMENTING TI SYSTEMS

As mentioned before, ITSPs are defined here as companies created and owned by a mother company, and their main mission is to offer a wide range of technological services to create value for their holding company. A company may decide to have its own technology service provider because it brings many benefits such as: (1) no risk of supplier dependency; (2) no risks of information leakage and possible losses of technology-based core competencies; and (3) no risk of hindering a firm's ability to capture the full potential for new product ideas. In general terms, the TI processes in ITSPs may cover two different 'levels': (1) the TI system is meant to serve only the ITSP; and (2) the TI system is expected to serve the ITSP itself and internal customers (that is, other units or groups that also belong to the mother company).

In this specific case an ITSP in the chemical and pharmaceutical area was taken as reference to discover what particular characteristics should be taken into account when choosing appropriate TI tools. Several aspects that make the TI activities more challenging were defined. In this section the most critical ones are briefly described.

There is a High Diversity of Areas to Watch

When the system is meant to serve only the ITSP, it is expected to be a continuous and systematic watching process of technological developments focused on the strategic areas of the ITSP. However, it faces additional challenges because unlike other companies where the intelligence gathered is meant to provide input for a specific set of research fields, ITSPs are usually serving a wide range of clients from different sectors (especially when the offered services are useful for a large variety of industries). In the specific case of an ITSP in the chemical area, their services may be required by clients in the agriculture, manufacturing, construction, materials and health care sectors, and so on. Thus, even though the monitoring system is meant to serve only the ITSP, the number and variety of technologies to scan and monitor tend to be high. Usually, companies don't have the financial and human resources to keep abreast of technology developments in a high diversity of areas.

In the case where the TI system is expected to serve not only the ITSP itself but also internal customers (also part of the holding), as is the case for the ITSP of reference, it may be necessary to add an 'ad hoc' TI service to the systematic TI system mentioned before. This ad hoc TI system is meant to respond on demand and fast to the client's requirements, and can be working in different sectors. In this case the diversity of areas to watch is even larger and the implementation of a TI system is more challenging.

TI Tends to be Driven by Customer Needs rather than Strategic Interest

The fact that ITSPs usually work as project-based businesses may cause more attention to be paid to short- and medium-term technology customer needs, often at the expense of other projects of high importance for the company in the long term. This also has an impact on the TI system as the activities may end up being only concerned with looking for solutions to immediate customer problems, rather than innovation opportunities, and being only steered by the customer's needs, rather than by the strategic interests. In this scenario it is also likely that the balance between bottom-up and top-down intelligence flows is lost, as the monitoring activities become strongly oriented to provide clients' need-based intelligence to the top management, in some cases to the detriment of the top-down strategy-oriented intelligence. Thus, in ITSPs it is a challenge to keep TI activities linked with strategic processes, especially when the ITSP also serve internal clients, as not only one but several strategies need to be taken into account.

Ad Hoc and Reactive Way of Working to Respond to Clients' Requirements

TI is one of those activities where many people within a company are indirectly involved (from top managers to scientists and researchers). Regardless of the company's centralized, decentralized or ad hoc approach to TI, there are always employees within the company who, as part of their job, need to keep themselves updated about what is going on in their field, and information gathering is for them a continuous activity performed in parallel with other daily activities. Thus, intelligence activities are usually widely spread in the company, cutting across many units and processes, and many pieces of intelligence are held by employees all over the company. This fact is likely to cause problems of overlapping and effort duplication in TI activities, which may easily be reinforced by the 'on-demand', project-based working mode in ITSPs. Different projects are carried out by different groups at the same time, and decisions have to be made sometimes in a reactive and ad hoc way to satisfy the client's requirements in a timely manner, losing the chance to achieve synergies. Table 5.3 summarizes the challenges mentioned.

Thus, in the urgency to satisfy the requirements of customers, the ITSP may end up involved in projects scattered over many fields which are not necessarily strategic, and the organization may start to spend more resources in the 'look for the customer solution' mode at the expense of innovation projects. Communication among groups may become weak, and then wasted resources due to lack of synergies is evident.

TOOLS ANALYSED AND SELECTED

This section does not seek to provide a comprehensive list and description of the available TI tools, but to show some examples and a general overview of some of them which are considered useful for TI systems in ITSPs facing challenges like the ones mentioned earlier. As was said earlier, the choice of methods is influenced by many variables, and is subject to the specific challenges of a company. Besides, some tools may require the fulfilment of certain preconditons such as an open-minded culture, high top management commitment, a certain amount of human and financial resources available, and so on. Thus, the tools shown next are the ones

Table 5.3 Some challenges faced by the ITSP when implementing TI systems

Why TI?	Customer need-driven
	Strategy-driven
What is TI for?	Customer projects – many areas: 'look for the solution' approach
	Innovation projects – strategic projects: 'anticipation' approach
How?	Ad hoc approach – TI is gathered mostly on demand
	'You need to know approach' – bottom-up No strong links among internal and external projects Informal communication (overlapping and effort duplication)
	Regular basis approach
	'I need to know approach' – top-down (link to or alignment with strategy) Commonalities and synergies among internal and external projects Formal communication (better use of resources)

that were considered appropriate in the specific context of the analysed ITSP. Only the selected tools will be described in more detail, the others will only be mentioned.

Bearing in mind the challenges described in the previous section, several TI tools were analysed. It was observed that not all of them cover the whole TI process (for example, some of them focus on data acquisition and analysis) and are not sufficiently strategy related and communication oriented. For this reason, besides monitoring and scanning tools to detect emerging technologies, it was necessary to incorporate other types of tools to avoid the potential issues mentioned before. It was determined that three types of tools should be analysed:

- Monitoring and scanning tools (as defined initially): to identify emerging (high-potential) technologies in a timely manner. Both quantitative and qualitative tools were studied.

Table 5.4 Proposed tools to support TI systems in the ITSP

Potential issues in the ITSP	Focus on short- and medium-term technology needs of the customers Projects are scattered all over many fields Ad hoc approach
Potential challenges for TI systems	Difficult to know what to monitor Weak alignment with strategy Lack of synergies
Potential actions	Identify focus areas and gaps Common understanding of path to follow: more direction and focus for team members Stronger link between strategic planning and implementation, and technology development Systematize and formalize the process
Potential tools	TI tools Strategic planning tools TI frameworks

- Strategic planning tools: to help select the focus areas and keep a balance between the customer needs and the ITSP's long-term strategic vision. Tools that help to keep the TI and strategic processes aligned were preferred.
- TI frameworks: to formalize the TI process, facilitate communication among groups and use in-house intelligence effectively. Attention was focused on participative, workshop-based tools that allow carrying out a structured communication process based on the client's needs and the strategic vision.

Table 5.4 summarizes the tools chosen.

As can be noted, to have a TI system able to respond to internal and customer needs requires more than tools to collect and analyse data. There are many challenges, such as: keeping strategy driven and aligned; managing overlapping and effort duplication; having a good link with the 'top' and the 'bottom' of the organization; being able to integrate systematic and ad hoc TI approaches; communicating intelligence effectively, and so on, which need to be taken into account when designing TI systems and choosing the tools that will support them. The following section focuses on the tools analysed and selected.

Strategic Planning Tools

Roadmapping

In the case of strategic planning tools the focus was on roadmapping as there was a clear need to link TI activities to strategy processes from the very beginning of the TI implementation. 'Technology roadmapping represents a powerful technique for supporting technology management and planning, especially for exploring and communicating the dynamic linkages between technological resources, organizational objectives and the changing environment' (Phaal, n.d.).

Roadmapping has become the method of choice for linking strategic and technology planning (Rohrbeck, 2012), among other things, because it is a flexible tool that can be adapted to the specific needs of a company and it can be used for different purposes: product planning, service and capability planning, strategic planning, long-range planning, knowledge asset planning, programme planning, process planning, integration planning (Figure 5.1).

For this specific case roadmapping was chosen because it focuses on the development of a vision of the future (markets, business, products, technologies, and so on), and therefore gaps and intelligence needs can be

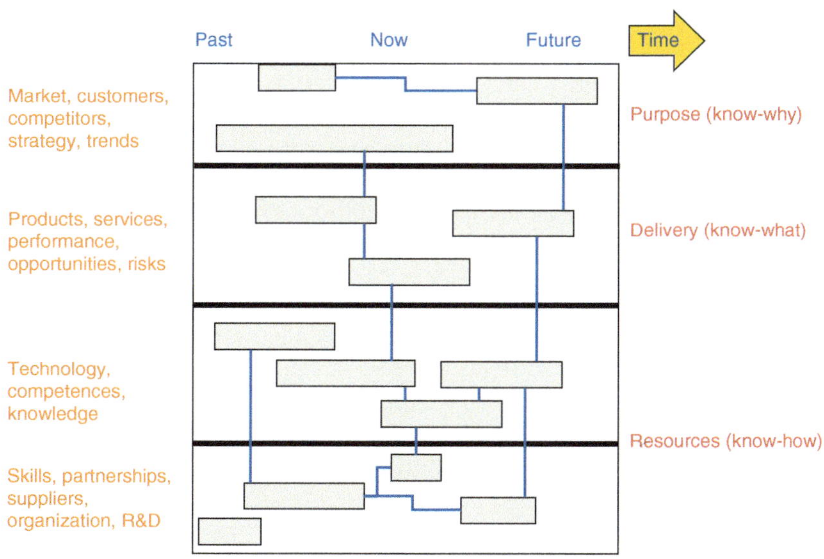

Source: Phaal (n.d.).

Figure 5.1 Roadmapping

identified, and strategic options can be visualized more clearly. Besides, this is a good tool to link different intelligence activities and streams (technology, market, competitors). Thus, this tool is useful to guide the TI system, help link strategic planning and technology developments, provide focus and direction to the team members, and coordinate the group's efforts. It is important to mention that it is not only the final roadmap itself which brings these benefits, but the roadmap building process and the communication and interaction it fosters.

There are many methodologies to create a roadmap. In this case, and given that participative tools were more necessary, special attention was given to workshop-based methodologies. The Fraunhofer Institute for Systems and Innovation Research (FhG-ISI) has defined a roadmapping methodology which combines three basic steps: (1) technology overview; (2) technology and market analysis; and (3) roadmapping workshops. The first phase is research on the technology and it is based mainly on expert interviews and literature review. The second step involves market analysis, and for this databases, market surveys and secondary analysis are used. The third stage is when the roadmap is actually built and consists of a series of workshops where experts from the academy and the industry level get together (Reiss and Thielmann, 2012).

The University of Cambridge has created the 'T-Plan' methodology which comprises four facilitated workshops. The first three are focused on the three key layers of the roadmap (market or business, product or service, and technology), and the final one brings the layers together on a time basis to construct the chart (Phaal, n.d.).

The University of Nottingham has developed an integrated company technology roadmapping tool called the 'strategic technology alignment roadmapping' or STAR system. This methodology provides software to support all steps from data collection to evaluation and ranking of R&D projects (Arman and Foden, 2010).

The 'opportunity landscape' tool, developed by ETH Zürich, was also considered as a potential tool to use. It was chosen mainly because it is a useful tool to identify strategic technology areas and technologies, and has the potential to identify technology focus areas and gaps, and clarify intelligence needs, roles and responsibilities. This tool also allows the technology and strategic processes to stay linked (Savioz and Blum, 2002).

TI Frameworks and Tools

By researching TI frameworks and tools, it was evident that many of the qualitative tools developed to either monitor or detect emerging technologies also provide a framework to formalize TI processes; for this reason, both TI frameworks and monitoring and scanning tools are described together below.

Forecasting innovation pathways

Within the field of future-oriented technology analysis, Robinson et al. (2011) have proposed a methodology called 'forecasting innovation pathways' (FIP), which was selected because it is useful for new and emerging technologies (that is, where applications are still very speculative or with just some applications in the market) and provides a clear step sequence to carry out TI activities.

In FIP, indicators of developments in the field of technology are determined and explored to identify path-dependencies. The pathways, based on the dynamics of the present, may be broad and tentative but they are useful to identify endogenous futures ('futures embedded in the present which can give indications and insights into the transition from present into future'; Robinson et al., 2011). The definition of those indicators is a challenge, though, and the approach requires the collection of large amounts of data, from many and often heterogeneous sources, sufficient to determine the pathways.

The FIP framework includes four stages: (1) understand; (2) profile and link; (3) project and assess; and (4) report. Stage 1 serves 'to understand a technology's level of maturation and the contextual dynamics that could affect its potential innovation pathways' (Robinson et al., 2011). Some tools proposed by the FIP creators are: 'technology readiness level', which is a measure used by some United States government agencies and many of the world's major companies (and agencies) to assess the maturity of evolving technologies; and the technology delivery system, which is appropriate for emerging technologies by distinguishing the businesses to develop the innovation and take it to market, and the key contextual factors affecting the success of that innovation process. Stage 2 is meant to profile R&D (activities and actors) and link it to potential applications. The tools proposed for this are bibliometric analyses, social networking and trend analyses. Stage 3 includes the exploration of possible futures, the technology assessment and the technology impact. And finally, stage 4 consists of integration and communication. These four stages are divided again into steps (Table 5.5).

It should be noted that the 'Engage experts' step ideally occurs throughout all the process.

Other analysed tools include an indicator-based tool, developed by the Chamber University in the USA, to monitor technologies which are already known to a company (Nyberg and Palmgren, 2011), and the TRIZ

Table 5.5 Steps and tools proposed in the 'forecasting innovation pathways' approach

Step	Description	Candidate tools
Characterize the technology's nature	Description of the technology and what it does Technology development stage analysis	Technology readiness levels
Model the technology delivery system	Contextual forces and factors analysis Push to pull enterprise analysis	Socio-technical mapping Techno-economic networks
Profile R&D	Bibliometric analysis Top actors and topics analysis R&D disciplinary analysis	Tech mining Publication analyses Patent analyses Social network analyses Trend analyses
Profile innovation actors and activities	Enterprise cooperation analysis	Cross-charting matrix
Determine potential applications	Competing technologies analysis Competitor, risk and market analysis Commercialization pathway exploration and assessment	Data mining on co-authoring among top corporations Platform profiling
Lay out alternative innovation pathways	Consolidate intelligence from previous steps Promising innovation paths and contingencies analysis	Multi-path mapping (first round) Visualizations such as innovation chain
Explore innovation components	Analyse promising innovation paths within key selection environments	Interviews with multiple experts Multi-actor workshops
Technology assessment	'Winning' potential innovation pathways impact assessment Alternative technologies comparative impact assessment	Multi-path mapping (second round) Innovation chain Technology comparison matrix
Synthesize and report	Iterative review Communicate findings (use multiple media; interactive as possible)	Papers Standards reports
Engage experts	Support to the analyses in all previous steps	Interviews Questionnaires Workshops Co-author papers

Source: Robinson et al. (2011).

(theory of inventive problem-solving)-based 'technology intelligence', a method to identify competing technologies and forecast their development (Schuh and Grawatsch, 2003).

FhG-ISI tool

The FhG-ISI is developing a decision support tool to identify the emergence of technologies using tech-mining techniques. The main characteristics of this methodology are: (1) the emergence of a technology is considered a relative concept and therefore it should be measured by comparing technologies (rather than identifying the emerging phase of a technology in comparison to other development phases of the same technology); (2) the emergence of a technology is identified by the co-occurrence of stronger scientific, research and inventive activity; the proxies used are publications[1] for the former and patents for the later; and (3) the share ratio is used as indicator to identify emerging technologies as it measures the average growth of a given technology relative to the overall average growth of the reference sample (Pero, 2013). The possible results can be classified in a 2 × 2 matrix as shown in Table 5.6.

This methodology is based on publication and patent data, therefore a time lag has to be taken into account due to the administrative procedures. Indeed, 'the strength of this method is to rely on accurate data to assess the statistical significance of technology developments and patterns' (Pero, 2013). It was selected, among other things, because it allows tangible evidence to support decisions to be obtained quickly. Table 5.7 shows the data collection steps of this methodology.

Table 5.6 Emergence typology

Technology status during period p		Technological Sharp Ratio	
		Negative	Positive
Scientific Sharp Ratio	Positive	Basic Strong growth of publications; Weak growth/decline of patents	Emerging Strong growth of publications; Strong growth of patents
	Negative	Appearing Weak growth/decline of publications; Weak growth/decline of patents	Applied Weak growth/decline of publications; Strong growth of patents

Source: Pero (2013).

Table 5.7 Steps and tools proposed by the FhG-ISI approach

Step	Description	Candidate tools
Define the technology set to analyse	Narrow down the scope of the analysis and define level of analysis	Literature review and expert reviews
Link the technology to its research and inventive activities	Define adequate keywords to measure both the scientific and technological activity of the technologies selected	Keyword strategy based on specialized literature and expert input
Collect data	Execute queries in databases	Patent and publication databases
Validate	Verify a randomly selected subset of the extracted data	Yamane formulae (to select a representative subset)
Determine emergence of the technologies selected	Compute the scientific and technological Sharpe ratio	Sharpe ratio, scripts

Implementation

Once the potential tools were selected, it was necessary to define which should be tested first by implementing a pilot test. Taking into account that the strategic planning tools and the frameworks to formalize the TI process require significant commitment and support from top management and possibly more resources than are available, it was necessary to start on a small scale. Thus, it was decided to test the scanning tool first. The FhG-ISI tool would provide tangible evidence and data which could be transformed easily into actionable intelligence to convince management. Besides, it could be implemented easily with the staff available and the help of the project partner FhG-ISI.

The pilot test is currently being carried out and has started by looking and testing for free and low-cost publication and patent databases. The results are being compared with the results obtained from commercial databases. The goal is to identify a combination of tools that offer good cost-effectiveness for the ITSP. The fields chosen for the tests are nanotechnology and biotechnology. The support of experts from both the FhG-ISI and the ITSP has been essential.

CONCLUSION

TI activities are essential in industry sectors such as the pharmaceutical and chemical sector where the product cycles are long and the development costs are very high. To successfully innovate, changes in markets and technologies need to be identified early. However, even though TI is getting more popular and is receiving more attention from academics and industry practitioners, it seems that its implementation continues to be a big challenge for companies. The TI activities are usually not systematized or formalized, and intelligence depends on what employees find out in available data sources which are then communicated informally. The tools used to support these processes are limited (if any) and therefore the strategic planning process is usually disconnected from any TI activities.

Technology developments need to be continuously monitored to create awareness of potential changes, which may represent business opportunities or threats for a company. Thus, companies need to start to evolve towards a continuous TI approach. There are qualitative and quantitative tools that can be used to support the TI processes, and the options increase as there are also different ways to adapt and combine them.

This chapter contributes to the literature on technology intelligence by describing the tool box selected to support the TI implementation in an ITSP. Businesses of this kind face particular challenges when implementing TI systems due to their nature and particular way of operating. By analysing a specific ITSP, several of these challenges were identified: how to select areas and projects to be monitored, while keeping a balance between customer needs and its own strategic vision; how to keep the TI and strategic processes aligned; and how to be able to reduce effort duplication and overlapping. If not taken into account, these challenges can become barriers which can significantly hinder TI activities.

By analysing different IT tools it was also concluded that it would be difficult to find one or a set of TI tools that helps with all the specific challenges that the ITSP may face. Hence, the focus of attention should not be only on scanning and monitoring tools as initially defined, but also on complementary tools that are outside the boundaries of TI. Interaction and communication are two additional and important aspects for choosing the tools, as the goal is to have a tool kit that, besides providing actionable intelligence, is also sufficiently strategy-related and communication-oriented.

Finally, the tools chosen included: (1) a strategic planning tool, which is useful for strategy alignment purposes and determining strategic options; (2) a method to formalize the TI process created for emerging technologies; and (3) a scanning tool to perceive weak signals based on patent and publication analysis. It is expected that this combination of tools will allow the ITSP, besides detecting threats and opportunities, to make faster and more informed decisions, to convince top and middle management of the importance of this process and keep it aligned with the strategic processes, and to foster communication among different groups and units.

NOTE

1. In journals, trade journals, book series and conference materials that have an International Standard Serial Number assigned to them, as well as conference papers (Pero, 2013).

REFERENCES

Ahlgrimm, Michael (2008). External knowledge acquisition and transfer from innovation clusters to central R&D unit: the mediating role of R&D listening posts as technological gatekeepers. http://lnu.diva-portal.org/smash/record.jsf?pid=diva2:158336 (accessed May 2012).

Arman, H. and Foden, J. (2010). Combining methods in the technology intelligence process: application in an aerospace manufacturing firm. *R&D Management*, 40, 2.

Behkami, N. and Daim, T. (2012). Research forecasting for Health Information Technology (HIT), using technology intelligence. *Technological Forecasting and Social Change*, 79, 498–508.

Caldwell, B., Wang, E., Ghosh, S., Kim, C. and Rayalu, R. (2005). Forecasting multiple generations of technology evolution: challenges and possible solutions. *International Journal of Technology Intelligence and Planning*, 1(2), 131–149.

Kerr, C.I.V, Mortara, L., Phaal, R. and Probert, D.R. (2006). A conceptual model for technology intelligence. *International Journal of Technology and Planning*, 1(2), 73–93.

Lichtenthaler, E. (2006). Technology intelligence: identification of technological opportunites and threats by firms. *International Journal of Technology Intelligence and Planning*, 2(3), 289–323.

Mortara, L., Kerr, C., Probert, D. and Phaal, R. (2007). *Technology Intelligence: Identifying Threats and Opportunities from New Technologies*. Cambridge: Cambridge University Press.

Nyberg, A. and Palmgren, S. (2011). Using indicator for technology monitoring. Thesis, Chalmers University of Technology.

Pero, Mickael (2013). Identifying emerging technologies: an application to nanotechnology. Unpublished paper.

Phaal, Robert (n.d). Technology roadmapping. http://www.unido.org/fileadmin/import/16963_TechnologyRoadmapping.pdf.

Reiss, T. and Thielmann, A. (2012). Roadmapping und Monitoring zu Energiespeichernfür die Elektromobilität. Presentation.

Robinson, D., Huang, L., Guo, Y. and Porter, A. (2011). Forecasting Innovation

Pathways (FIP) for new and emerging science and technologies. Unpublished paper.
Rohrbeck, Rene (2006). Harnessing a network of experts for competitive advantage. Presented at the 4th Seminar on Project and Innovation. http://www.rene-rohrbeck.de/documents/Rohrbeck_(2006)_Technology-Scouting_Presentation.pdf.
Runge, Wolfgang (2006). *Innovation, Research and Technology Intelligence in the Chemical Industry*. Stuttgart: Fraunhofer IRB.
Savioz, P. and Blum, M. (2002). Strategic forecast tool for SMEs: how the opportunity landscape interacts with business strategy to anticipate technological trends. *Technovation*, 22, 91–100..
Schuh, G. and Grawatsch, M. (2003). TRIZ-based technology intelligence. Presented at the European TRIZ Association meeting TRIZ Futures 2003.

PART II

Collaboration and networking in managing emerging technologies

INTRODUCTION

Emerging technologies addressing grand societal challenges increasingly stem from collaboration within and across diverse networks and communities in Europe and worldwide (see Chapter 1 of the book). We endeavour here to unravel the socio-economic impact of such emerging technologies for all kinds of stakeholders in order to manage them more efficiently and effectively in their early critical stages of development. This part of the book increases our understanding in terms of key challenges, in particular, underlying knowledge bases and industrial architectures of collaboration and networking, highlighting the importance of these levers for managing the emergence of such enabling technologies for a sustainable and competitive future. The four chapters respectively present results on health care-related technologies, with a particular focus on particle therapies for cancer stemming from high-energy physics and biophysics-related communities; nanotechnology in a small innovative national economy; renewable energy for addressing environmental pollution and sustainability; and last but not least, research infrastructures in materials science, from metallurgy to nanotechnology, within a large European economy also facilitating international collaboration.

Following previous studies in collaboration and networking across different institutional spheres, key technology sectors, disciplinary areas and related scientific and technological communities (Etzkowitz and Leydesdorff, 2000; Murray, 2002; Assimakopoulos, 2003, 2007; Breschi et al., 2003; Powell et al., 2005; Gittelman, 2007; Ahrweiler, 2010; Lissoni et al., 2011; Leydesdorff et al., 2013) the chapters here highlight the importance of underlying knowledge vectors and heterogeneous institutional, disciplinary and industrial architectures. As knowledge vectors reflect the increasing interconnectedness of scientific and technological communities, chapter contributors highlight academic publications and industrial patents, as patterns of collaborative endeavours in highly networked communities of co-publishing and co-patenting individuals and organizations. The authors deploy both quantitative and qualitative methods, including network and statistical analyses, coupled with qualitative data derived from in-depth case studies, observations and interviews with key stakeholders in academia, industry and government in Europe and worldwide.

Chapter 6, by Carafa, Assimakopoulos and Parker, discusses the case of particle therapy for cancer, drawing on bibliometric data, co-publication and co-citation networks among the high-energy physics community and related scientific and technological communities in biophysics, software and hardware engineering, as well as radiology and oncology. A striking finding here is the very long time horizon of more than 40 years of developing the underlying knowledge base spanning a spectrum of several dozens of disciplines and subject categories from particle physics to biomedicine. The idea of particle therapy for cancer was originated in the USA at the Lawrence Berkeley Laboratory (LBL) in the late 1940s and was initially developed at Harvard University and the Massachusetts General Hospital in the 1950s. But it took off later on in Europe, in particular in Switzerland and Germany, after the 1970s. The very first trials on humans were developed in the German national lab of heavy ions and biophysics in Darmstadt, now the Helmholtz Centre for Heavy Ion Research (GSI), and were transferred to the historical faculty of medicine in Heidelberg and its state-of-the-art University Hospital facilities (HIT, http://www.heidelberg-university-hospital.com/World-Class-Oncology-at-Heidelberg-University-Hospital.217.0.html), with support from the local government and the Siemens health care business unit. In parallel, there has been much advanced research and development in Japan, through the National Institute of Radiological Research, Tsukuba University and Hitachi for the commercialization of particle therapies mainly in Japan, as well as in the USA and worldwide.

Chapter 7 by Schrempf and Dolgova presents original findings of co-patenting nanotechnology collaboration networks in Ireland from the late 1970s to the early 2010s. They discuss how these patterns of knowledge diffusion develop in interdisciplinary rather than multidisciplinary networks, so that policy makers can make meaningful interventions for fostering knowledge propagation across different fields over time. They also help practitioners in enabling knowledge commercialization across many industry sectors that apply nanotechnology, for example in information and communication technology (ICT) or nanotechnology application areas. There is still however a long way to go before there is conclusive evidence about how to measure unambiguously the integration of nanotechnology in different industries and associated technological communities. The average path length and geodesic distances of different patent classes highlight the need for making investments in certain connections that bridge nanotechnology and different application areas generating the highest potential for innovation and socio-economic impact.

Chapter 8 by Gao also focuses on Ireland. She studies the multidisciplinary emerging collaborative architecture of the renewable energy clusters and associated policy instruments for enabling the emergence of the sector in Ireland. The role of the European Commission (EC) funding is duly acknowledged in the analysis of the empirical findings, as well as the role of the national policies in community building across 15 research clusters in nine technological areas of interest. Key actors are identified through a social network analysis of the collaboration graph, and unsurprisingly large public companies, utilities and universities are identified as the key drivers for innovation in the renewable and clean energy clusters. Young entrepreneurial firms do not seem able to cross the 'valley of death' and the few surviving after a few years depend on their sustained collaboration with the bigger players. Collaborative projects also depend on the intentional brokerage of the regional cluster agencies managing the funding and implementation of central government policy. Gao contributes to our understanding of how policy instruments may (or may not) facilitate the emergence of sustained innovation ecosystems. This is particularly true in sectors that are characterized by slow return on investment, the monopolistic role of key players in energy production, and the slow transition of the energy production and distribution system to a future of clean and renewable energy.

Last but not least, in Chapter 9, Pero explores the central role of research infrastructures (RIs) in the evolution of the materials science network in Italy in the past few decades, as well as their significance for fostering international collaborations. The chapter discusses in particular the brokerage role of RIs in the interorganizational cluster network evolution, shaping materials science for such diverse areas of interest as nanotechnology and metallurgy. The main findings of social network analysis, of a set of publications from 1984 to 2004 in the Scopus database, provide insights into the different brokerage roles that RIs provide. Furthermore a statistical analysis shows that RIs fulfil their role with respect to supporting multiple organizations and clusters in reaching out and strengthening international collaborations.

Dimitris G. Assimakopoulos

REFERENCES

Ahrweiler, P., 2010, *Innovation in Complex Social Systems*, Oxford: Routledge.

Assimakopoulos, D. (ed.), 2003, 'Collaboration and Networking', *International Journal of Technology Management*, 25(1–2), 1–199.

Assimakopoulos, D., 2007, *Technological Communities and Networks: Triggers and Drivers for Innovation*, Oxford: Routledge.

Breschi, S., Lissoni, F. and Malerba, F., 2003, 'Knowledge-relatedness in firm technological diversification', *Research Policy*, 32 (1), 69–87.

Etzkowitz, H. and Leydesdorff, L., 2000, 'The dynamics of innovation: from national systems and "Mode 2" to a triple helix of university–industry–government relations', *Research Policy*, 29, 109–123.

Gittelman, M., 2007, 'Does geography matter for science-based firms? Epistemic communities and the geography of research and patenting in biotechnology', *Organization Science*, 18 (4), 724–741.

Leydesdorff, L., Carley, S. and Rafols, I., 2013, 'Global maps of science based on the new Web-of-Science categories', *Scientometrics*, 94 (2), 589–593.

Lissoni, F., Llerena, P. and Sanditov, B., 2011, 'Small worlds in networks of inventors and the role of science: an analysis of France', Working Papers of BETA 2011-18, Bureau d'Economie Théorique et Appliquée, UDS, Strasbourg.

Murray, F., 2002, 'Innovation as co-evolution of scientific and technological networks: exploring tissue engineering', *Research Policy*, 31 (8–9), 1389–1403.

Powell, W.W., White, D.R., Koput, K.W. and Owen-Smith, J., 2005, 'Network dynamics and field evolution: the growth of inter-organizational collaboration in the life sciences', *American Journal of Sociology*, 110 (4), 1132–1205.

6. Network evolution at the science–technology overlap in the triple helix of particle therapy of cancer

Andrea Carafa, Dimitris G. Assimakopoulos and Andrew Parker

INTRODUCTION

The social processes which allow for scientific and technical progress have largely drawn the attention of scholars of science and technical change, and policy makers, for many years. International scientific collaboration has increased due to a number of factors internal and external to science (Wagner and Leydesdorff, 2005). Science and technology developments are intertwined (Price, 1984) and can be contextualized in the 'triple helix' of industry–university–government relations (Etzkowitz and Leydesdorff, 2000). Scientific and technological networks are distinctive, co-evolve and overlap to an extent (Murray, 2002) and 'mode 2' knowledge production involves greater transdisciplinarity and collaboration between sites (Gibbons et al., 1994). Overall this points to the underlying relational nature of innovation, which deserves researchers' attention (Edquist, 1997; Etzkowitz and Leydesdorff, 1997).

Nowadays researchers are puzzled about innovation network dynamics. Generally speaking, this is because a given relationship may change over time (for example, from weak to strong), as well as a whole structure of relationships. When these changes in relationships and structures are a result of a coherent social process, this is termed network evolution.

This chapter deals with the social processes of scientific and technical progress and the respective network evolution by using, as a starting point, Murray's (2002) contribution on scientific and technological networks' co-mingling. She acknowledges that:

> the old view that 'science' was an exogenous and self-contained process has been replaced by a growing awareness that science may be, to a considerable extent, endogenous. Moreover much work suggests that science and technology

> may in fact co-evolve and that the nature of such interaction may be much more bi-directional than was originally thought. (Murray, 2002: 1389)

How do science and technology co-evolve in an interdisciplinary setting across the triple helix? Murray (2002) quotes Nelson's statement (1995: 63) that scientific progress may in part 'reside in the connections between science and technology'. If science and technology overlap, does the view of scientific progress based on communities (Kuhn, 1970 [1962]; Crane, 1972) still explain the changes in relationships and structure of scientific collaboration? For the sake of clarity this question may be reformulated in at least two different ways: Is the underlying social process of building science still the same in a world in which companies may participate in the scientific community life which has grown internationally, and in which scientists may enter the technology sphere (for example, via advising)? How do social networks of scientists evolve throughout the shaping of intertwined science and emerging technologies within the triple helix?

Murray (2002) explored a number of avenues of science–technology overlap: co-publishing and citation, co-patenting, consulting, advising, movement of human capital from academia to industry, proximity to 'star' scientists, licensing and company founding. A consideration is that most of them can be studied via social network analyses: co-publishing and citation, co-patenting, consulting, advising, licensing and company founding may be studied as different (multiplex) networks (Hanneman and Riddle, 2005: Ch. 16), which are labelled as 'avenue networks' in this chapter. Movement of human capital can be studied by looking at the organizational affiliation attribute. Proximity to 'star' scientists may be studied not only geographically but also via network position analysis and measures of distance.

Based on a previous conference proceedings paper (Carafa et al., 2011), this chapter explores in more depth the evolution of one avenue network, co-authorship, by means of bibliometric data and network analyses coupled with semi-structured interviews. The other avenues are studied through semi-structured interviews. Further data collection is necessary to study the other avenue networks. The role of 'scientific instrumentalities' (Price, 1984) emerged in the semi-structured interviews, and it is considered in the bibliometric data gathering.

This work proposes a theoretical framework which integrates the discourses on the science–technology overlap (Murray, 2002) and the 'duality' of people and groups (Breiger, 1974) within the 'triple helix' (Etzkowitz and Leydesdorff, 2000). Different actors and groups shape new technologies (Blosch and Preece, 2000) which may be regarded as emerging in configurations of disciplines, skills and potential users (Leydesdorff and Rafols, 2010). If science and technology overlap, these actors and groups may belong to scientific and/or technological communities, while being affiliated to organizations which can be positioned in the triple helix. The framework spans the individual to the systemic level, including different layers and aspects of the relational nature of innovation. A network perspective is applicable at each level (people, groups and organizations, and triple helix) and between levels to study their dualities.

The empirical study focuses on 'particle therapy', a new paradigm of cancer therapy, having two competing streams using different particles: protons and carbon ions. Proton and carbon-ion therapies have been emerging at the cutting edge of cancer treatment as they offer remarkable advantages compared to conventional X-ray therapy of cancer which is widespread in the world. Yet, several barriers (technical and non-technical) need to be overcome for these technologies to be adopted for conventional clinical therapy. This affects both science and technology developments.

As remarked by Baba et al. (2009), different bodies of literature studied university–industry research and development (R&D) collaborations by focusing on different industries. This is important because: (1) innovation processes may vary among industries depending on their knowledge base (Asheim and Coenen, 2005; Asheim and Gertler, 2005; Moodysson et al., 2008); and (2) public research plays a different role among industries (Pavitt, 1984). Accordingly, a study on particle therapy in the health care sector may offer new insights, given its interdisciplinary nature and the public research focus on such scientific and technological developments.

THEORETICAL FRAMEWORK

Murray's (2002) contribution showed that distinctive scientific and technological networks exist and overlap to an extent. Building on Murray's work, this chapter seeks to contextualize her argument explicitly within the triple helix of industry–university–government interactions. Below are three subsections on the community-based view of science and technology, the science–technology overlap, and the 'chain of dualities' from people to the triple helix.

Scientific and Technological Communities

The social structures of science and technology are different (Dasgupta and David, 1994). This is the basis of Murray's distinction (2002: 1390) between scientific and technological networks:

> Science . . . is characterized by publication, supported by a priority-based reward system and exists predominantly (but not exclusively) in research universities. This is in contrast to the world of technology in which ideas are produced for economic ends and encoded in patents and other modes of protection to facilitate appropriability. This simplified distinction provides a starting point from which to explore how the individual scientists, scientific and technical communities and their institutions shape the co-evolution and co-production of new ideas.

A stream of authors (Price, 1963; Kuhn, 1962, 1970; etc) have focused on the social organization of science based on communities. By building upon Kuhn's key contribution (1970 [1962]), Constant (1984) and Assimakopoulos (1997, 2007) have focused on the social organization of technological practices based on communities. Crane (1972) conceptualizes the evolution of scientific knowledge as growing along an S-shaped curve. Dosi (1982) conceptualizes technological progress similarly as evolving in an S-shaped logistic growth curve. Performance is limited and thus individuals and organizations explore alternative approaches to the problem they focus on. While these studies focused on the scientific or technological communities only, Murray's contribution is on their co-evolution.

Kuhn (1970 [1962]) argues that communities of scientists are the social locus of scientific knowledge. In each community, scientists create and follow a paradigm. This gives specific exemplars and has a narrow and specific usage which is associated with the evolution of scientific knowledge. For Kuhn, a paradigm is a set of beliefs regarding several aspects: the object of research, methods, concept of science, theories, relevant issues, and the values and meanings of a discipline (Bonet et al., 2009). Kuhn argues that scientific progress is not continuous and that two periods alternate over time, that is, 'periods of normal science' and 'periods of scientific revolutions'. In periods of normal science research is undertaken in the scientific community and faith in a paradigm is necessary for people to engage in it. From this, optimism becomes rooted in a scientific community and new theories are created. Periods of scientific revolutions are those in which a paradigm shift occurs. In periods of scientific revolutions consensus is difficult to reach. When anomalies appear which cause the existing paradigm to be questioned, and there is a rejection of the old paradigm in favour of a new one, this implies the abandonment of the old community for a new one (Kuhn, 1970 [1962]; Bonet et al., 2009).

Based on Kuhn's argument, Constant (1984) draws a parallel and defines technological communities. He argues that the communities of technology practitioners are the social locus of technological knowledge. These individuals create and follow a technological tradition of practice associated with the evolution of a particular technology. But differently from scientific paradigms, technological traditions have not only a strong cognitive dimension but also a strong socio-cultural dimension. The cognitive dimension includes scientific theory, methods, hardware and software, implying both tacit and explicit knowledge. The socio-cultural dimension includes communication structures, values and beliefs which are the glue of the community. The key point of contact between the scientific and technological communities can be expected to be cognitive.

By comparing scientific revolution (Kuhn, 1970 [1962]) and technological evolution (Constant, 1984; Assimakopoulos, 2007) it is possible to highlight a key difference as regards the evolution of these different communities. In scientific communities the rejection of an old paradigm for a new one implies the abandonment of one community for the new one. In technological communities there is instead no implied rejection of old traditions in favour of the new ones: multiple memberships to different communities are possible.

Crane (1972: 22) argues that scientific knowledge grows in a kind of a diffusion process (see Figure 6.1). In turn the growth of publications in any scientific field follows the logistic growth curve depicted for the diffusion of innovations. When a paradigm appears there is little or no social organization in the community. In periods of normal science, there are groups of collaborators and/or an invisible college. When major problems are solved or anomalies appear in the paradigm, there is increasing specialization and controversy within the community. And when there is exhaustion or crisis of the paradigm, a decline in membership can be observed.

Science–Technology Overlap and Co-evolution

Murray (2002) points to the concept of co-evolution of scientific and technological networks and quotes Nelson's statement (1995: 63) that scientific progress may in part 'reside in the connections between science and technology'. She shows that science and technology networks are distinctive but co-evolve and overlap to an extent, providing a novel methodology with which to study this empirically. Her contribution explores a number of avenues of science–technology overlap: co-publishing and citation, co-patenting, consulting, advising, movement of human capital from academia to industry, proximity to 'star' scientists, licensing and company founding.

Regarding co-publishing and citation, she found that few corporate individuals participate in science, differently from what was found by Henderson and Cockburn (1994), Arora and Gambardella (1994) and Liebeskind et al. (1996), who focused on how company ties to scientific

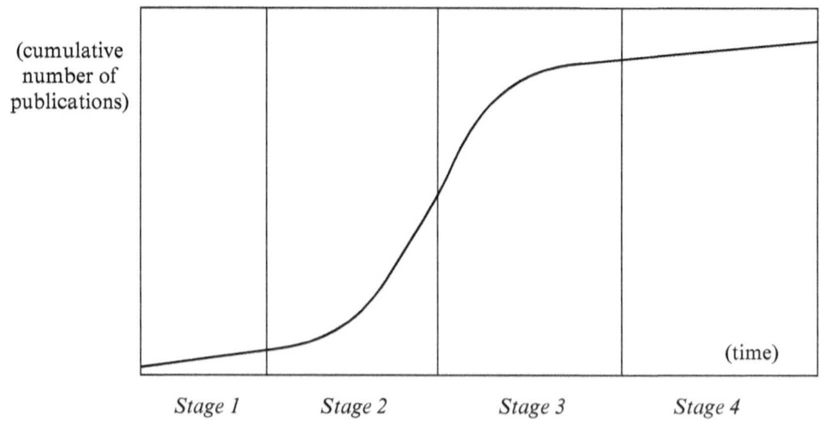

	Stage 1	Stage 2	Stage 3	Stage 4
CHARACTERISTICS OF KNOWLEDGE	Paradigm appears	Normal science	Solution of major problems	Exhaustion
			Anomolies appear	Crisis
CHARACTERISTICS OF SCIENTIFIC COMMUNITIES	Little or no social organization	Groups of collaborators and an invisible college	Increasing specialization	Decline in membership
			Increasing controversy	Decline in membership

Source: Crane (1972: 172).

Figure 6.1 Crane's view on the evolution of scientific knowledge and communities

communities affect the company performance, especially in terms of technological progress for new and science-based technology. The latter group of contributions showed that in times of a technological paradigm shift associated to one in science, a firm's publishing is crucial for a successful transition to the new technological paradigm. Instead, Murray (2002) found that overlap occurs more importantly via licensing, consulting, advising (Scientific Advisory Boards membership) and company founding by scientists. She points out that academic scientists actively seek to establish different links with firms, including patenting. Regarding the latter, it is important to add when and where companies are founded. In this regard, Zucker et al. (1998) found that most productive 'star' bioscientists played a central role in where and when new biotech firms were formed.

An additional element may be considered. As remarked by Leydesdorff and Meyer (2007: 2), Price (1984: 6) proposed the concept of scientific 'instrumentalities' (for example, Galileo's telescope used for momentum discoveries) that mediate between science and technology developments in periods of 'normal science' and 'normal technology'. Price (1984: 13) stressed that instrumentalities – unlike instruments – could also set the ground for binding scientists and engineers in invisible colleges by making a new methodological and technical option available to them. A new instrumentality may produce a scientific breakthrough or paradigm shift, and the possibility of making new technological products available that were not around previously. Government laboratories play an important role in providing new instrumentalities (Rosenberg, 1992) and methodologies (Salter and Martin, 2001). And these can be further developed to adapt to commercial needs (OTA, 1995).

Interestingly Breschi and Catalini (2010) explore the co-evolution of science and technology and find that author-inventors who bridge the boundaries between science and technology domains are crucial for allowing connectivity. These gatekeepers can be expected to be important in science or technological networks, yet in only one of these networks and not in both. Breschi and Catalini also point out that European corporate scientists tend to be less present at the intersection of scientific and technological networks compared to their US counterparts.

Gittelman and Kogut (2003) show how crucial ties with open science through boundary-spanning 'gatekeepers' are for biotech companies. But having bridging ties (that is, spanning organizational boundaries) is not conducive to innovation per se, as they need to be Simmelian (strong and sticky) ties (Tortoriello and Krackhardt, 2010). Zucker et al. (1998, 2002) show that the most successful biotech companies are co-authoring with university professors. In these contributions, people in a certain position (boundary-spanning gatekeepers) in a network of individuals are crucial for producing an outcome for their company. This company outcome – for example, technical advances – may affect its ties with other organizations and people in these organizations, such as other universities and academics.

From People to the Triple Helix: A Chain of Dualities

When science and technology overlap (Murray, 2002), the different actors and groups which shape new technologies (Blosch and Preece, 2000) may belong to scientific and/or technological communities, while being

Perspective	Illustration	Key references	Key points
Community–Social organization	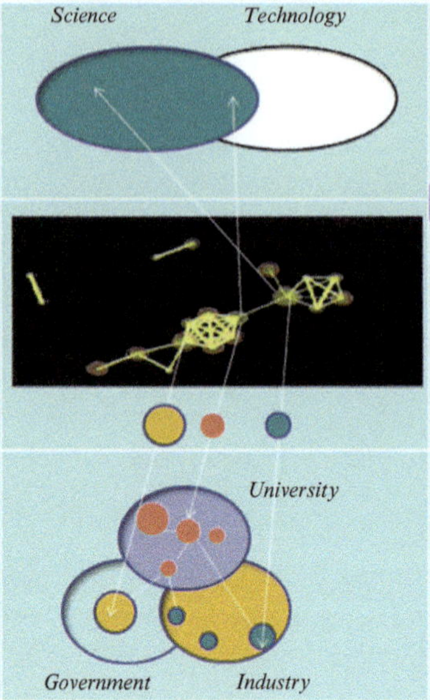	Scientific communities (Kuhn, 1970 [1962]; Crane, 1972) Tech communities (Constant, 1987: Assimakopoulos, 2007; Dosi, 1982)	Parallel discourses
		Overlap of Science & Technology networks (Murray, 2002)	
Networks of: a) Individuals • Scientific authors • Inventors		Social networks • 'Duality' (Breiger, 1974: Lazega et al., 2008)	SN theory, used data on researchers
b) Organizations		University–industry collaboration	Using SNA as a method
Innovation system		Triple helix (Etzkowitz and Leydesdorff, 1997, 2000)	Suited to a SNA/ bibliometric analysis
		Triple helix indicators (e.g. Park and Leydesdorff, 2010)	Based on bibliometric data

Figure 6.2 The 'chain of dualities' framework

affiliated to organizations which can be positioned in the triple helix of university–industry–government relations. By focusing on actors it is possible to move from people to their corresponding groups, organizations and the triple helix. It is also possible to move along a 'chain of dualities': people and groups, groups and organizations, organizations and helices (see Figure 6.2). This chain of dualities may be used to look at the overlap between science and technology from both a static and a dynamic viewpoint.

Breiger (1974) initiated the research tradition on the duality of people and groups, and Lazega et al. (2008) recently contributed to it. They looked at multilevel networks of superposed and partially connected interdependencies between actors in different networks, that is, interpersonal and interorganizational. Lazega et al. created a method of structural linked design which articulates these different levels. First, the analysis of the whole networks at each level is carried out. Then, the two levels are put in relation to each other by looking at the organizational affiliation of individuals and how each pairs correspond in the two levels, as in bipartite networks. This way, they can look at centrality measures in both levels and identify small and big fish (cancer researchers) and small and big ponds (labs). Finally they look at (scientific) outcomes measures to identify catching strategies of small fish. They argue that this adds a new dimension to the sociological study of outcomes, meso-level phenomena (for example, opportunity structures), and macro-level phenomena such as social inequality.

The outcomes obtained by an organization may affect the resources obtained by another organization in the same network or outside it (for example, a successful university lab may obtain more resources over time and, since public budgets are shrinking, other labs may see their pond shrinking). An illustrative example can be used: competition for funding research on HIV cures vis-á-vis particle therapy for cancer. If a university lab is successful in its HIV cure, research leading to a new cure which is highly cost-effective, then the respective government may be confronted with increasing funding for one stream or another to the benefits of industry and society at large. This links the duality of people and groups with the triple helix of university–industry–government relations.

Three pillars are at the basis of the proposed framework: a community-based view of science and technology; the science–technology overlap; and a chain of dualities from people to the triple helix. The framework spans from the individual to the systemic level of analysis, and includes different layers and aspects of the relational nature of innovation. Social network analyses are suited to bringing these together in a single framework. A network perspective is applicable at each level (people, groups and organizations, and triple helix) and between levels to study their dualities.

One may recall a classical interrogative in social theory: does position in structure (network) determine the actor outcome, or vice versa? By adding the time variable, which position at a given point in time determines which outcome at a given point in time, and vice versa? Now, which position of an actor determines which outcome in an intergroup or interorganizational structure, and vice versa (fish–pond duality)? And which position of an organization determines which outcome in the triple helix (pond–helix duality)? By adding the time variable, when are these positions determining these outcomes, or vice versa?

These two dualities may be renamed by using astronomy metaphors which researchers can refer to in their empirical contributions. First, a star–constellation duality would include scientists and their research groups. The most cited and prolific scientists are big stars and have a position in the constellation structure. Young scientists (small stars) are more peripheral but gravitate around big stars. Their research groups may be bigger or smaller (in terms of resources, expertise and number of scientists) and may collaborate with other groups or not; the size and structure of the constellation depends on these differences. Yet, these groups belong to an organization which can be positioned in the triple helix galaxy (constellation–galaxy duality). Gravitation of a node around another might not imply a tie, but tie formation may be observed over time.

Such a framework might be linked to the research efforts in developing new science, technology and innovation indicators. Lepori et al. (2008) stress that changes in indicators of science, technology and innovation are linked to shifts in the research and innovation system, as well as to the transition from a centralized to a distributed intelligence. They call for a change of rationale from an input–output framework to a positioning indicators framework, focusing on flows and linkages between actors in the innovation system. Thinking of a chain of dualities, and looking at network positions data (for example, centrality measures) as well as outcomes, may allow their call to be addressed.

As shown in Figure 6.2, such a framework may be used to map these literature bodies more comprehensively. One simple consideration can be made by referring to some relevant references in these research streams: on the one hand, bibliometric and social network analyses are used to study, respectively, the science–technology overlap (Murray, 2002) and the evolution of scientific collaboration (Barabási et al., 2002); on the other hand, social network theory is built by using data on scientists and technological practitioners (Lazega et al., 2008; Tortoriello and Krackhardt, 2010). This is what may make these different literature bodies informative of each other. The framework can further include a section on methodologies used, industries studied, and so on.

METHODOLOGY AND DATA SOURCES

First, semi-structured interviews were carried out with leading experts in order to explore and refine the scope of the study. They also pointed to key secondary data such as reports and scientific publications that were worth analysing.

Second, co-authorship and co-citation networks of scientific papers for both routes were studied. The corresponding data sources are the Science Citation Index Expanded and Conference Proceedings Citation Index – Science in the Web of Science. In the first interviews, the interviewees were asked what they considered as the relevant search strings to be used for the bibliometric data gathering. A small online survey was administered to them to give time to reflect and list relevant strings.

Two datasets were generated covering mostly the clinical side and including 1920 and 440 records for route 1 and 2 respectively.[1] Then two other datasets of 1634 and 1024 records were generated for route 1 and 2, respectively, and covering the spillovers from accelerator science to the other disciplines. Further details on data gathering are provided in the subsection below. With the latter datasets, the network evolution visualization and analysis proceeded by ten-year slices (five slices for proton, and two for carbon ion), based on the publication date between 1965 and 2011 corresponding to the data retrieved.

Multilevel networks were studied: interpersonal, and interorganizational, international (being superimposable and interdependent). The analysis of networks was done at each level, and the levels put in relation by looking at the organizational affiliation and geographical location of individuals. This entails pairs (for example, individual and organization), as in bipartite networks. The analysis looked at outcomes measures (scientific records) and network position to identify small and big stars (scientists), and their constellation of collaborators. It also looked at the corresponding organizations' galaxies (each interorganizational network component). The structure and size of each component, and the helices to which the organizations correspond, was studied. The international networks were analysed. Each level is informative of the other one.

Finally, semi-structured interviews with leading experts (identified through the bibliometric analyses and interviews' snowball sampling) were carried out with the aim of triangulating the findings obtained in the first two steps of the research design. Twelve interviews were carried out with key scientists who have been active in the US, Europe and Japan.

The semi-structured interviews covered the history and characteristics of the particle therapy community, including the different disciplines involved in this research, their ways of collaborating for scientific and technical advances, and how these are brought to society and the economy. The role of star scientists and spillovers from one discipline to another were investigated too. The interviewees were asked generic questions which were gradually brought to focus on certain specific elements. They were left room to highlight aspects they considered important, as well as to name colleagues to be interviewed (snowball sampling).

To a minor extent the analysis covered reports and scientific papers, and direct observation of: (1) an international scientific workshop in which the core of the particle therapy community united (including mainly scientists from university, government labs and companies); and (2) the functioning of research facilities as well as commercially oriented clinical facilities.

Bibliometric Data-Gathering Approach

Step 1: Retrieval of two general datasets

A formal 'nominalist' search was based on search strings collected by asking experts in the field of particle therapy. These strings were used to generate two initial datasets, that is, one per route, which included more than 100 000 records each. These datasets also included irrelevant data and therefore the following steps were taken in order to get rid of such data.

Step 2: Refining by document type (articles and proceedings papers)

Both published articles and conference papers were included in the dataset as they represent the outputs of research to be disseminated. This search limitation allowed for excluding other irrelevant records from the datasets. Although journal publications and proceedings papers may be considered as overlapping, both were included for the following reason. Conference papers seem to be particularly relevant, as semi-structured interviews have highlighted the important role played by certain conferences in the field. Several star scientists stressed that important decisions are taken at conferences, and that some proceedings papers have been key for future developments, becoming highly cited.

Step 3: Combination with an 'organizations' search string (considering the scientific instrumentalities)

Both datasets were combined with a new search for organizations. These organizations were again identified together with particle therapy experts and include all major centres of research and technological development (RTD) in the area of proton and carbon-ion therapy. These centres are universities, government labs and companies; they include both physics and clinical research (including university hospitals).

The companies involved can easily be listed by experts in the fields as they are limited in number worldwide. This is due to the high level of investments, risk-bearing and expertise required for participating in the RTD of particle therapy facilities. Likewise, the government labs and universities involved in such research can easily be traced into the data for one key reason: there are a limited number of accelerator facilities (clinical and non-clinical) worldwide where such research could be carried out, and this is due to their high construction costs (hundreds of millions of dollars). Overall, this is consistent with the concept of scientific instrumentalities which mediate between science and technology developments (see the theory section).

Step 4: Reiteration of previous steps to include and exclude records
Some key terms were excluded and new key terms and organizations were included in the search strings, and the differences with previous datasets were analysed. Not only was number of records monitored, but also the presence of records from different disciplines. In this reiterative process, the opinion of experts was asked.

EMPIRICAL STUDY

This study started by discussing with experts in accelerator physics the different scientific developments which followed the improvements in accelerator science and allowed for relevant technological applications. While some developments related to industrial facilities, particle therapy for cancer is interesting for several reasons: mainly its interdisciplinarity, its international dimension, and the fact that the new clinical applications developed are associated with a new paradigm of cancer therapy. Below are detailed findings on the evolution of particle therapy.

Two Routes of Particle Therapy

The evolution of scientific knowledge on 'particle therapies' for cancer has been intertwined with the development of cutting-edge technologies based on protons and carbon ions since the 1950s and 1990s, respectively. Both proton and carbon-ion routes are at the intersection of physics, biomedical science, clinical and general medicine and chemistry, and resulted from collaboration among several universities, companies, government laboratories and agencies mainly in the USA (for example, Loma Linda University, LBL Berkley,[2] Varian), Japan (for example, Chiba University, NIRS, Hitachi) and Europe (for example, CERN, GSI, Siemens).

Particle therapy (also known as hadrontherapy)[3] offers one key strength vis-à-vis conventional radiation therapy for cancer (that is, based on beams of X-rays). In particle as well as in conventional radiation therapy, beams are delivered to the patient to destroy tumour cells. Inevitably, the surrounding healthy tissues also receive some radiation doses. The damage is lower with charged protons and carbon ions than with X-rays. Both protons and carbon ions offer higher precision and energy when irradiating the target region. The 'Bragg peak' distinguishes the particle therapy paradigm: the charged particle's energy deposition reaches this peak before stopping, allowing tissue penetration with low diffusion (Dosanjh et al., 2007; Enlight, 2010).

Route 1 technologies (that is, based on protons[4]) are nowadays maturing after more than 50 years of research and technology development (RTD). Route 2 technologies (that is, based on carbon ions) are emerging after two decades of RTD, yielding potentially dramatic improvements vis-à-vis route 1 but also higher uncertainty about their biological effects. In the USA, route 1 has been developed on a large scale, while route 2 has not been backed with the necessary support by companies and government, having not been approved by the corresponding US agency within the US Department of Health and Human Services, the Food and Drug Administration (FDA). The US research involvement in route 2 RTD has therefore shifted significantly to collaborations with European actors.

Interviews and papers (Dosanjh et al., 2007; Enlight, 2010) point to the origin of the proton route as being back in 1946 in the USA. After the Second World War, major research projects on ion beams were started at Berkeley Lab, where the physicist Ernest Lawrence asked Robert Wilson, a former student of his, to collaborate. Wilson was at Harvard at that time and proposed proton therapy for cancer by exploiting the properties of the 'Bragg peak'. The first patients were treated in the 1950s by using non-dedicated accelerators. At that time only a few parts of the body could be treated due to the low power of accelerators, and proton therapy was only at its outset. It was in the late 1970s that proton therapy started becoming viable for clinical applications thanks to the advances in accelerators, medical imaging and computing. This links back to the concept of scientific instrumentalities which allowed mediation between science and technology and bound together different experts from different disciplines. Proton facilities entered medical clinics only in the early 1990s. Nowadays there are about 30 proton centres in the world.

Research on the carbon-ion route originated in the USA in the late 1980s and migrated to Europe and Japan in the 1990s after the closure of the Bevalac accelerator in Berkeley, California. The most significant contributions in Germany and Japan were achieved under the leadership of two scientists who had previously carried out research at Berkeley. In 1994 a first dedicated facility was operational at the Heavy Ion Medical Accelerator (HIMAC) in Japan, and in 1997 the Helmholtz Centre for Heavy Ion Research (GSI) in Germany treated the first patient with carbon ions. In 2009 a first facility which could use both protons and carbon ions started treating patients in Germany (the Heidelberg Ion Therapy Centre, HIT). Other facilities of this kind have been or are being constructed in Europe and Japan.

In the RTD work of both routes, some landmark phases can be identified:

- The development of scientific knowledge on particle therapy which results in publications and the application of this knowledge, protected through patents. This phase requires adaptation of non-dedicated facilities, or the design, construction and testing of new facilities. The relationship with firms is limited to commissioning parts of the designed facilities.
- Transfer of expertise to industry for the construction of commercial facilities to be possible. This tends to happen with the commissioning of the construction of research-oriented facilities designed by government labs or university scientists, a phase in which the interaction between industry, university and public research organizations (PROs) is key.
- Initial commercialization of clinical facilities (construction and operation) limited to further clinical trials.
- The clinical trials are carried out to acquire the statistical evidence required by medical doctors for the acceptance of these technologies for standard clinical practice. This produces further publications, patents, standards and protocols for the use of machines and therapy administration.
- Wider commercialization of the technology is possible as the technology is considered to be 'ready' by medical doctors, based on extensive statistical evidence.

These phases are reiterative and overlap over time, underlying a non-linear innovation process. This seems to be in line with previous contributions. Indeed, government laboratories play an important role in providing new instrumentalities (Rosenberg, 1992) and methodologies (Salter and Martin, 2001). And these can be further developed to adapt to commercial needs (OTA, 1995).

First Bibliometric Analysis: From the Clinical and Biomedical Side

The first data gathering includes 1920 records for route 1, and 440 records for route 2. It includes scientific progress especially within the clinical and medical side. The records that were picked by the search are classified by ISI subject category, and their respective disciplines identified by using Leydesdorff and Rafols's (2009) global map of science based on the ISI subject categories. It can be observed from Table 6.1 that the total number of records broken down by subject and discipline (2153 and 511) is higher than the total number of records gathered (1920 and 440). This is because the records may include different subject areas when they are interdisciplinary. The top two disciplines are within clinical science and biomedicine,

Table 6.1 Records by top subject category and discipline retrieved from the Science Citation Index Expanded and Conference Proceedings Citation Index – Science

Proton (1920)		Carbon ion (440)	
ISI Subject	Discipline	ISI Subject	Discipline
Radiology, nuclear medicine and medical imaging (988)	Clinical science	Radiology, nuclear medicine and medical imaging (209)	Clinical science
Oncology (537)	Biomedical science	Oncology (129)	Biomedical science
Nuclear science and technology (265)	Physics	Nuclear science and technology (70)	Physics
Engineering, biomedical (206)	Clinical science	Biology (54)	Biomedical science
Clinical neurology (157)	Neuroscience	Instruments and instrumentation (49)	Material sciences

Proton (1920)			Carbon ion (440)		
Discipline	Records No.	% Records	Discipline	Records No.	% Records
Clinical science	1194	55.46	Clinical science	209	40.90
Biomedical science	537	24.94	Biomedical science	183	35.81
Physics	265	12.31	Physics	70	13.70
Neuroscience	157	7.29	Material sciences	49	9.59
Total	2153	100.00		511	100.00
Top 2		80.40			76.71
Top 3		92.71			90.41

and account for 80.40 and 76.71 per cent of the scientific records. Physics is the third top discipline in both proton and carbon-ion routes.

This corresponds to the early stages of the particle therapy community social organization. The yearly peaks of 60 publications in carbon ions and 198 in protons can be compared with the peak in conventional radiation therapy publications, that is, 420 records (see Figure 6.3). A significant

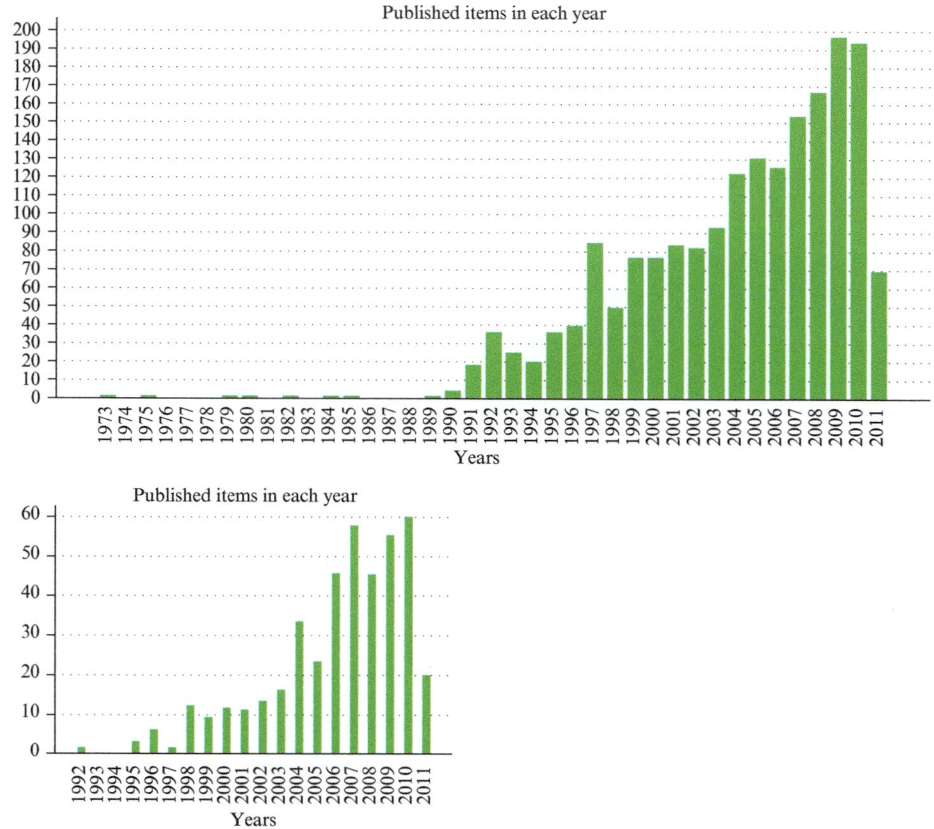

Figure 6.3 *Number of papers per year retrieved from the Science Citation Index Expanded and the Conference Proceedings Citation Index – Science, 1973–2011*

increase in authoring activity is registered since the early 1990s for both routes. This is consistent with the fact that proton facilities entered clinics in the 1990s and that the first carbon-ion accelerator facilities were set up in the 1990s.

Regarding key authors by records count (see Table 6.2), a number of names are found to correspond to top cited authors, top co-cited in co-citation networks, and nominated by interviewees. For route 1, the overlapping names across these data are in particular the medical doctors J.D. Slater and J.M. Slater (Loma Linda University), medical doctor Suit from MGC (USA), and Eros Pedroni from PSI (CH). For route 2, the overlapping names across these data are Tsujii, Kraft, Kanai, Debus, Sholz, Kato, Durante and Kramer. Both Kraft and Tsujii did research in California, where the route 2 idea originated, and then brought it to Germany and Japan where it developed extensively. Recall that the carbon-ion route was not approved by the FDA and hence it could not develop further in the USA.

Regarding the institutions (see Table 6.3), the first observation is that universities are predominant in terms of published contributions. A few hospitals (usually linked to universities, for example MGC linked to Harvard Medical School) and clinics also appear as they contributed to clinical research. There are no companies observed in the top 40 organizations (nor in the top 100) for route 1 research. In route 2, one company is found in the top ten. Accelerator Engineering Corp. is based in Chiba, Japan and was involved in the HIMAC project. Its corporate scientists Noda, Sato and Takada appear in the bibliometric analyses (even in the second one) as important contributors. Note that Chiba University is a major contributor to route 2 research. Therefore spatial proximity between scientists seems relevant in this case.

Second Analysis: From Accelerator Science to Clinical Therapy

Let us recall that particle therapy was first proposed by the physicist Robert Wilson. He proposed the use of protons in 1946 and the therapy entered clinics only in the 1990s. During the 1970s important developments in accelerators, medical imaging and computing allowed for significant developments of proton therapy. Since the first analysis was centred more on the biomedical and clinical science, a second data gathering was carried out to be more inclusive of the other disciplines that also contributed to particle therapy, that is, physics, general medicine, chemistry, engineering and other subject areas. A more representative dataset was gathered which includes the progress done during the 1990s too.

The dataset includes 1634 records for route 1, and 1028 records for route 2. The evolution of the corresponding co-authorship networks by

Table 6.2 Top 40 authors based on records count in the Science Citation Index Expanded and Conference Proceedings Citation Index – Science

R1: Proton route

Top 5	Top 10	Top 15	Top 20	Top 25	Top 30	Top 35	Top 40
Lomax, A.J. (57)	Newhauser, W.D. (38)	Akine, Y. (29)	Habrand, J.L. (24)	Titt, U. (20)	Ferrand, R. (18)	Mazal, A. (17)	Egger, E. (16)
Paganetti, H. (48)	Cuttone, G. (34)	Goiten, M (29)	Goiten, G. (23)	Bortefeld, T. (19)	Gragoudas, E.S. (18)	Miller, D.W. (17)	Georg, D. (16)
Slater, J.D. (48)	Tokuuye, R. (32)	Sugahara, S. (28)	Glimelius, B. (22)	Loeffler, J.S. (19)	Liebsch, N.J. (18)	Suit, H.D. (17)	Grideley, D.S. (16)
Slater, J.M. (46)	Mohan, R. (30)	Hug, E.B. (25)	Schneider, U. (21)	Olko, P. (19)	Noel, G. (18)	Chauvel, P. (16)	Grusell, E. (16)
Munzenrider, J.E. (38)	Pedroni, E. (30)	Cirrone, G. (25)	Ohara, K. (20)	Deshardins, L. (18)	Delaney, T.F. (17)	Delacroix, S. (16)	Mirkovic, D. (16)

R2: Carbon ion

Top 5	Top 10	Top 15	Top 20	Top 25	Top 30	Top 35	Top 40
Tsujii, H. (69)	Ando, K. (26)	Miyamoto, T. (23)	Yanagi, T. (19)	Baba, M. (16)	Koike, S. (14)	Iwata, Y. (12)	Kato, S. (11)
Yamada, S. (41)	Furusawa, Y. (26)	Kraft, G. (22)	Yoshikawa, K. (19)	Haberer, T. (16)	Drentje, A.G. (13)	Matsufuji, N. (12)	Ogawa, H. (11)
Kamada, T. (37)	Jakel, O. (26)	Murumatsu, M. (22)	Kato, H. (18)	Scholz, M. (16)	Ohno, T. (13)	Nikoghosyan, A. (12)	Suzuki, M. (11)
Kanai, T. (30)	Kandatsu, S. (23)	Mizoe, J.E. (21)	Mizoe, J. (18)	Enghardt, W. (15)	Schardt, D. (13)	Sato, Y. (12)	Durante, M. (10)
Debus, J. (28)	Kitigawa, A. (23)	Nakano, T. (19)	Yamamote, N. (17)	Kramer, M. (15)	Schulz-Ertner, D. (13)	Zhang, H. (12)	Fujisawa, T. (10)

Table 6.3 Top 25 institutions based on records count in the Science Citation Index Expanded and Conference Proceedings Citation Index – Science

R1: Proton				
Top 5	Top 10	Top 15	Top 20	Top 25
Harvard Univ (175)	Univ Tsukuba (65)	Natl Inst Radiol Sci (33)	Indiana Univ (22)	Massachussets Eye Ear Infirm (19)
Massachussets Gen Hosp (146)	Inst Nazl Fis Nucl (58)	Univ Florida (32)	German Canc Res Ctr (21)	Univ Heidelberg (19)
Loma Linda Univ (110)	Univ Calif San Francisco (48)	Karolinska Inst (29)	Deutsch Kzentrum (20)	Univ Wollongong (19)
Paul Scherrer Inst (102)	Inst Curie (34)	National Canc Ctr (28)	Univ Uppsala Hosp (20)	Univ Uppsala (18)
Univ Texas MD Anderson Canc Ctr (81)	Univ Texas (34)	Univ Milan (23)	Fox Chase Canc Ctr (19)	Inst Gustav Roussy (17)
R2: Carbon ion				
Top 5	Top 10	Top 15	Top 20	Top 25
Japan NIRS (164)	Gunma Univ (65)	Forschungszentrum Rossendorf EV (8)	Univ Lyon (7)	Inst Natl Fis Nucl (5)
GSI Darmstard (55)	Univ Heidelberg (21)	Kyoto Univ (8)	Kitasato Uni (6)	Karolinska Inst (5)
Chiba Univ (35)	Accelerator Engn Corp Chiba (13)	Kyushu Univ (7)	Nara Med Univ (6)	Nagoya Univ (5)
German Cancer Res Ctr (30)	Jaoan Atom Energy Agcy (9)	Osaka Univ (7)	Res Ctr Hosp Charged Particle Therapy (6)	Tech Univ Darmstadt (5)
Chinese Acad Sci (23)	Tech Univ Dresden (9)	Tokyo Inst Tech (7)	Toyo Univ (19)	Univ Groningen (5)

using ten-year time slices was analysed. This is technically semi-statics, and is used to visualize the evolution of scientific collaboration which may be meaningful in every decade. The analysis focused on the scientific papers authored by the top 250 cited actors in each decade, due to technical limitations of the software used. Table 6.4 shows the total number of nodes (authors) in each route. The number of ties and density of the co-authorship networks is documented by ten-year time slice.

It can be noted that during the last two decades, when route 2 records are picked, the number of nodes increases in route 2 and decreases in route 1. This makes it worth exploring the social networks of scientists in more depth for each route during the last two decades. Before doing so, the evolution of the networks per each route is visualized (see Appendix 6.1 and 6.2). This visualization permits following the position of nodes in the structure of co-authorship over time: individuals, organizations and countries.

The focus is on proton first. Regarding the interpersonal co-authorship networks, this seems the most complex to understand. Authors Kato and Takeda are pointed out in route 1 and their authorship activity (node size) and position is followed.

Regarding the interorganizational networks of co-authorship, it can be observed that for route 1, Hitachi Ltd is the key company in terms of scientific contributions. It can be observed that it goes from being a big contributor with two ties in slice 1, to being a smaller contributor with a much higher number of ties and a central position in slice 2. In slice 3 the Hitachi component becomes bigger and Ibaraki University becomes the biggest contributor of the component, while Hitachi's contribution is decreased but still significant. A small Japanese company is also present in the network (see red star) and can be still visualized in the last time slice. The nature of the Japanese system may explain this variation in terms of firms' involvement in the scientific community. Indeed, the system is strongly based on collaboration between industry, university and national labs. In any case, this is compatible with the view that technology acquisition by firms requires long-term investment in knowledge acquisition. And this long-term investment involves firms in co-publishing activities and participation in the scientific community.

CERN (the European Organization for Nuclear Research, in Switzerland) and Triumf (Canada) are the key government labs appearing in the visualizations, with CERN's component growing bigger. In the 1990s Loma Linda University (USA) appears in the network, giving a significant initial contribution, as expected from the interviews. In the next decade it continued to contribute, even though its interorganizational collaborations decreased. Yet, the interviews and archival analysis show

Table 6.4 Size, number of ties, and density of the co-authorship network by time slice

R1: Proton route

Time slices	Total number of authors	Top 250 nodes	Ties	Density	Total number of organizations	Top 250 organizations	Ties
1965–1974	12	12	25	0.3788	5	5	2
1975–1984	516	250	2721	0.0874	42	42	19
1985–1994	1261	250	6056	0.1946	126	126	73
1995–2004	2744	250	1496	0.0482	283	250	170
2005–2011	1608	250	6212	0.1996	158	158	108

R2: Carbon-ion route

10-year slices	Total number of nodes	Top 250 nodes	Ties	Density	Total number of organizations	Top 250 organizations	Ties
1995–2004	1026	250	2466	0.0792	123	123	86
2005–2011	2253	250	5906	0.1898	245	245	183

Note: The last slice in each route includes seven years instead of ten due to limitations in the data available.

that there are at least two key people behind Loma Linda's contribution, Doctors J.M. and J.D. Slater.

Regarding route 2, Noda, Matsui, Metral and Giovannozzi appear as key contributors in interesting network positions. Further analysis is necessary on the interpersonal networks, looking at network positions measures. They can also be better understood by analysing the corresponding interorganizational and international networks, as follows. The key organizations in terms of scientific contribution are Chiba University (Japan), GSI Darmstadt (Germany, government lab), Hitachi Ltd (Japan), and Ibaraki University (Japan). They belong to different components and a few ties are found across components from slice 1 to 2, though some are lost and other ties are established. The Japanese components have higher connectivity between university, government labs and industry, compared to the USA and Europe. By looking at the organizations which appear in the visualizations, it seems that there is an overlap between organizations involved in proton and carbon ion research, compared to the first analysis. The interview data makes us think that while clinical contributions are more on one route or the other, contributions regarding instrumentalities such as imaging and accelerators can show an overlap across both routes. It seems interesting to check this further. One way of looking at it is to analyse the overlap between co-authorship networks in both routes at both the interpersonal and the interorganizational level.

Regarding countries, these visualizations may be considered a simple way of looking at how national systems of innovation are featured by international collaborations and how this affects their outcomes in terms of research contribution. Japan stands out as having less international co-authorship ties, and yet the highest scientific contributions in both routes. As some interviewees stressed, Japan exhibits high coordination in research collaboration between industry, universities and national laboratories. These may be linked to the *kairetsu* culture, and be the result of its industrial and research policies. In route 1, Switzerland is the second-biggest contributor; and in route 2, Germany. In route 2, higher international connectivity is found between European and US actors and this is expected to be linked to the lack of FDA approval of route 2 in the USA, which led to the international collaboration option.

Findings from Semi-Structured Interviews

Particle therapy paradigm and the community

Mainly the interviews, but also archival analysis of scientific papers and reports, show that the particle therapy paradigm lies in the unique physical and radiobiological properties of protons and carbon ions, compared to

X-rays. Yet protons and carbon ions are also distinguished based on their different physical and biological properties.

These two streams are somehow competing but also cooperating (coopetition). An international coordination initiative (Enlight) and conferences bring both streams together. This entails communication structures to: (1) share scientific advances with the community; and (2) achieve standards of carrying out clinical trials and evidence collection. The Enlight network was created in 2002 at CERN in order to bring together clinicians, physicists, biologists and engineers with an interest in particle therapy. There is heterogeneity of subcommunities (disciplines) with particle therapy, and the corresponding challenge is their alignment. A couple of specialized international conferences now reunite the world's community of scientists, industrialists and practitioners interested in particle therapy.

RTD for both routes has involved different disciplines. While these people previously belonged to their respective disciplinary communities, they have formed the nucleus of a new scientific interdisciplinary community of particle therapy. They currently maintain multiple affiliations in pre-existing and emerging communities of scientific and technological practices. Yet, this may apply to a different extent in different disciplines, and further data and analyses are required to map it. So far it seems that when science is built across disciplines, multiple affiliations are possible. The presence of 'scientific instrumentalities' (for example, new accelerators; positron emission tomography or PET imaging) is binding this new community by making new research options available. This has allowed testing of different designs, scanning methods and different doses for different types of tumours, for example.

Avenues of science–technology overlap

Overlap between the scientific and technological communities is found across the triple helix. As shown in the previous section this overlap occurs via co-authorship; this chapter's findings are different from Murray's (2002) and similar to those of Henderson and Cockburn (1994), Arora and Gambardella (1994) and Liebeskind et al. (1996). In times in which a technological paradigm shift is associated to one in science, publishing by a firm is key for a successful transition to the new technological paradigm. Licensing relations between firms and government labs or universities are found, with cases of licensing of a full portfolio of patents. Advising of companies via scientific advisory boards and informal advising is found. Consulting and co-patenting relations are found, with an important role played by scientists in both government labs and universities. Circulation of human capital within and from government labs and universities to industry is also found.

Zucker et al. (1998) found that most productive 'star' bioscientists played a central role in where and when new biotech firms were formed. Murray (2002) found that scientists played a role in founding firms for the commercialization of intellectual property. But extensive evidence of company founding is not found. This seems to be linked mainly to the high investment size necessary to develop and/or purchase particle therapy facilities; investment which could only be provided by big companies and government. However, 'star scientists' played an important role in: (1) where and when new clinical facilities were built for further clinical trials; (2) attracting necessary funding from governments and industry for the construction of clinical facilities; (3) fostering collaboration with industry for the transfer of expertise, and the construction and development of clinical facilities; and (4) creating a community of scientists and practitioners around particle therapy.

Companies bear the risks and invest only, or especially, in one stream of developments (for example, Varian in protons, Siemens in carbon ions), with some exceptions for the most recent carbon-ion facilities being able to use protons too.[5] When they do not publish, they are found to be present and to interact at conferences, where presentations and exchange of information are vehicles of knowledge flows. This is in line with Murray's findings.

The nature of user needs

Baba et al. (2009) argue that a 'two-way' interaction model between universities (U) and industry (I) – both U–I and I–U – seems to be the most appropriate to study innovation dynamics in the advanced materials industry. Scientific knowledge can be expected to flow from universities to firms, while knowledge of the market and demand by companies flows to universities. Companies have knowledge about user needs in order to carry out RTD work (Maine and Garnsey 2006). Yet, in this study the use of particle therapy devices requires the interaction of technicians, radiologists and other medical doctors, thus making user needs a complex topic.

The role of clinical trials

There seems to be a discrepancy between the physicists who consider the technology as ready, and the medical doctors who consider the technology as not yet ready for standard clinical practice and point to the need for collecting further statistical evidence from clinical trials. In the end, what makes the difference for companies are the needs of medical doctors, statistical evidence included. Firms that did not invest in these technologies had two main motives: (1) their development requires big investments (both private and public); and more importantly (2) clinicians do not consider the technologies to be ready, and technology acceptance for standard clinical practice depends on the results that statistical evidence will show. This is a process which takes decades, such as in drug testing, but with higher complexity (for example, combinations of treatments, type of cancers and their body location have to be tested). Thus, the inclusion of such technologies into their business portfolio has a high-risk profile. While in biotech and some related pharma fields universities develop new ideas which are captured rapidly by industry (Cohen et al., 2002), in this study this process is slower and the role of public research is fundamental.

CONCLUSIONS, IMPLICATIONS AND FUTURE RESEARCH DIRECTIONS

Overlap is found between the scientific and technological communities across the triple helix and internationally: more across the triple helix (and less internationally) for Japan, vis-à-vis the US and Europe. Japan is also the highest contributor in terms of scientific publications. The science–technology overlap occurs via co-authorship of scientific papers, consulting, patenting, advice, circulation of scientists and proximity to star scientists. There is no extensive evidence of company founding, but evidence that star scientists played a key role in getting new clinical and research-based facilities built.

While scientists previously belonged to their respective disciplinary communities they have formed the nucleus of a new interdisciplinary community of particle therapy. This entails heterogeneity of the community in terms of disciplines, yet scientific instrumentalities bind it together. The two research streams in the community are both competing and cooperating (coopetition).

Users are clinical practitioners and researchers; they are also heterogeneous in terms of discipline and tradition of practice, and their needs are a complex topic which is difficult to master fully. Clinical trials play a big role in opening the markets and decrease the risk borne by companies that invested in particle therapy technology.

The role of government lab and university helices is central in terms of scientific knowledge production as well as the development of applications. Government plays an important role in terms of funding research and the construction of new facilities, and approving the clinical use of such therapies for trials, with major impacts in terms of whether the research stream develops or not. Industry plays an important role in terms of risk-taking and investing heavily in the development of cutting-edge technologies.

Even if science and technology overlap at an international level, the old view of scientific progress based on communities still explains the changes in relationships and structure of scientific collaboration over time. This view still holds, but further research will address whether the timing (speed) of scientific and technological development is affected, and how.

This research may develop further in order to: (1) advance a typology of scientists from a social network analysis viewpoint; (2) identify both 'hidden stars' and 'rising stars'; (3) link these roles and the evolution of science and technology developments; (4) investigate the central role of specific institutions in the evolving triple helix of particle therapy worldwide; (5) gather network data on other 'avenue networks' to study their multiplexity; and (6) look at to what extent and when different authors in different disciplines cite each other or move to co-authoring.

Understanding the triple helix of emerging technologies from a network evolution viewpoint may be relevant to inform public funding and private investments in emerging technologies. This may be a potentially complementary approach to cost-effectiveness. The aspect of gatekeepers spanning boundaries (organizational and disciplinary in this case) deserves further research based on the data available. A better understanding of bridging may inform industrial and research policy-making as well as company strategies in order to foster ties between industry and open science, and support the role of key scientists and the different institutions in the triple helix. These aspects may also have implications for new science, technology and innovation indicators. Lepori et al. (2008) stress that changes in such indicators are linked to shifts in the research and innovation system, as well as to the transition from a centralized to a distributed intelligence. They call for a change of rationale from an input–output framework to a positioning indicators framework, focusing on flows and linkages between actors in the innovation system. Thinking of a chain of dualities and looking at network positions data (for example, centrality measures) as well as outcomes may allow their call to be addressed.

ACKNOWLEDGEMENTS

This research was funded by the European Commission (Marie Curie ITN Action), CERN (the European Organization for Nuclear Research, Switzerland) and GSI (the Helmholtz Centre for Heavy Ion Research, Germany). Sincere thanks are addressed to the many scientists and experts who contributed to this research by providing access to data and sharing their experience. This chapter is based on an earlier paper (Carafa et al., 2011) presented at the Triple Helix IX International Conference held at Stanford University, USA, in 2011.

NOTES

1. We number the routes in order to better communicate the historical sequence of proton and carbon-ion routes. The aim is to ease the reading as we acknowledge the complexity of this particle therapy study.
2. A US Department of Energy National Laboratory operated by the University of California.
3. In this chapter only the term 'particle therapy' will be used, for the sake of simplification.
4. We number the routes in order to better communicate the historical sequence of proton and carbon-ion routes. The aim is to ease the reading as we acknowledge the complexity of the particle therapy case.
5. Carbon-ion therapy facilities are more expensive than proton therapy facilities.

REFERENCES

Arora, A. and Gambardella, A. (1994). Evaluating technological information and utilizing it: scientific knowledge, technological capability and external linkages in biotechnology. *Journal of Economic Behavior and Organization*, 24 (1), 91–115.

Asheim, B.T. and Coenen, L. (2005). Knowledge bases and regional innovation systems: comparing Nordic clusters. *Research Policy*, 34 (8), 1173–1190.

Asheim, B.T. and Gertler, M.S. (2005). The geography of innovation: regional innovation systems. In Mowery, F.J.D. and Nelson, R. (eds), *The Oxford Handbook of Innovation*. Oxford University Press, Oxford, pp. 291–317.

Assimakopoulos, D. (1997). The Greek GIS Community. PhD thesis, Department of Town and Regional Planning, University of Sheffield.

Assimakopoulos, D. (2007). *Technological Communities and Networks*. London: Routledge.

Baba, Y., Shichijo, N. and Sedita, S.R. (2009), How do collaborations with universities affect firms' innovative performance? The role of 'Pasteur scientists' in the advanced materials field. *Research Policy*, 38 (5), 756–764.

Barabási, A.L., Jeong, H., Néda, Z., Ravasz, E., Schubert A. and Vicsek, T. (2002). Evolution of the social network of scientific collaborations. *Physica A*, 311, 590–614.

Blosch, M. and Preece, D. (2000). Framing work through a socio-technical ensemble: the case of Butler Co. *Technology Analysis and Strategic Management*, 12 (1), 91–102.

Bonet, E., Jensen H.S., Sauquet, A. (2009), Course handouts on Epistemology of Science, Vol. I. ESADE Business School, Barcelona.

Breiger, R.L. (1974). The duality of persons and groups. *Social Forces*, 53, 181–190.

Breschi, S. and Catalini, C. (2010), Tracing the links between science and technology: an exploratorary analysis of scientists' and inventors' networks. *Research Policy*, 39, 14–26.

Carafa, A., Assimakopoulos, D. and Parker, A. (2011). Network evolution in the triple helix of particle therapies: insights from bibliometric and interview data. Paper presented at Triple Helix IX, 2011, Stanford University, USA.

Cohen, W.M., Goto, A., Nagata, A., Nelson, R.R. and Walsh, J.P. (2002). R&D spillovers, patents and the incentives to innovate in Japan and the United States. *Research Policy*, 31, 1349–1367.

Constant, E.W. (1984). Communities and hierarchies: structure in the practice of science and technology. In R. Laudan (ed.), *The Nature of Technological Knowledge: Are Models of Scientific Change Relevant?* Dordrecht: Reidel, pp. 27–46.

Constant, E.W. (1987). The social locus of technological practice: community, system, or organisation? In W.E. Bijker, T. Hughes and T.J. Pinch (eds), *The Social Construction of Technological Systems*, Cambridge, MA: MIT Press, pp. 223–242.

Crane, D. (1972). *Invisible Colleges: Diffusion of Scientific Knowledge*, Chicago, IL: University of Chicago Press.

Dasgupta, P. and David, P. (1994). Toward a new economics of science. *Research Policy*, 23 (5), 487–521.

Dosanjh, M., Hoffmann, H.F. and Magrin, G. (2007). Status of hadron therapy in Europe and the role of ENLIGHT. *Nuclear Instruments and Methods in Physics Research Section A: Accelerators, Spectrometers, Detectors and Associated Equipment*, 571 (1–2), 1 February, pp. 191–194; Proceedings of the 1st International Conference on Molecular Imaging Technology – EuroMedIm 2006.

Dosi, G. (1982). Technological paradigms and technological trajectories: a suggested interpretation of the determinants and directions of technical change. *Research Policy*, 11 (3), 147–162.

Edquist, C. (ed.) (1997). *Systems of Innovation: Technologies, Institutions and Organizations*. London: Pinter/Cassell.

Enlight (2010). Hadrontherapy: history and current status. European Network for Light Ion Hadron Therapy. http://enlight.web.cern.ch/ (accessed 12 March 2011).

Etzkowitz, H. and Leydesdorff, L. (eds) (1997). *Universities in the Global Economy: A Triple Helix of University–Industry–Government Relations*, London: Pinter/Cassell.

Etzkowitz, H. and Leydesdorff, L. (2000). The dynamics of innovation: from national systems and 'Mode 2' to a triple helix of university–industry–government relations. *Research Policy*, 29, 109–123.

Gibbons, M., Limoges, C., Nowotny, H., Schwartzman, S., Scott, P. and Trow, M. (1994). *The New Production of Knowledge: The Dynamics of Science and Research in Contemporary Societies*. London: Sage.

Gittelman, M. and Kogut, B. (2003). Does good science lead to valuable knowledge? Biotechnology firms and the evolutionary logic of citation patterns. *Management Science*, 49 (4), 366–382.

Hanneman, R.A. and Riddle, M. (2005). Introduction to social network methods. Riverside, CA: University of California, Riverside. Available at http://faculty.ucr.edu/~hanneman/.

Henderson, R. and Cockburn, I. (1994). Measuring competence? Exploring firm effects in pharmaceutical research. *Strategic Management Journal*, 15 (Special Issue), 63–75.

Kuhn, T.S. (1970 [1962]), *The Structure of Scientific Revolutions*, 2nd edn. Chicago, IL: University of Chicago Press.

Lazega, E., Jourda, M.-T., Mounier, L. and Stofer, R. (2008). Catching up with big fish in the big pond? Multi-level network analysis through linked design. *Social Networks*, 30 (2), 159–176.

Lepori, B., Barré, R. and Filliatreau, G. (2008). New perspectives and challenges for the design and production of S&T indicators. *Research Evaluation*, 17 (1), 33–44.

Leydesdorff, L. and Meyer, M. (2007). The scientometrics of a triple helix of university–industry–government relations (introduction to the topical issue), *Scientometrics*, 70 (2), 207–222.

Leydesdorff, L. and Rafols, I. (2009). A global map of science based on the ISI subject categories. *Journal of the American Society for Information Science and Technology*, 60 (2), 348–362.

Leydesdorff, L. and Rafols, I. (2010). How do emerging technologies conquer the world? An exploration of patterns of diffusion and network formation. Digital Libraries (cs.DL), Social and Information Networks (cs.SI), Cornell University.

Liebeskind, J., Oliver, A.L., Zucker, L. and Brewer, M. (1996). Social networks, learning and flexibility: sourcing scientific knowledge in new biotechnology firms. *Organization Science*, 7 (4), 428–443.

Maine, E. and Garnsey, E. (2006). Commercializing generic technology: the case of advanced materials ventures. *Research Policy*, 35 (3), 375–393.

Moodysson, J., Coenen, L. and Asheim, B. (2008). Explaining spatial patterns of innovation: analytical and synthetic modes of knowledge creation in the medicon valley life science cluster. *Environment and Planning A*, 40 (5), 1040–1056.

Murray, F. (2002). Innovation as co-evolution of science and technological networks: exploring tissue engineering. *Research Policy*, 31 (8–9), 1389–1403.

Nelson, R.R. (1995). Recent evolutionary theorizing about economic change. *Journal of Economic Literature*, 33 (1), 48–90.

OTA (US Office of Technology Assessment) (1995). *Innovation and Commercialization of Emerging Technologies*. Washington, DC: US Government Printing Office.

Park, H. and Leydesdorff, L. (2010). Longitudinal trends in networks of university–industry–government relations in South Korea: the role of programmatic incentives. *Research Policy*, 39 (5), 640–649.

Pavitt, K. (1984). Sectoral patterns of technical change – towards a taxonomy and a theory. *Research Policy*, 13 (6), 343–373.

Price, D.J. de Solla (1963). *Little Science Big Science*. New York: Columbia University Press.

Price, D.J. de Solla (1984). The science/technology relationship, the craft of experimental science, and policy for the improvement of high technology innovation. *Research Policy*, 13, 3–20.

Rosenberg, N. (1992). Scientific instrumentation university research. *Research Policy*, 21, 381–390.

Salter, A.J. and Martin, B.R. (2001). The economic benefits of publicly funded basic research: a critical review. *Research Policy*, 30 (3), 509–532.

Tortoriello, M. and Krackhardt, D. (2010). Activating cross-boundary knowledge: the role of Simmelian ties in the generation of innovations. *Academy of Management Journal*, 53 (1), 167–181.

Wagner, C.S. and Leydesdorff, L. (2005). Network structure, self-organization, and the growth of international collaboration in science. *Research Policy*, 34 (10), 1608–1618.

Zucker, L., Darby, M.R. and Brewer, M.B. (1998). Intellectual human capital and the birth of US biotechnology enterprises. *American Economic Review*, 88 (1), 290–336.

Zucker, L.G., Darby, M.R. and Armstrong, J.S. (2002). Commercializing knowledge: university science, knowledge capture, and firm performance in biotechnology. *Management Science*, 48 (1), 138–153.

APPENDIX 6.1: THE EVOLUTION OF SCIENTIFIC COLLABORATION – PROTON ROUTE (R1)

Evolution of the Interpersonal Network of Co-authorship

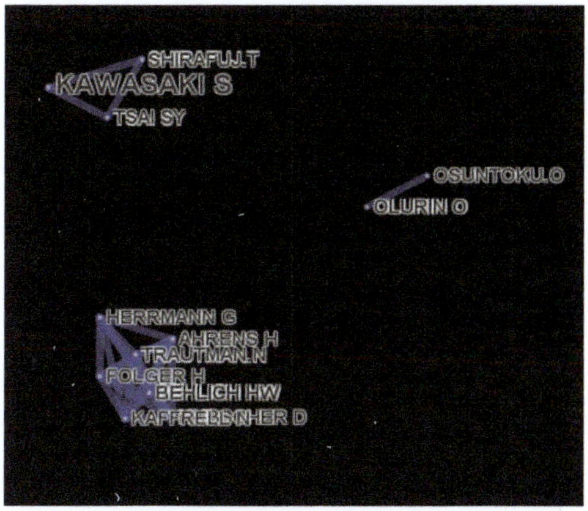

- Ties: co-authorship
- Nodes size: scientific contribution by author
- Ties size: frequency of co-authorship

Figure 6A.1a Evolution of the interpersonal network of co-authorship, 1965–1974

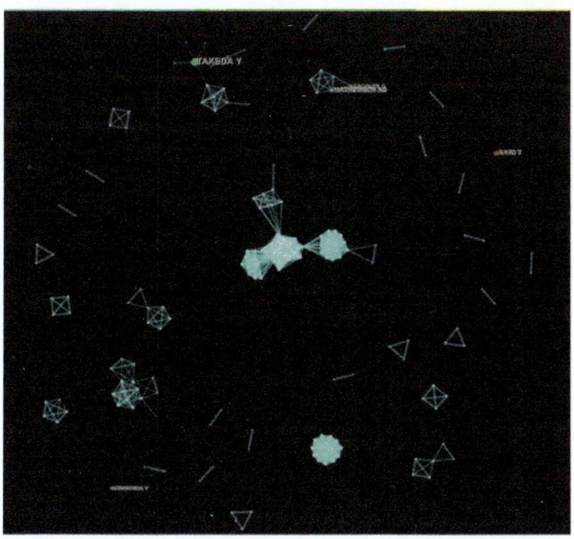

Figure 6A.1b Evolution of the interpersonal network of co-authorship, 1975–1984

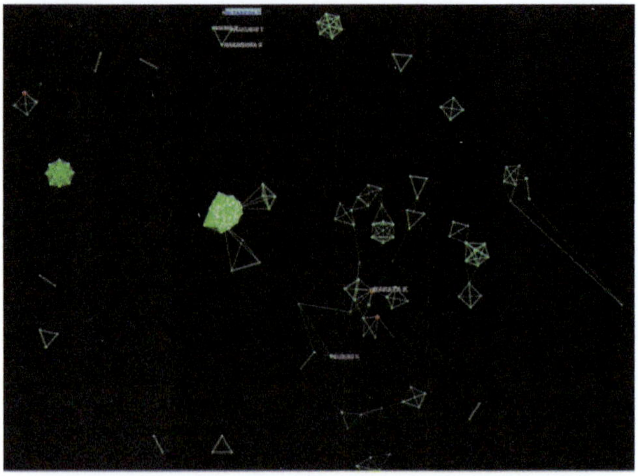

Figure 6A.1c Evolution of the interpersonal network of co-authorship, 1985–1994

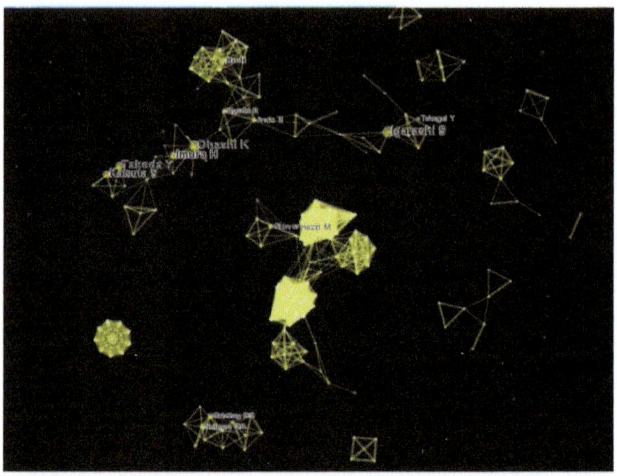

Figure 6A.1d Evolution of the interpersonal network of co-authorship, 1995–2004

Figure 6A.1e *Evolution of the interpersonal network of co-authorship, 2005–2011*

Evolution of the Interorganizational Network (Co-authorship)

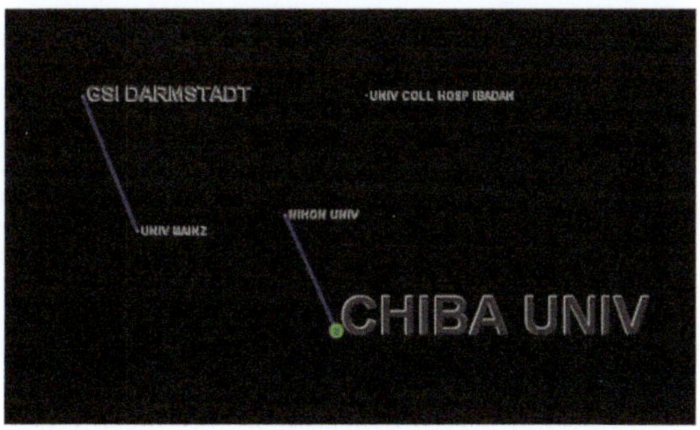

Figure 6A.2a *Evolution of the interorganizational network (co-authorship), 1965–1974*

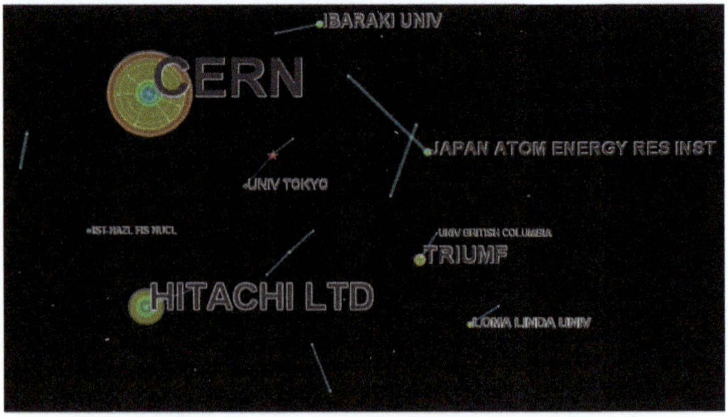

Figure 6A.2b *Evolution of the interorganizational network (co-authorship), 1975–1984*

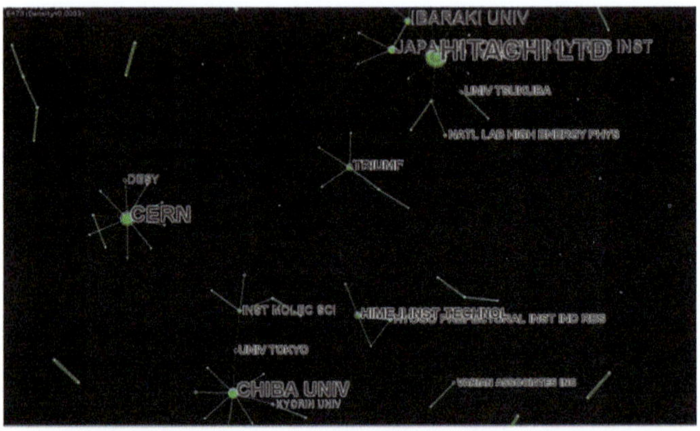

Figure 6A.2c *Evolution of the interorganizational network (co-authorship), 1985–1994*

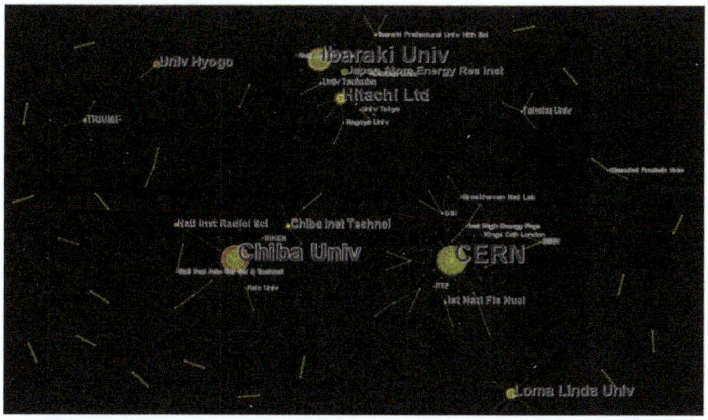

Figure 6A.2d Evolution of the interorganizational network (co-authorship), 1995–2004

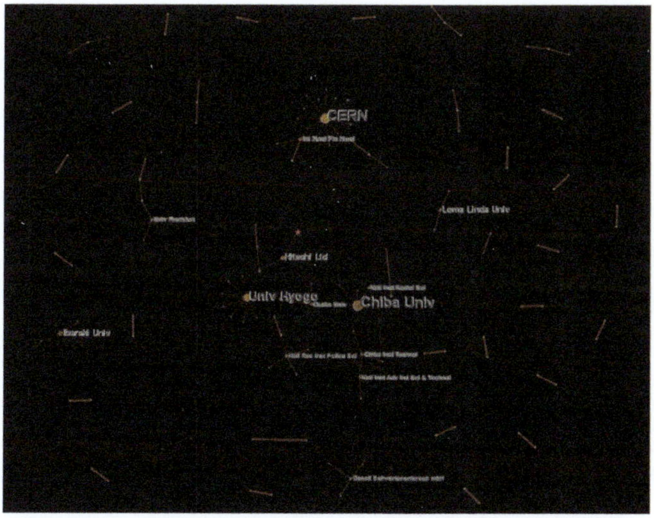

Figure 6A.2e Evolution of the interorganizational network (co-authorship), 2005–2011

Overall Contribution by Country

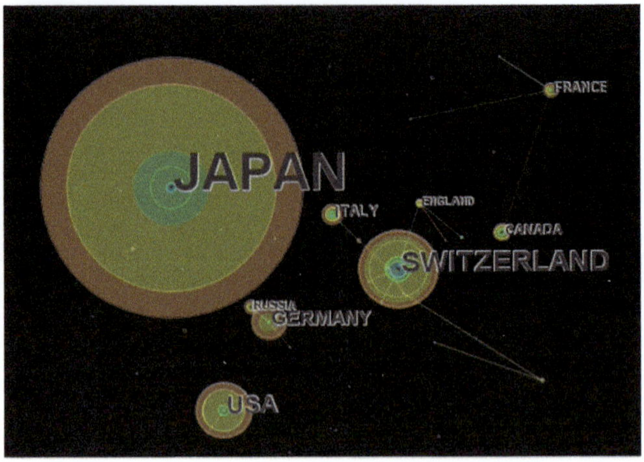

Figure 6A.3 Overall contribution by country (each concentric ring represents a decade)

APPENDIX 6.2: THE EVOLUTION OF SCIENTIFIC COLLABORATION – CARBON ION (R2)

Evolution of the Interpersonal Network of Co-authorship

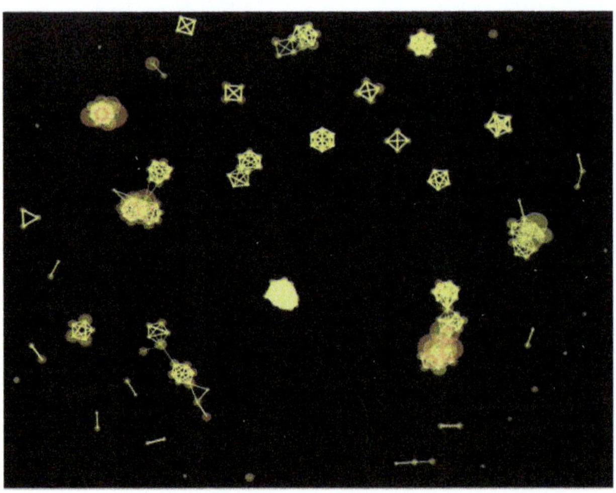

- Ties: co-authorship
- Nodes size: scientific contribution by author
- Ties size: frequency of co-authorship

Figure 6A.4a Evolution of the interpersonal network of co-authorship, 1995–2004

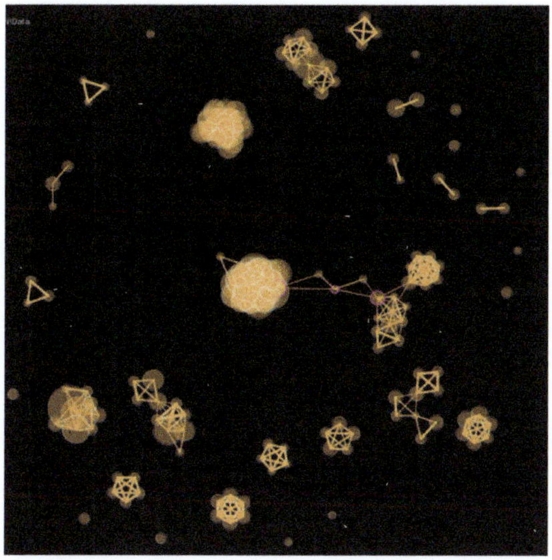

Figure 6A.4b Evolution of the interpersonal network of co-authorship, 2005–2011

Evolution of the Interorganizational Network (Co-authorship)

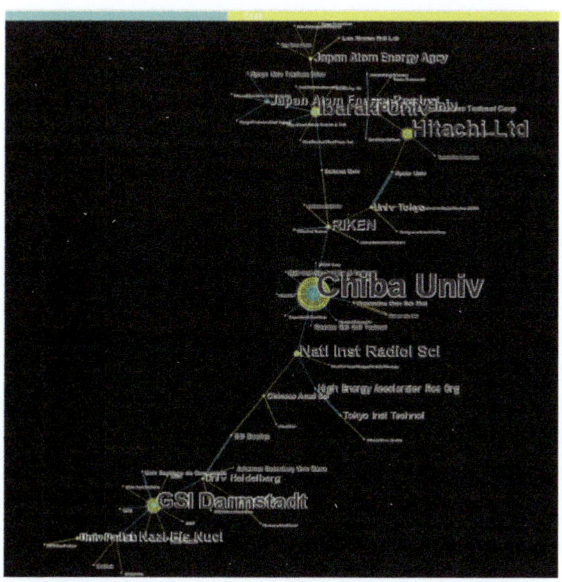

Figure 6A.5a Evolution of the interorganizational network (co-authorship): merged (see colours for each decade) 1995–2004, 2005–2011

Figure 6A.5b Evolution of the interorganizational network (co-authorship): overall contribution by country (each circle is one decade; e.g. a tree section)

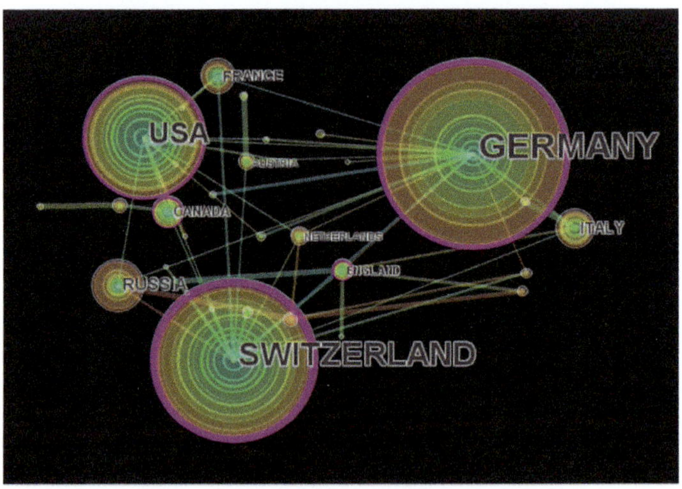

Figure 6A.5c Evolution of the interorganizational network (co-authorship): focus on the international collaborations

7. Nanotechnology in Ireland: an analysis of the patent co-classification network

Benjamin Schrempf and Evgenia Dolgova

INTRODUCTION

Nanotechnology – named as 'the key-technology [*sic*] of the 21st century' (Bhattacharya and Shilpa, 2011) – is often defined as technology to manipulate and control matter at the scale of 1–100 nanometres. Nanotechnology represents an interesting case for the scholars of emergent technologies, as it is applied across a variety of sectors, from information and communication technology (ICT) to biotechnology, and from the pharmaceutical industry to material science and manufacturing. Although nanotechnology as a scientific field substantially impacts upon the socio-economic development of the European economy, it is particularly important for the Republic of Ireland, as the country holds a revealed technological advantage in nanotechnology compared to other countries of Western Europe (OECD, 2009). At the same time the Republic of Ireland also hosts strong semiconductor, materials and biotechnology sectors (Forfás, 2010) that apply nanotechnology extensively.

Taking the importance of the nanotechnology sector in Ireland, an important question arises as to how its development can be sustained and fostered. What are the principles that drive the growth and evolution of nanotechnology? Although research has recently generated a lot of helpful insights into the scope, structure and dynamics of the nanotechnology field (Yu et al., 2010), the debate is still ongoing with regards to characteristic patterns of knowledge generation and diffusion in the nanotechnology sector. As patterns of knowledge diffusion are often considered major prerequisites for innovation, it is essential to understand how knowledge is generated, structured and propagated throughout the network. These processes are of particular importance for policy makers, as awareness of the knowledge diffusion patterns helps to design efficient policy measures and interventions for fostering knowledge transmission.

As nanotechnology is represented across different industries and scientific fields, it is essential to understand whether nanotechnology develops in an interdisciplinary or multidisciplinary (Miyazaki and Islam, 2007; Schummer, 2004) manner. Does nanotechnology exhibit an interdisciplinary nature, originating in one discipline and then spilling over to the related fields? Or does it follow a different pattern, taking root in multiple disciplines and then converging across fields over time? Knowing how nanotechnology propagates among different fields would help practitioners to target policy interventions more efficiently, which in turn would enable synergies in knowledge commercialization and innovation diffusion.

In this chapter we use nanotechnology and non-nanotechnology patents to represent the relevant knowledge base (KB) in the Republic of Ireland. Thus we map the co-classification network of all patents – those related to as well as unrelated to nanotechnology – that have at least one assignee in the Republic of Ireland. We chose Ireland as the country that pioneered multiple nanotechnology applications and hosts many industry sectors that apply nanotechnology. Therefore nanotechnology in Ireland graphically illustrates how the knowledge is structured in this emergent field. We contribute to the debate about the interdisciplinarity and multidisciplinarity of nanotechnology by providing a visual and descriptive network analysis of the empirical case of Ireland in order to be able to investigate how and where nanotechnology connects to other knowledge available to the economy.

THEORETICAL BACKGROUND: KNOWLEDGE IN PATENT NETWORKS

Knowledge base of a firm, region or nation constitutes a critical factor (Saviotti et al., 2005) for thriving in the modern knowledge-based economy (OECD, 1996). Although in early theories knowledge is represented as a simple and homogeneous input factor for the production function (see e.g. Griliches, 1979), recent evolutionary approaches (Krafft et al., 2011; Saviotti, 2007, 2009) treat knowledge as a system and differentiate between two important properties of knowledge: (1) its interpretative or retrieval structure; and (2) its co-relational structure. In this evolutionary framework (Saviotti, 2009), knowledge could be represented as a network: a structure of units (concepts, publications or patent classes) and links between these units (for example, the joint utilization of concepts or co-citations among patents or papers). The systemic approach allows us to represent how knowledge accumulates through the reconfiguration of available knowledge units and the emergence of new ones (Krafft et al., 2011). This approach could be applied to characterize the knowledge base of various systems: companies, industry sectors or nations. Proxies for the knowledge base of these systems can be found, for instance, in publications and citation patterns among them (Onel et al., 2011), as well as patents and their co-citations (Breschi et al., 2003). In our approach, we are focusing not on publications, but on patents and their co-classifications as previously conducted by other researchers; patent co-classification to analyse the knowledge base of sectors (Krafft et al., 2011), companies (Saviotti, 2009) and nations (Dolfsma and Leydesdorff, 2011).

To understand how policy makers could foster the development of the knowledge base for an emergent technology, we focus our investigation on the national innovation system (Nelson, 1993), taking Ireland as an example. The knowledge base of a national innovation system can be seen as the combination of individuals' knowledge, represented in patents (Saviotti et al., 2005). To understand how the knowledge base is structured, we analyse a network of patent co-classifications (Nesta and Saviotti, 2005), taking International Patent Classification (IPC) classes cited in a patent as nodes of the network, and co-occurrence of different classes on the same patent as edges (Krafft et al., 2011).

We can represent the knowledge incorporated in patents related to nanotechnology as a network, visualize it and apply statistical network analysis techniques to reveal how nanotechnology is related to other areas of knowledge. The following changes in the structure of the knowledge network characterize an emergent knowledge base (Krafft et al., 2011): (1) new nodes appear; (2) old nodes (old knowledge) dissolve; (3) new connections form between nodes or clusters (for example, links between new and old knowledge appear within the network); and (4) the position of nodes in the system changes. In these analyses, innovation is captured when new connections are formed between the knowledge units, or when the new nodes – patent classes – appear.

As nanotechnology has a broad range of applications across multiple fields, we are interested in how nanotechnology and other technologies are connected. A co-classification analysis reveals how nanotechnology is embedded within the broader knowledge (Igami, 2008), codified in patents. Nanotechnology literature addresses these interconnections in the interdisciplinarity–multidisciplinarity debate.

IS NANOTECHNOLOGY INTERDISCIPLINARY OR MULTIDISCIPLINARY?

It is essential to understand how new knowledge is embedded into existing scientific disciplines, as this will enable us to identify possible areas for policy intervention. Interdisciplinarity implies that nanotechnology connects previously separated fields and leads to convergence between them. Multidisciplinarity characterizes a mix of loosely connected, unintegrated scientific disciplines. In this sense a new technology serves as an 'amalgam of relatively traditional pockets' (Porter and Youtie, 2009: 1023).

Debate is still ongoing about whether nanotechnology exhibits an interdisciplinary or a multidisciplinary nature. Few researchers emphasize multidisciplinary characteristics of nanotechnology. Schummer (2004) investigates co-authorship networks in nanotechnology publications and concludes that nanotechnology is a multidisciplinary phenomenon: nanotechnology connects more or less unrelated disciplines such as nanophysics, nanochemistry, nano-electrical engineering or nanobiotechnology. According to Schummer (2004), nanotechnology does not show a higher degree of interdisciplinarity than other disciplines. Porter and Youtie (2009) corroborate the idea of multidisciplinarity by demonstrating that nano-related disciplines do not converge over time: the network links between nanotechnology and other disciplines remain relatively constant from 1991 to 2005. In their analysis, nanotechnology clusters around materials sciences, showing a high degree of disciplinary diversity or multidisciplinarity.

Similarly, Meyer (2007) points to the high degree of specialization among companies active in the field of nanotechnology: most companies focus on one specific nano-scale technology. This indicates that nanotechnology could be considered as a general-purpose technology, applied among – but not leading to convergence of – various industrial sectors.

Rafols and Meyer (2007) demonstrate that the degree of interdisciplinarity varies depending on the indicator chosen: interdisciplinarity is high in the subfield of nanobiotechnology if references or instruments from other scientific disciplines are used as an indicator; interdisciplinarity is low if the background of researchers and their affiliation are considered. From these studies, Huang et al. (2010) conclude that the question of interdisciplinarity very much depends on the definition of 'interdisciplinarity'.

Although the debate is still ongoing about whether or not nanotechnology fosters convergence among various scientific fields and existing technologies, Meyer (2007) observed that the subdiscipline of instrumentation connects nanotechnology clusters. The ICT sector – an established, mature technology field – developed many instruments for nanotechnology from the beginning, which raised questions about the origins of nanotechnology: is nanotech a radically new technology, or does it follow the developments in other sectors (Meyer, 2007)?

In this chapter we would like to address this question by examining the network structure between nanotechnology and related fields. We rely on the nanotechnology patent co-classification network to provide insights on how nanotechnology is embedded within other disciplines. We suggest that if nanotechnology is multidisciplinary, there would be sparse links between most knowledge areas and this should be reflected in the co-classification network. Some nanotech knowledge fields, such as instrumentation, would connect those formerly separated disciplines.

As Igami (2008) states, nanotechnology patent applications are distributed differently among the European Union (EU) countries. Due to path-dependence, knowledge in nanotechnology accumulates and clusters around established scientific fields. As Ireland has strong biotechnology and ICT sectors, we could expect high activity in these areas. Nanotechnology activity would be low in areas that are peripheral to the Irish knowledge base. As nanotechnology develops, new nodes would attach to established clusters, building upon already existing knowledge and competencies. This effect – first deepening and then widening specialization (Menz and Ott, 2011) – determines the dynamics of pathways between previously disconnected fields and gives a first indication of the type of logic that drives nanotechnology evolution.

We argue that nanotechnology would exhibit interdisciplinary network structures in patent co-classification networks, as shorter geodesic paths between different network clusters would be established:

Hypothesis: The average path length of the network should be significantly lower in a network with nanotechnology than in a network without nanotechnology.

METHODS AND DATA

Ireland as a Case Study

We chose Ireland as an example of a national innovation system for several main reasons. First, nanotechnology in Ireland represents a critical case: Ireland is a champion in the field of nanotechnology in Western Europe. According to the Organisation for Economic Co-operation and Development (OECD, 2009), it is the country with the highest revealed technological advantage in this area. Thus, in the case of Ireland we can learn how knowledge could be successfully integrated, and can better understand the processes behind the emergence and dissemination of nanotechnology within the nation's entire economy.

Second, nanotechnology has had a profound effect on two fundamental pillars of the Irish economy: the ICT and biotechnology sectors. Although nanotechnology originated in the field of microprocessor manufacturing, which is a cornerstone of Ireland's ICT industry (Choi and Mody, 2009), both ICT and biotech sectors host a variety of practical nano-applications.

Third, Ireland prioritized nanotechnology and some specific fields within it (materials, biotech and ICT) and is keen to understand which policy measures could be applied to amplify technological development in this area (Forfás, 2010). Therefore, Ireland is a particularly interesting case for technology policy assessment.

As patents are considered to be an accurate proxy for scientific and technological knowledge (Bhattacharya and Shilpa, 2011), we generated our dataset from the Delphion database that covers all patents filed at the United States Trademark and Patent office (USPTO), which holds a more detailed database than its European counterpart, the European Patent Office. We extracted all patents that have been assigned to at least one person or organization located in the Republic of Ireland (ROI) in the years 1978 to 2011.[1] We focus our investigation on knowledge of economic importance for Ireland: knowledge that could be used for innovation and production. Therefore, patents assigned to companies in the Republic of Ireland, and not the ones invented in Ireland, are considered.

Nanotechnology patents were identified by the Boolean keyword search for the term 'nano*' (see Hu et al., 2011). We chose a Boolean keyword search instead of the IPC classification Y01N (nanotechnology) to find patents related to nanotechnology, as in the rapidly expanding field of emergent technologies keyword search provides a more precise identification of nanotechnology patents (Bhattacharya and Shilpa, 2011). A separate IPC class for nanotechnology has been established only recently (Dolfsma and Leydesdorff, 2011). It has not been applied by the USPTO up until now. IPC classifications are nevertheless assigned to the data obtained by Boolean search, to help us to identify distinct clusters of knowledge and the sectors affected by nanotechnology (for example, measurement, materials, pharmaceuticals, semiconductors). IPC classes have been matched to industry sectors for approximately 30 per cent of all classes, including the most important ones by node degree.

The search returned 2953 patents assigned to at least one Irish person or organization. The dataset was then split into two subsets. The first subset includes all patents related to nanotechnology (225 patents that contain the term 'nano*' in their title or abstract; subsequently referred to as 'nano patents' – NP). The second subset contains patents unrelated to nanotechnology (2728 patents without a direct relationship to the term 'nano*'; further refered as 'other patents' – OP). Prior to further analysis and graph creation all items without IPC code information were deleted from the dataset (252 items).

Co-classification analysis has been widely used: to identify how knowledge relatedness explains firms' technological diversification (Breschi et al., 2003); to analyse the structure of the biotechnology knowledge (Krafft et al., 2011); and to investigate how Dutch and Indian innovation systems affect the nanotechnology knowledge base (Dolfsma and Leydesdorff, 2011). This method enables policy makers to grasp how the knowledge structure and the innovation system are interconnected, and how manufacturing structure – which depends on available knowledge – may evolve in the short run. Dolfsma and Leydesdorff argue that knowledge from one class would be transferred and applied more easily to another class if co-classification is found. Knowledge codified in patents and thus patent data can be used as an indicator for potential and actual knowledge flow within the respective system, for example a company, or a national innovation system of innovation network (Dolfsma and Leydesdorff, 2011). Using classes as nodes in a network is based on the assumption that these nodes represent discrete concepts, which otherwise would lead to a multilevel network. Since IPC classes are a hierarchical system with discrete divisions between the classes (Leydesdorff et al., 2012), these classes can be considered to be suitable for a unimodal network analysis.

The hypothesis was tested with three co-classification analyses, one for each dataset: a subset of nanotechnology patents (NP), a subset of all other patents (OP – Irish patents that do not mention nanotechnology), and the combined dataset. Table 7A.1 in the Appendix reveals the network statistics for all three datasets. The combined dataset allows us to visualize how the nanotechnology network connects to other fields, and to investigate the effects of the links which only come into existence with the combination of the two datasets, that is, the links which describe the connections between NP and OP. We used Gephi software (Bastian et al., 2009) to visualize and analyse patent networks.

Results

The share of nanotechnology patents in the Irish nanotechnology KB steadily increased from about 2 per cent in the year 1998 to about 8.4 per cent in 2010 (see Table 7A.1 and Figure 7A.1 in the Appendix). Within the same time span the share of new nanotechnology patents on all new patents ranged from 4.6 per cent in 1998 to around 20 per cent in 2009, with the absolute number of new nanotechnology patents remaining relatively stable from 2009 at approximately 30. Although Schmoch and Thielmann (2012) argue that nanotechnology evolution has reached a phase of stagnation, our data indicate that nanotechnology steadily increased in importance for the Irish economy and knowledge base over the decade.

The patent co-classifications map for all nanotechnology-related patents reveals the strong prevalence of organic fine chemistry and pharmaceuticals (coloured in red). A high level of detail is used to classify patents: sections, classes and subclasses. In Figures 7A.2 and 7A.3, the node size represents betweenness centrality (larger nodes have higher betweenness centrality). Nodes in yellow indicate electrical engineering (mainly ICT, including semiconductors), biotechnology nodes are in blue, and basic materials chemistry in green. The graph contains 334 components, having a low graph density of 0.022. The giant component consists of 3149 nodes and 34935 edges. This means that 67.82 per cent of all nodes and 91.84 per cent of all edges are included in the giant component.

In Figures 7A.3 and 7A.4 all edges related to a nanotechnology patent are coloured in red; all others are in blue. Visual inspection reveals that all edges in the network related to nanotechnology are connected to the giant component. This indicates the interdisciplinary nature of nanotechnology development, as newly added patent classes are directly related to classes that are already included in other patents. Figure 7A.2 shows all nanotechnology-related patent classes as a co-classification network. The network is much less coherent, however, not contradicting the notion of a rather path-dependent development as nanotechnology patents are attached to various areas in the knowledge base.

Both the OP as well as the combined network show similar average clustering coefficients (0.786 and 0.782, respectively). With 0.859 the average clustering coefficient of the NP network is higher, the difference however is not yet tested for significance. Average path length is highest for the OP network (5.945), 5.505 for the combined network, and 2.71 for the NP network. With only the average path length being lower in the combined network, the results provide support to our hypothesis, which suggested that the average path length in the combined networks should be lower than in the OP network. In that sense nanotechnology can be seen as technology which connects formerly unconnected areas of the network more directly due to its interdisciplinary nature.

However, further investigation into the evolution of betweenness centralities (see below) is needed to further strengthen the statement of interdisciplinarity and a possibly significant impact of nanotechnology on the average path length. The almost unchanged average clustering coefficient might be interpreted as nanotechnology introducing additional, equally clustered IPC class-cliques.

Figures 7A.3 and 7A.4 show the co-classification map of all patents having an assignee located in the ROI. The OP network has 4270 nodes (classes); the nanotechnology network has 712 classes. When combined, only 4643 classes are mentioned. Thus, there are 373 nanotechnology-specific classes, which are added because of their appearance on a nanotechnology patent.

Betweenness centralities are calculated for the detailed dataset of the combined and the nanotechnology network. The betweenness centrality measures how often a node 'is located on the shortest path (geodesic) between other nodes in the network' (Leydesdorff, 2007: 1304). It therefore also takes into account all indirect links of the respective node and is normalized by the number of all geodesic distances in the graph. In Figure 7A.5 the highest 500 values of the combined and the NN network are plotted, whereas for the nanotechnology network all values are included (191 values). In general, betweenness centrality can be taken as an indicator for interdisciplinarity (Leydesdorff, 2007). Future research could address how the betweenness centrality distributions evolve.

DISCUSSION AND CONCLUSION

With its widespread applicability and thus economic potential, nanotechnology is, on the one hand, of great interest for policy makers. For this reason it is important to understand which factors influence the evolution of this emerging technology and which policy instruments could be applied to foster knowledge integration for economic benefit. This applicability, on the other hand, also raises the question of how this new technology relates to already existing knowledge areas. Scholars have endeavoured to understand the developmental trajectories of nanotechnology by focusing on how nanotechnology is embedded within the adjacent disciplines, and looked for the multidisciplinary or interdisciplinary indicators of its development. We focused on the question of whether nanotechnology is more multidisciplinary, thus affecting other technologies in a rather uncorrelated way, or more interdisciplinary, bringing existing areas of knowledge together. For our case study we chose Ireland and the patents in the Irish patent portfolio, as it already has a competitive edge in the field of nanotechnology and its applications. Evidence suggests that nanotech develops incrementally, in an interdisciplinary manner. With a network representation of the knowledge available to Irish companies in the form of patents, we looked at whether the nanotechnology network in Ireland also spans several unrelated technology areas. Our analysis indicates that nanotechnology loosely connects various fields of knowledge, providing some support for nanotechnology being of an interdisciplinary nature.

Further questions remain, however. Evolutionary analyses of the networks may reveal whether a multidisciplinary character within the networks is a constant or a temporary phenomenon. In other words, does it change from earlier stages of the network evolution to later ones? Interdisciplinarity in this sense might be only a recent phenomenon, that could gain in importance over time. This would imply that even more newly established connections would bridge nanotechnology with various application areas. If so, fostering the establishment of these interdisciplinary bridges may hold great potential for innovation: novel patent class combinations could be promising areas for policy interventions. A citation analysis would provide more detailed insights: patents often cite from other technology areas without being classified in these. Citation mapping might unveil direct links between various technologies (Igami, 2008).

The importance lies in identifying which nanotechnologies have the highest impact and fit best into the already existing industrial knowledge network. This is to help policy makers to allocate resources accordingly, and to focus on the commercialization of certain nanotechnologies, as is being proposed in the Irish Nanotechnology Commercialisation Framework (Forfás, 2010).

NOTE

1. Although the number of patents granted in one year may be affected, for example, by the availability of human resources at the respective patent office, aggregation among multiple years smoothes these differences and provides an insight into the overall structure of the network.

REFERENCES

Bastian, M., Heymann, S. and Jacomy, M. (2009). Gephi: an open source software for exploring and manipulating networks. International AAAI Conference on Weblogs and Social Media.

Bhattacharya, S. and Shilpa, M. (2011). Mapping nanotechnology research and innovation in India. *DESIDOC Journal of Library and Information Technology*, 31(5), 349–358.

Breschi, S., Lissoni, F. and Malerba, F. (2003). Knowledge-relatedness in firm technological diversification. *Research Policy*, 32(1), 69–87.

Choi, H. and Mody, C.C.M. (2009). The long history of molecular electronics: microelectronics origins of nanotechnology. *Social Studies of Science*, 39(1), 11–50.

Dolfsma, W. and Leydesdorff, L. (2011). Innovation systems as patent networks: The Netherlands, India and nanotech. *Innovation-Management, Policy and Practice*, 13(3), 311–326.

Forfás (2010). Ireland's nanotechnology commercialisation framework 2010–2014. http://www.forfas.ie/media/forfas310810-nanotech_commercialisation_framework_2010-2014.pdf (accessed 30 November 2014).
Griliches, Z. (1979). Issues in assessing the contribution of research and development to productivity growth. *Bell Journal of Economics*, 10(1), 92–116.
Hu, G., Carley, S. and Tang, L. (2011). Visualizing nanotechnology research in Canada: evidence from publication activities, 1990–2009. *Journal of Technology Transfer*, 37(4), 550–562.
Huang, C., Notten, A. and Rasters, N. (2010). Nanoscience and technology publications and patents: a review of social science studies and search strategies. *Journal of Technology Transfer*, 36(2), 145–172.
Igami, M. (2008). Exploration of the evolution of nanotechnology via mapping of patent applications. *Scientometrics*, 77(2), 289–308.
Krafft, J., Quatraro, F. and Saviotti, P.P. (2011). The knowledge-base evolution in biotechnology: a social network analysis. *Economics of Innovation and New Technology*, 20(5), 445–475.
Leydesdorff, L. (2007). Betweenness centrality as an indicator of the interdisciplinarity of scientific journals. *Journal of the American Society for Information*, 58(9), 1303–1319.
Leydesdorff, L., Kushnir, D. and Rafols, I. (2012). Interactive overlay maps for US patent (USPTO) data based on International Patent Classification (IPC). *Scientometrics*, 98, 1583–1599.
Menz, N. and Ott, I. (2011). On the role of general purpose technologies within the Marshall–Jacobs controversy: the case of nanotechnologies. Karlsruhe Institute of Technology, Working Paper Series in Economics, No 18.
Meyer, M. (2007). What do we know about innovation in nanotechnology? Some propositions about an emerging field between hype and path-dependency. *Scientometrics*, 70(3), 779–810.
Miyazaki, K. and Islam, N. (2007). Nanotechnology systems of innovation – an analysis of industry and academia research activities. *Technovation*, 27(11), 661–675.
Nelson, R.R. (1993). *National Innovation Systems: A Comparative Analysis*. New York: Oxford University Press.
Nesta, L. and Saviotti, P.P. (2005). Coherence of the knowledge base and firm innovative performance: evidence from the US pharmaceutical industry. *Journal of Industrial Economics*, 53(1), 123–142.
OECD (1996). The knowledge-based economy. *General Distribution*. http://www.oecd.org/sti/sci-tech/1913021.pdf (accessed 30 November 2014).
OECD (2009). OECD Science, Technology and Industry Scoreboard 2009. OECD Publishing. http://dx.doi.org/10.1787/sti_scoreboard-2009-en (accessed 30 November 2014).
Onel, S., Zeid, A. and Kamarthi, S. (2011). The structure and analysis of nanotechnology co-author and citation networks. *Scientometrics*, 89(1), 119–138.
Porter, A.L. and Youtie, J. (2009). How interdisciplinary is nanotechnology? *Journal of Nanoparticle Research: An Interdisciplinary Forum for Nanoscale Science and Technology*, 11(5), 1023–1041.
Rafols, I. and Meyer, M. (2007). How cross-disciplinary is bionanotechnology? Explorations in the specialty of molecular motors. *Scientometrics*, 70(3), 663–650.
Saviotti, P.P. (2007). On the dynamics of generation and utilisation of knowledge:

the local character of knowledge. *Structural Change and Economic Dynamics*, 18(4), 387–408.
Saviotti, P.P. (2009). Knowledge networks: structure and dynamics. In Pyka, A. and Scharnhorst, A. (eds), *Innovation Networks: New Approaches in Modelling and Analyzing*. Dordrecht: Springer.
Saviotti, P.P., De Looze, M.-A. and Maupertuis, M.A. (2005). Knowledge dynamics, firm strategy, mergers and acquisitions in the biotechnology based sectors. *Economics of Innovation and New Technology*, 14(1–2), 103–124.
Schmoch, U. and Thielmann, A. (2012). Cyclical long-term development of complex technologies: premature expectations in nanotechnology? Fraunhofer ISI Discussion Papers, Innovation Systems and Policy Analysis, No 31.
Schummer, J. (2004). Multidisciplinarity, interdisciplinarity, and patterns of research collaboration in nanoscience and nanotechnology. *Scientometrics*, 59(3), 425–465.
Yu, G., Wang, M.-Y. and Yu, D.-R. (2010). Characterizing knowledge diffusion of nanoscience and nanotechnology by citation analysis. *Scientometrics*, 84(1), 81–97.

APPENDIX

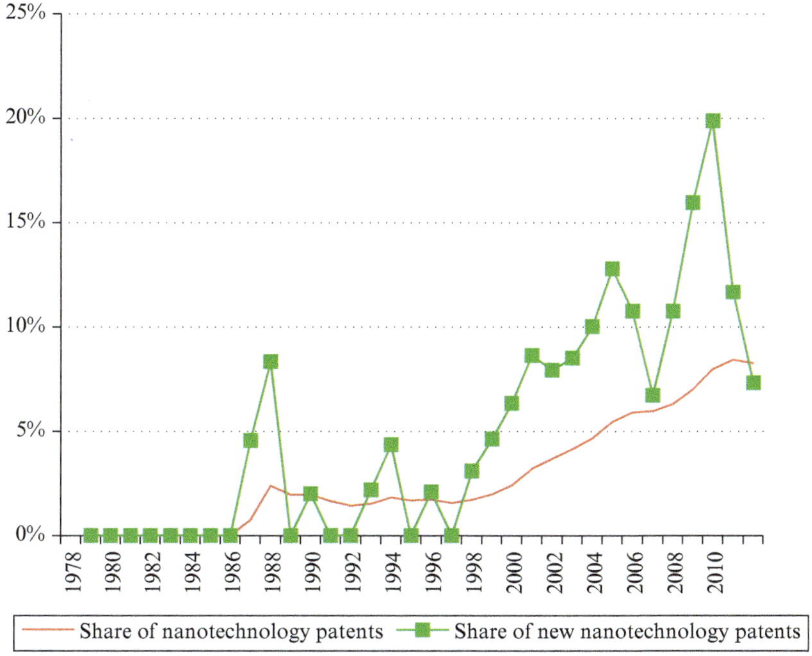

Figure 7A.1 Annual and cumulative share of nanotechnology patents (%)

Table 7A.1 Descriptive network statistics

	Nano patent network	Other patents network	Combined network
Nodes	711	4270	4643
Edges	6997	31369	37734
Av. degree	9.841	7.346	8.127
Av. clustering coefficient	0.859	0.786	0.782
Av. path length	2.710	5.945	5.505
Graph density	0.026	0.003	0.003

Figure 7A.2 Co-classification map of nanotechnology patents in Ireland

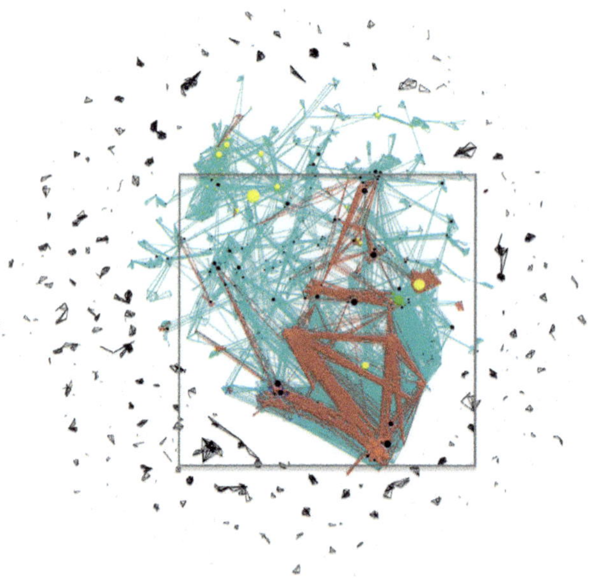

Note: Node colours: yellow: electrical engineering; blue: biotech; green: basic materials chemistry; red: organic chemistry and pharmaceuticals.

Figure 7A.3 All patents co-classification map

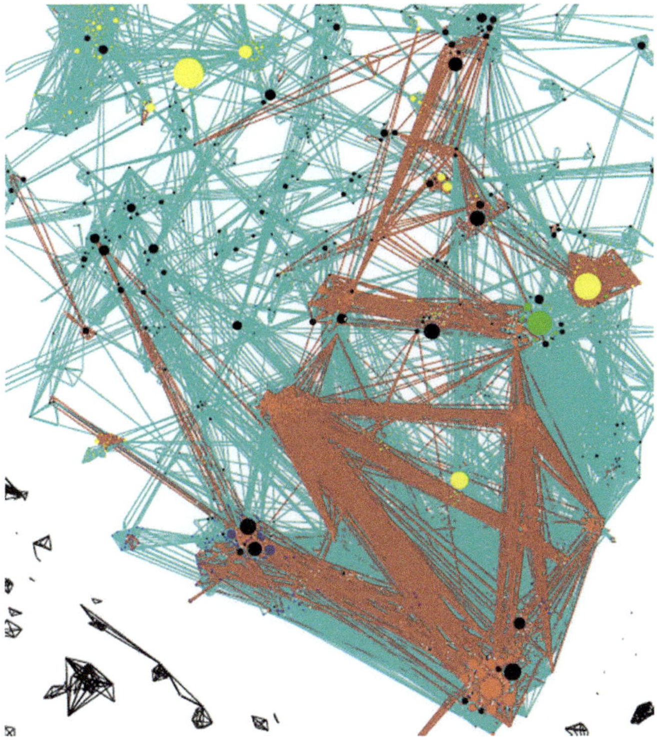

Figure 7A.4 Detail of Figure 7A.3

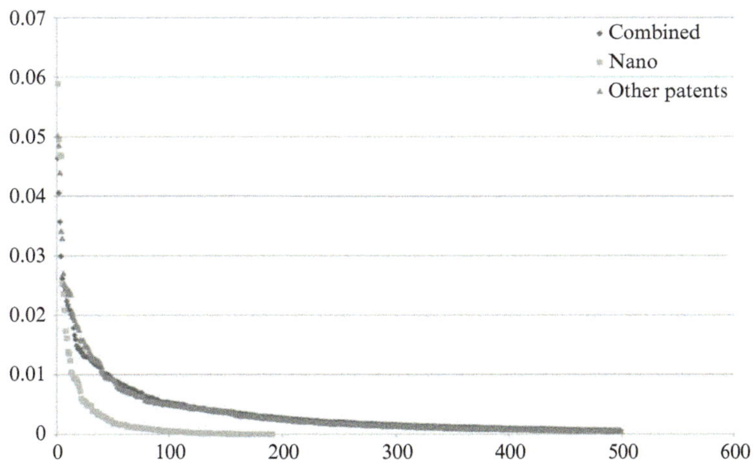

Figure 7A.5 Betweenness centrality distributions

8. Research clusters and emerging industrial architecture of the Irish renewable energy sector

Bei Gao

INTRODUCTION

Facing the future scarcity of fossil fuels, severe environmental pollution and the climate change caused through combustion of conventional energy sources, it is widely acknowledged that a substantial launch of renewable energy technologies is taking place (OECD IEA, 2007). Sustainable and competitively priced supplies of energy are fundamental for a country's economic development and international competitiveness. In order to make cleaner energy supplies sustainable and affordable, robust control of the energy industry by governments is not uncommon.

Renewable energy (RE) technology is a multidisciplinary field with an expanding knowledge base and widely dispersed sources of expertise. High development costs are associated with long time lags in investment return (Cooke, 2002). Therefore, in the RE field, innovation takes place in networks of learning. Companies and research institutions engage in research and development (R&D) networks to facilitate a broad range of knowledge exchange and cope with high uncertainty. Due to the specific environmental target of RE technology, the benefit of RE technology tends to accrue to social well-being rather than to innovators and early adopters of such technology. Market forces offer little incentive for RE development, thus as an alternative, policy initiatives and public funds for research, development and demonstration provide the first impetus for developing RE technology (Popp et al., 2010). These characteristics of RE technology make it an industry highly dependent upon the political system, institutional factors and innovation environments in a national context.

The purpose of this research is to investigate the performance characteristics and development opportunities of the RE innovation network, and to look deep into the question of how policy can foster the development of the RE innovation network. This chapter presents the first findings, illustrating emerging research clusters as well as industrial structure change, and exploring the role of R&D funding and policy instruments on the formation of RE research clusters as well as industrial structure.

The chapter is organized in the following way. It first presents the theoretical background of complex innovation systems, including national, sectoral, technological innovation systems, as well as evolutionary economics perspectives. The following section presents selected datasets and the methodology of the research. The chapter then shows the Irish policy initiatives and instruments in recent years in conjunction with the efforts made at a European scale. The structure of research clusters is visualized with social network analysis tools, and a discussion of the relation between funding and research clustering activities is provided. Policy instruments and emerging industrial architecture are presented. The last section summarizes preliminary findings and provides an outlook regarding future work.

THEORETICAL BACKGROUND

Complex Innovation Systems

Different approaches in Systems of Innovation (SI)

The development of innovation is a complex and interactive process based on learning among organizations (Dosi and Nelson, 2010). Firms do not innovate in isolation, but rather in systems of innovation that include 'important economic, social, political, organisational, institutional, and other factors that influence the development, diffusion, and use of innovations' (Edquist, 1997). The main emphasis of a systems approach to innovation was initially on national innovation systems (Freeman, 1987; Lundvall, 1988; Nelson, 1993; Edquist, 1997). Since the 1990s, new concepts of systemic characteristics of innovation, concerning other levels of the economy than the nation state, have been developed. Technological systems (Carlsson and Stankiewicz, 1991) focus on knowledge or competence flows, and attach more importance to technology diffusion and utilization. Regional systems of innovation deal with innovative regions within countries' geographical boundaries, or including parts of different countries (Cooke et. al., 1997; Maskell and Malmberg, 1999; Asheim, 1999). Sectoral innovation systems (Breschi and Malerba, 1997) emphasize firms active in making products and generating technologies within specific technological fields or industries. Intersectoral diffusion of knowledge and co-evolution of national and sectoral or technological systems are considered as the primary determinants of the international competitiveness of an industry (Hemert and Nijkamp, 2010).

Components and functions of SI

Interdependent organizations (firms, research institutions, governmental agencies, and so on), and institutions that constrain or stimulate the organizations to conduct innovation, are considered to be crucial components in all versions of the innovation systems. The institutions not only shape and are shaped by the actions of heterogeneous organizations, but also affect how the organizations perform the functions. The overall functions of an innovation system were summarized by Johnson and Jacobsson (2000): the creation and diffusion of new knowledge, the guidance of the direction of the technology search process, the supply of resources, the creation of positive external economies, and the formation of markets. System failure occurs when there is an absence of certain functions or a deficiency in the functioning of the system.

Relevance to the RE sector

RE development faces severe problems in resource supply and market formation, due to the dominance of the incumbent conventional energy suppliers and environmental externalities such as pollution costs. The state and relevant policies play a crucial role in the energy domain in providing initial impetus to RE technology R&D and boosting its industrial competence. However, a critique of the system of innovation approach is that it lacks a component examining the role of the state (Edquist, 2001). Based on the analytical framework provided by sectoral and national innovation systems, the present chapter highlights the importance of the state, referring to government initiatives, policy instruments and monitoring agencies promoting RE technology creation and implementation. Special attention is paid to firms, since competition and collaboration in R&D and market activities are essential components that link innovation and economic growth (Breschi and Malerba, 1997).

Evolutionary Economics

Path-dependency on the system level

The system of innovation approach takes a long-term historical perspective and considers innovation processes to be evolutionary. Innovation systems are characterized by non-linearity and non-equilibrium interactions between the components within them (Dosi and Nelson, 2010). In the learning and decision-making process of an innovation system, a strong tendency of path-dependency has been identified. David (2001) defined the term 'path-dependency' as 'a dynamic property of allocative processes', which captures the idea that history matters. Three causes for path-dependency in the process of technological change were specified: (1) technical interrelatedness of system components; (2) switching costs; and (3) increasing returns to scale (David, 1985). On a system level, related industries tend to congregate geographically for economical resources (Krugman, 1991), with imperfect information and limited foresight. Positive feedbacks among agents' choices lead to the permanence of lock-in to a specific path of allocation (Puffert, 2004). Increasing returns arise when a particular technology has specialized products or service and a large number of buyers. Positive network externalities increase the benefit for existing users while the total number of users rises (Arthur, 1989, 1994). As a result, switching costs are even higher.

For the energy sector, fuel choices in the past, related industries such as construction and electricity distribution, and decisions taken on research investments are interdependent and mutually reinforcing. External costs associated with combustion of fossil fuels, such as environmental and health costs, are not fully internalized into their price (Department of Public Enterprise Ireland, 2002). Under such conditions, customers are not willing to pay higher prices for cleaner energy. Challenges remain because of the high switching cost in terms of infrastructure development, the monopoly of conventional energy producers and the difficulties in achieving broader acceptance from users. Energy system transition is largely affected by the coordination of diverse policies, such as innovation policies promoting research and commercialization of scientific results, policies enhancing the competence of the RE industry, as well as environmental and other regulations. Thus, the implications of an evolutionary perspective are fundamental for the understanding of technology diffusion processes of the RE industry as well as for the analysis of relevant policies.

Path-dependency on the organization level

On the organizational level, path-dependent features can be identified from the cumulativeness of absorptive capacity and its effect on expectations formation. Absorptive capacity is defined by Cohen and Daniel (1990) as 'the ability of a firm to recognize the value of new, external information, assimilate it, and apply it to commercial ends [which] is critical to its innovative capabilities'. An organization's prior related knowledge and its diversity of background directly affect the formation of its absorptive capacity. Firms that conduct their own R&D are more sensitive and better able to grasp externally available knowledge. Insufficient investment in an area of expertise in the early stages is likely to prohibit the potential development of a technical capability in that domain. However, most innovations result from 'borrowing' rather than invention (March and Simon, 1958): learning from other organizations tends to have an accumulative effect on a firm's competence. Copying and learning from others are very common in the RE sector, due to the high technology development costs and long lag of investment return. Tacit elements of knowledge (Lam, 2000) are difficult to transmit simply by instructions. Tacit knowledge is often arranged and operated in dissimilar ways, based on path-dependent routines for a specific technique, across organizations in the same industry. These routines are formed by encoding inferences from history and changes in response to direct organizational experience, seen as a learning-by-doing (Arrow, 1962) process. Learning itself can be viewed as knowledge through which organizations develop competence (Levitt and March, 1988). Consequently, it is necessary to investigate the collaboration among innovative firms and their clustering activities with other associations, in order to understand the dynamic of the innovation process of an emerging technology.

DATA AND METHODOLOGY

Empirical Study

The empirical study started with mapping the relevant policy initiatives and instruments. Then three sets of data were collected: (1) R&D funding and research clustering (SEAI, 2011); (2) the population of enterprises and other organizations working in the renewable energy field; and (3) a trend of wind energy firms founded over the previous ten years.

R&D funding calculated the total funding for renewable energy infrastructure building, research network formation, education and research projects from 2004 to 2010. The initial dataset of research clusters consists of 96 organizations made up of universities, research institutions, governmental bodies and industrial enterprises, which form 15 clusters in nine technological categories. In order to identify the connection between technology type and the role of relevant actors in the innovation network, a visualization of data is applied with social network analysis (SNA) tools.

With the purpose of showing the current industrial structure of the RE sector, a list of active players in the renewable energy field containing roughly 300 firms and non-profit organizations was collected from different sources. The search process started from the LinkedIn website with a keyword search and a snowball sampling method. Search strings were based on a list of keywords generated by the World Intellectual Property Organization for alternative energy. RE firms that are listed by Energy

Ireland (the biggest energy forum in Ireland) and Enterprise Ireland (a government organization responsible for helping Irish enterprises start, grow and innovate) were then added to the dataset. Each firm's website was examined to identify whether it should be included in this study, depending on its main products or services. Only firms whose business is directly related to energy generation from renewable sources were chosen. The dataset contains the name and type of organization, location, year founded, technology specialty, value proposition, contact information and relation with other organizations if applicable.

Additionally, company registration data on Irish firms under the NACE Code 40.1 'Production and distribution of electricity' were accessed. As wind energy has the greatest potential and largest installed capacity among various types of renewable energy sources in Ireland, all 149 wind energy firms founded in the previous ten years were selected to illustrate a trend of firms entering the market responding to relevant policies.

Social Network Analysis

For technology-intensive industries, with a large knowledge base and discrete expertise, innovation takes place in networks of learning (Ahrweiler, 2010). Interorganizational innovation networks and the relevant structural features provide explanations for how innovation happens. Research on social interactions facilitating knowledge diffusion and technology advance has increased substantially in the last two decades (Powell et al., 1996; Smith and Ho, 2006; Ahrweiler, 2010).

A network is defined by Brass et al. (2004) as 'a set of nodes and the set of ties representing some relationship, or lack of relationship, between the nodes'. From an evolutionary perspective, changes in network structures occur as organizations enter or exit a network, and construct or dissolve relationships with other agents in the network (Glückler, 2007). The chance and ability to access and join knowledge- and learning-intensive relations determine the socio-economic position of organizations (David and Foray, 1995).

The relationships among nodes are expressed by the ties that can be directed or undirected. Degree of a node considers the number of links incident with each node in a network graph. It is a principal measure of centrality of a node, represented by the following equation:

$$D(i) = \sum_{j \in I \neq i} x_{ij}$$

Degree distribution is often used as an indicator of whether tie formation in a network is equiprobable for all or preferential to already connected nodes (Powell, 2005).

A node is considered as a gatekeeper in a graph when other nodes need to reach each other through this node. Betweenness measures the extent to which a node is placed among other nodes. A gatekeeper usually has high-value in-betweenness. It is at the boundary of a subgroup in a network, controls access of outsiders to the group, and plays a significant role in information and knowledge flow:

$$B(i) = \sum_{\substack{j \neq i \neq k \in I \\ j \neq k}} \frac{\sigma_{jk}(i)}{\sigma_{jk}}$$

Density is one of the most commonly used measures to illustrate the cohesion of a network. The maximum number of ties that can be present in a network is $n(n - 1)/2$. Density refers to the actual ratio of the ties present to the maximum possible:

$$\Delta = \frac{2l}{n(n - 1)}$$

The toolkit used to implement SNA study is Gephi software, which is open source software suitable for the visualization and exploration of complex networks. Various measures such as network degree, density, betweenness and closeness (Wasserman and Faust, 1994) are accessible in Gephi.

RESULTS

Ireland is positioning itself as a key actor in the export market of environmental goods and services (Government of Ireland, 2008). On the one hand, Ireland's limited indigenous energy resourses makes it greatly exposed to the fluctuation of global energy prices, and there is an emergent demand to reduce the high reliance on imported fossil fuels. On the other hand, large amounts of renewable resources for energy generation are potentially available: for example the potential of wind power is 5000 megawatts[1] which is sufficient to fulfil national peak electricity demand. Policy assistance and institutional frameworks capable of handling the complexity of energy transition are significant for promoting security of the energy supply. In this section, important policy initiatives and policy instruments for promoting renewable energy are presented. The relationship between public intervention and research clustering activities as well as the industrial architecture transformation are shown.

Policy Initiatives and Research Clusters

Policy initiatives

The Irish government has consistently supported renewable energy development through its policies, together with Europe-wide targets. European Union (EU) Council Decision 2002/358/EC set Ireland's target to limit its emissions to 13 per cent above its 1990 levels (OECD IEA, 2007). Great attention has been paid by the Irish government to renewable energy since the passing into law of the Sustainable Energy Act (Government of Ireland, 2002). The Sustainable Energy Act set up ambitious targets of 30 per cent renewable energy in electricity consumption by 2020, which was a milestone for the development of the RE sector. This was followed by an Energy Policy Green Paper in 2006, 'Towards a Sustainable Energy Future for Ireland' (Government of Ireland, 2006), which included consultations on all aspects of national energy policy since the 1970s. The government's White Paper on Energy Policy (Government of Ireland, 2007) was launched in 2007, which referred to the whole category of energy, announced national renewables targets and described the energy policy framework out to 2020. A policy document titled 'Building Ireland's Smart Economy' in 2008 (Government of Ireland, 2008) claimed that 40 per cent renewable electrical energy supply is technically feasible. The EU Directive 2009/28/EC, 2009 Revised Programme for Government (Government of Ireland, 2009) set a 40 per cent renewable electricity target and reinforced ocean energy policy after the 2006 Ocean Energy Strategy (Department of Communication, Marine and Natural Resources, 2006). In 2010, Ireland's National Renewable Energy Action Plan (Government of Ireland, 2010) set a new range of renewables objectives, with a special focus on offshore wind resources (Department of Communications, Energy and Natural Resources, 2010).

Research funding and clusters

As the major sponsor of renewable energy R&D, the Irish government provides more than 77 per cent of funding for universities and research institutes. A large proportion comes from the European Union, consisting of 11 per cent of the total funds. Ireland has successfully participated in 18 funded energy projects in FP7 and received €10.8 million from the European Commission since 2007, as a competitive participant in the upcoming FP7 calls in electrical grid, marine technology and energy efficiency, for example. Furthermore, industrial corporate bodies supply 2 per cent of funding to R&D (SEAI, 2011).

A breakdown of funding by technology category between 2004 and 2010 includes 21 per cent for energy efficiency and 18 per cent for bioenergy.

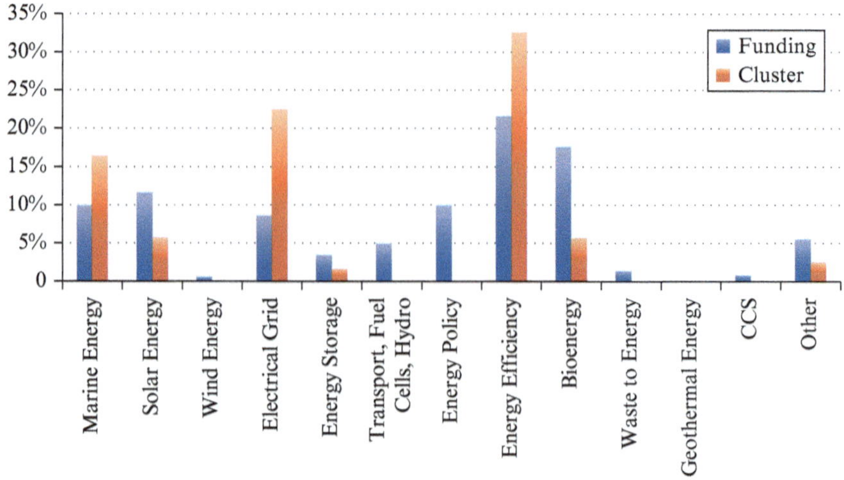

Figure 8.1 R&D funding by energy technology category 2004–2010 versus clusters size

Other fields of significant activity are solar energy (13 per cent), marine energy (10 per cent), energy policy (10 per cent), and the electrical grid (9 per cent), as shown in Figure 8.1. The second series of columns shows the respective research cluster size (adjusted according to the number of participants) in each technology category.

Comparing the two series, cluster size is generally coherent with the funding ratio to each category. Nevertheless, there are obvious differences in the energy policy area and transport, fuel cells and hydrogen. Firstly, for energy policy research, the Sustainable Energy Authority of Ireland, Forfás, the Environmental Protection Agency and a number of universities devote themselves to research on energy policies. However, the lack of official collaboration is the main reason explaining the inconsistency of relatively high energy policy funding and lack of evidence on research clusters. Secondly, transport, fuel cells and hydrogen, wind energy and carbon capture and storage (CCS) are in a more developed stage rather than fundamental research. Therefore, research activities are correspondingly less due to the product life cycle phase (Levitt, 1965). Thirdly, for marine energy and electrical grid, Figure 8.1 shows a very positive result of research clustering. Marine energy research is profiting from the policy orientation of recent years (Department of Communication, Marine and Natural Resources, 2006). Meanwhile, similarly for both marine energy and the electrical grid, the well-developed marine and information and

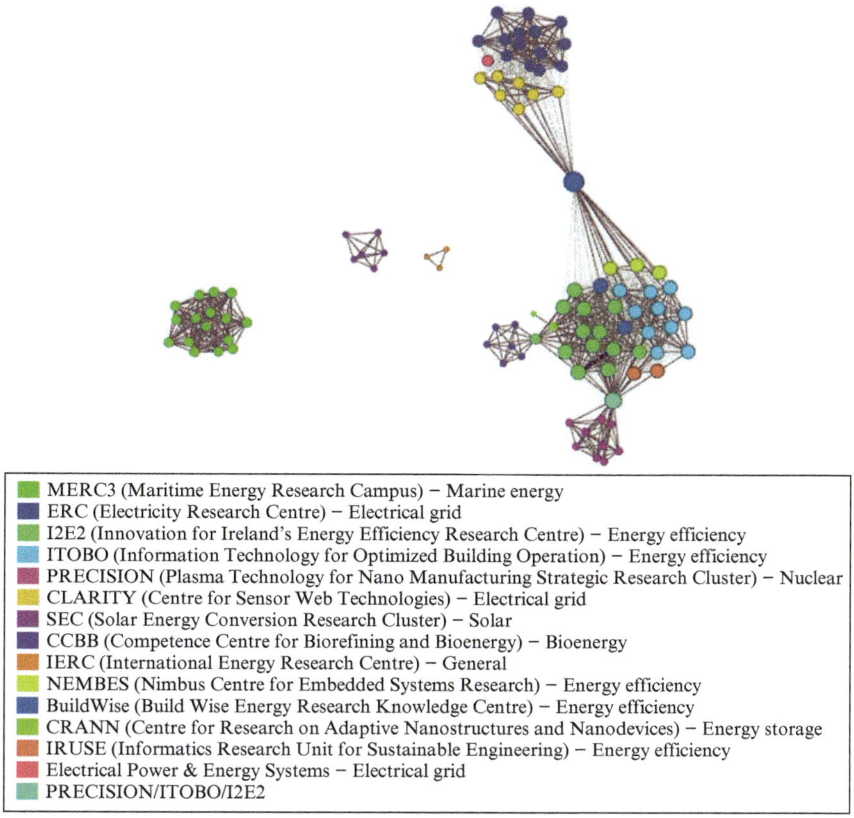

Figure 8.2 *Irish renewable energy research clusters*

communication technology (ICT) collaboration networks easily form new cooperations in related research fields.

In addition to RE capacity building, community building is always a central concern of Irish policy. Figure 8.2 demonstrates the network structure of RE research clusters. This network is based on the data on research clusters and centres from the Sustainable Energy Authority of Ireland (2011), generated using the Gephi software package. It shows 15 research clusters in nine technological areas. Nodes present the organizations involved in one or more research projects. The weight of ties is adjusted according to participation in the same project (weight: 3) and the same technology categories (weight: 2).

A giant component in the network consists of the energy efficiency, energy storage, electrical grid, bioenergy and nuclear clusters. Marine, solar energy and the International Energy Research Centre (IERC) group that facilitates RE expansion in general are independent sub-networks in the entire structure. Density of the network calculated as 0.206 indicates a relatively loose structure. A lack of integration of different technologies in the R&D collaboration network may lead to difficulties in the commercialization phase, when the renewable generators connect to the electric distribution system. Meanwhile, the average path length of 2.074 guarantees efficient information flow through the relative short path within each sub-network.

Organizations' proximity promotes common ground and interaction. Gatekeepers linking different research fields are essential for the successful development of a cluster and the promotion of new technology. Information and knowledge diffusion among clusters is enabled by the gatekeepers between two clusters. The electrical efficiency cluster is the largest research cluster and occupies the central position in the giant component, connecting with the electrical grid, bioenergy and nuclear clusters respectively via the gatekeepers of the Tyndall National Institute (University College Cork), the STOKES Institution (University of Limerick), and Intel. The three gatekeepers and TrinityHaus (Trinity College Dublin) hold top ranking on globe centrality, listed in Table 8.1. The Tyndall National Institute and Intel are also the top two in degree calculation, which indicates a local centrality of the nodes in a graph. The size of nodes was scaled to a standardized network degree in the entire network, representing the degree of connectivity among the organizations. Not surprisingly, the central nodes are multinational enterprises and universities with a broad-ranging knowledge base and high absorptive capacity. A tendency of preferential attachment by incumbency, and preferences for diversity, can be seen from the degree distribution.

Considering the organizational type, besides 10 per cent governmental agencies, 46 per cent of participants within the network are universities, and the other 44 per cent are industrial partners. This ratio demonstrates a high level of industrial contribution in research, and implies that renewable

Table 8.1 Betweenness centrality, top four organizations

Organization	Technology category	Value
Tyndall National Institute (University College Cork)	Energy efficiency	0.237
Intel	Energy efficiency	0.123
STOKES (University of Limerick)	Energy efficiency	0.086
TrinityHaus (Trinity College Dublin)	Energy storage	0.015

energy technology has reached an industry-driven stage of innovation, in which the implementation of research results is critical.

Policy Interventions and Emerging Industrial Architecture

Policy instruments

Although Ireland has been keen on building its competitiveness in renewable energy, challenges lie ahead for the implementation of RE technologies. High development costs and long time lags in the return on investment provide little incentive to private investment. Meanwhile, the free-rider problem is one of the primary reasons why the huge potential capacity of renewable energy remains unreached (Wiser and Pickle, 1997). The disadvantage is further reinforced by the dynamic increasing returns of production until the clean energy market achieves a critical mass. Renewable energy is not able to successfully compete with conventional energy in an open market due to the unique features of the industry. Multilevel policy intervention is indispensable for stimulating the implementation of RE technologies.

National energy policies have been influenced by developments on the international level, particularly on the EU level. Major support mechanisms commonly used in European countries include competitive bidding (tendering) schemes, feed-in tariff schemes, green credits trading and green pricing schemes. A competitive tendering scheme was introduced in Ireland with the launch of the Alternative Energy Requirement Programme in 1996 (Government of Ireland, 1996). Though the tendering scheme brought low costs of renewable energy plants, developers bid too low to be profitable from the projects. Failure occurred in achieving planning permission, known as the 'winner's curse' (OECD IEA, 2007). Apart from the low price, another reason for the failure is the tight time frame of the projects. In 2003, tax relief was partially or fully granted to energy produced from renewable sources, according to Directive 2003/96/EC (European Commission, 2003). After that, the Renewable Energy Feed in Tariff scheme (REFITs) was introduced in 2006. Furthermore, direct capital grants issued by the state to assist developers in sourcing their remaining capital requirements were in place for a long period from the 1990s, and significant amounts of grants were realized on a yearly basis from 2002.

The success of policy instruments greatly depends on the respective framework conditions in the country and the specific style of promotion models used (Reiche and Bechberger, 2004). Given that the Irish government appears not to be interested in increasing the number of small independent operators, REFITs suit the Irish context to minimize commercial

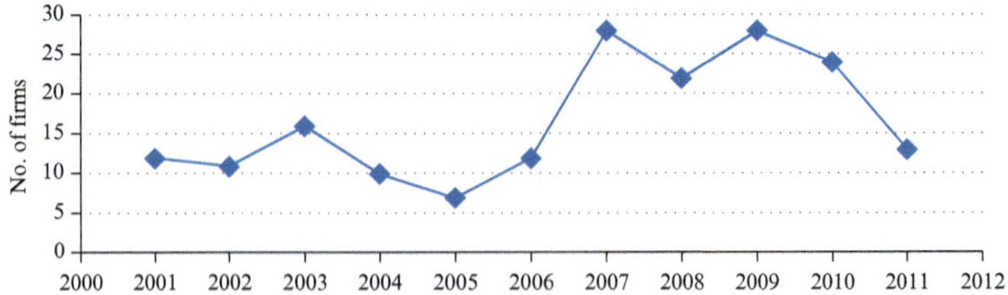

Figure 8.3 New wind energy firms, 2001–2011

risk by supporting the capital cost for renewable installations. Figure 8.3 illustrates increased numbers of wind energy firms established over the ten years to 2011. Fast-growing numbers of firms are shown in 2002, and then a boom period can be seen from 2006, responding to the introduction of REFITs. After the peak in 2007, development slowed down when the financial crisis hit the country.

Industrial architecture

Regarding the industrial structure of the renewable energy sector, although state companies for power generation and distribution, and multinational equipment suppliers, still dominate the market, a large number of small and medium-sized enterprises (SMEs) providing energy consultancy and services, and developing small-scale equipment, have become more and more active. Non-profit organizations (NPOs) working in wind farm community fund raising or providing wind energy usage training to local citizens, as well as venture capital, also play an important role. Table 8.2 illustrates the classification and technological specialities of the emerging firms. Entrepreneurs include experts with years of experience in the energy industry, researchers in universities, and active users of or investors

Table 8.2 Renewable energy industrial architecture

Specialty	Type	Source	Technologies type
Equipment/ services	SME	University spin-offs Branches of multinational enterprises	Wind, solar, efficiency Biomass, ocean
Consultancy	SME	University spin-offs Entrepreneurs	General, wind, efficiency
Community	NPO	Entrepreneurs	Wind, ocean, solar
Investment	VC	Financial firms	General

in community renewable energy. Branches of multinational companies have entered the renewable energy market from other fields, including civil engineering, mechanical and electrical engineering, natural resource management, and conventional energy enterprises. Not specializing in one type of technology, many firms are involved in both small-scale wind and solar energy, or even provide a wide range of services within wind, solar, biomass and energy efficiency. Besides wind energy, fast-growing technology implementations are around small-scale biomass and solar power for end-users, due to the implementation of a series of schemes such as BioHeat Alternative Heat, Greener Homes, and ReHeat Deployment schemes (SEAI, 2012) in the past few years.

Although a number of SMEs entered the RE market in the ten years to 2011, some of them were not able to survive for longer than a two- or three-year period. Young firms receive funding and support from governmental agencies and other public bodies, as they contribute to public goods, not only in terms of R&D outcomes and tangible innovation, but also in terms of contributions to social welfare and long-term societal challenges related to climate change and sustainability. However, these funds, which are meant by policy to foster private actors to help with societal issues, do not help a considerable number of young firms through the 'valley of death'. An overproportionally high exit rate for young firms was observed. In addition, large firms may draw back from their investment and engagement in early phases, when policy support is inconsistent.

It is common that large energy generation companies outsource project planning, site analysis and other types of consultancy or services to SMEs. Such SMEs having formal contracts with large state-owned companies tend to be long-lived. Many cases show that participating in collaborate R&D work with others has a positive effect on a firm's longevity. For instance, university spin-offs are able to carry out research in university laboratories at little cost to get through the 'valley of death'.

CONCLUSIONS AND FUTURE WORK

The structure of research clusters shows a relatively coherent relationship between R&D funding and research activities in the Irish renewable energy sector. Public support in the form of government-funded R&D projects triggers the formation of collaboration networks in research. Both the performance of an innovation system and the organizations within it benefit from knowledge exchange within collaboration networks. Thus, the outputs of collaborative projects need to be measured beyond tangible scientific results. It is important for policy makers to understand how organizations choose their partners and how knowledge diffuses among the networks, in order to facilitate research activities in a dynamic way. Furthermore, integration of various projects in the early stages requires public monitoring agencies to play a role in linking players in different research fields.

Due to the monopoly of conventional energy sources, the costs of externalities of pollution, the imperfection of information flows, and extremely slow return on investment, the RE industry faces an extra-complex process of innovation. The RE industry is characterized by a high significance of regulatory drivers, and demands multilevel policy intervention within a national innovation system. Unguided energy markets are unlikely to deliver renewable energy targets. A significant lag between technological inventions and implementation was found. Technology-push policies focusing on early-stage R&D and market engagement projects do not provide enough incentive for a massive implementation of renewable energy. A high exit rate of SMEs can be seen as a result of high uncertainty in the RE market and inconsistency in policy support. Creating a supportive environment for firms to carry out sustainable energy targets on different stages is essential for the implementation of RE technologies.

This chapter provides a limited contribution on the way towards an integrated framework for the analysis of the energy system transition and the impact of relevant policies. In the next step of our research, public interventions will be classified into categories including policies aiming for knowledge creation, R&D cooperation, commercialization, competence for industry, and regulations on environmental and other aspects. The impact of policy instruments will be analysed, using semi-structured interviews with firms, NPOs and research institutions working in RE fields. The performance and interrelation of organizations in the industrial architecture will be mapped out via surveys and examined via SNA tools; expected outcomes consist of the reconfiguration of the existing industry architecture influenced by innovation policy and the interaction of different players. In further phases of the study, an analytical framework will be developed and compared with other existing analytical frameworks.

NOTE

1. Available at http://news.cnet.com/8301-11128_3-9888020-54.html.

REFERENCES

Ahrweiler, P. (ed.). 2010. *Innovation in Complex Social Systems*. Routledge: London.

Arrow, K.J. 1962. The economic implications of learning by doing. *Review of Economic Studies* 29 (3): 155–173.

Arthur, W.B. 1989. Competing technologies, increasing returns, and lock-in by historical events. *Economic Journal* 99 (394): 116–131.

Arthur, W.B. 1994. *Increasing Returns and Path Dependence in the Economy*. Ann Arbor, MI: University of Michigan Press.

Asheim, B.T. 1999. Innovation, social capital and regional clusters: on the importance of cooperation, interactive learning and localised knowledge in learning economies. Paper presented at the Regional Studies Association International Conference on Regional Potentials in an Integrating Europe, University of the Basque Country, Bilbao, Spain, 18–21 September.

Brass, D.J., Galaskiewicz, J., Greve, H.R. and Tsai, W. 2004. Taking stock of networks and organisations: a multilevel perspective. *Academy of Management Journal* 47 (6): 795–817.

Breschi, S. and Malerba, F. 1997. Sectoral innovation systems: technological regimes, Schumpeterian dynamics, and spatial boundaries. In Edquist, C. (ed.), *Systems of Innovation: Technologies, Institutions, and Organisations*, London, UK and Washington, DC, USA: Pinter, pp. 130–156.

Carlsson, B. and Stankiewicz, R. 1991. On the nature, function and composition of technological systems. *Journal of Evolutionary Economics* 1: 93–118.

Cohen, W.M. and Daniel, A.L. 1990. Absorptive capacity: a new perspective on learning and innovation. *American Science Quarterly* 35: 128–152.

Cooke, P. 2002. Biotechnology clusters as regional, sectoral innovation systems. *International Regional Science Review* 25 (1): 8–37.

Cooke, P., Gomez Uranga, M. and Etxebarria, G. 1997. Regional systems of innovation: institutional and organisational dimensions. *Research Policy* 26: 475–491.

David, P.A. 1985. Clio and the economics of QWERTY, *American Economic Review* 75: 332–337.

David, P.A. 2001. Path dependence, its critics and the quest for 'historical economics'. In Garrouste, P. and Ioannides, S. (eds), *Evolution and Path Dependence in Economic Ideas: Past and Present*. Cheltenham, UK and Northampton, MA, USA: Edward Elgar Publishing, pp. 15–40.

David, P.A. and Foray, D. 1995. Assessing and expanding the science and technology knowledge base. STI Review 16.

Department of Communication, Marine and Natural Resources. 2006. Ocean Energy Strategy. Dublin.

Department of Communications, Energy and Natural Resources. 2010. Draft Offshore Renewable Energy Development Plan. Dublin.

Department of Public Enterprise Ireland. 2002. European Commission Green Paper: 'Towards a European strategy for the security of supply'. Response by the Department of Public Enterprise Ireland. Dublin.

Dosi, G. and Nelson, R.R. 2010. Technical change and industrial dynamics as evolutionary processes. In Hall, B.H. and Rosenberg, N. (eds), *Handbook of Innovation*, Vol. 1. Amsterdam: North-Holland, pp. 51–127.

Edquist, C. 1997. Systems of innovation approaches – their emergence and characteristics. In Edquist, C. (ed.), *Systems of Innovation: Technologies, Institutions and Organisations*. London: Pinter Publishers/ Cassell Academic, pp. 1–35.
Edquist, C. 2001. The system of innovation approach and innovation policy: an account of the state of the art. Lead paper. Nelson Winter Conference, Aalborg.
European Commission. 2003. Council Directive 2003/96/EC of 27 October 2003 restructuring the Community framework for the taxation of energy products and electricity. Brussels.
Freeman, C. 1987. *Technology Policy and Economic Performance: Lessons from Japan*. London: Pinter.
Glückler, J. 2007. Economic geography and the evolution of networks, *Journal of Economic Geography* 7 (5): 619–634.
Government of Ireland. 1996. *Renewable Energy – A Strategy for the Future*. Dublin.
Government of Ireland. 2002. *Green Paper on Sustainable Energy and the National Climate Change Strategy*. Dublin.
Government of Ireland. 2006. *Energy Policy Green Paper 'Towards a Sustainable Energy Future for Ireland'*. Dublin.
Government of Ireland. 2007. *Government's White Paper on Energy Policy*. Dublin.
Government of Ireland. 2008. *Building Ireland's Smart Economy: A Framework for Sustainable Economic Renewal*. Dublin.
Government of Ireland. 2009. *Programme for Government*. Dublin.
Government of Ireland. 2010. *Ireland's National Renewable Energy Action Plan*. Dublin.
Hemert, P. and Nijkamp, P. 2010. Knowledge investments, business R&D and innovativeness of countries: a qualitative meta-analytic comparison. *Technological Forecasting and Social Change* 77 (3): 369–384.
Johnson, A. and Jacobsson, S. 2000. The emergence of a growth industry: a comparative analysis of the German, Dutch and Swedish wind turbine industries. *Proceedings of the Schumpeter Conference 2000*.
Krugman, P. 1991. History and industry location: the case of the manufacturing belt. *American Economic Review* 81 (2): 80–83.
Lam, A. (2000). Tacit knowledge, organizational learning and societal institutions: an integrated framework. *Organization Studies* 21(3): 487–513.
Levitt, B. and March, J. 1988. Organizational learning. *Annual Review of Sociology* 14: 319–340.
Levitt, T. 1965. Exploit the product life cycle. *Harvard Business Review* 43: 81–94.
Lundvall, B.A. 1988. Innovation as an interactive process: from user–producer interaction to national systems of innovation. In Dosi, G., Freeman, C., Nelson, R. and Soete, L. (eds), *Technical Change and Economic Theory*. London: Pinter Publishers, p. 338.
March, J.D. and Simon, H.A. 1958. *Organisations*. New York: Wiley.
Maskell, P. and Malmberg, A. 1999. The competitiveness of firms and regions: 'ubiquitification' and the importance of localized learning. *European Urban and Regional Studies* 6: 9–25.
Nelson, R.R. 1993. *National Innovation Systems: A Comparative Study*. Oxford: Oxford University Press.
OECD International Energy Agency (OECD IEA). 2007. Energy policies of IEA Countries Ireland 2007 review.

Popp, D., Newell, R.G. and Jaffe, A.B. 2010. Energy, the environment, and technological change. In Hall, B.H. and Rosenberg, N. (eds), *Handbook of the Economics of Innovation*, Vol. 2. Amsterdam: Elsevier, pp. 873–937.

Powell, W.W. 2005. Networks of innovators. In Powell, W.W. and Grodal, S. (eds), *The Oxford Handbook of Innovation*. Oxford: Oxford University Press, pp. 56–85.

Powell, W.W., Koput, K.W. and Smith-Doerr, L. 1996. Interorganisational collaboration and the locus of innovation: networks of learning in biotechnology. *Administrative Science Quarterly* 41(1): 116.

Puffert, D.J. 2004. Path dependence, network form, and technological change. In Guinnane, T.W., Sundstrom, W.A. and Whatley, W.C. (eds), *History Matters*, Stanford, CA: Stanford University Press, pp. 63–95.

Reiche, D. and Bechberger, M. 2004. Policy differences in the promotion of renewable energies in the EU member states. *Energy Policy* 32 (7): 843–849.

Smith, H. and Ho, K.W. 2006. Measuring the performance of Oxfordshire's spin-off companies. *Research Policy* 35 (10): 1554–1568.

Sustainable Energy Authority of Ireland (SEAI). 2011. Energy research in Ireland 2004–2010.

Sustainable Energy Authority of Ireland (SEAI). 2012. Energy in Ireland 1990–2011.

Wasserman, S. and Faust, K. 1994. *Social Network Analysis: Methods and Applications*. New York: Cambridge University Press.

Wiser, R. and Pickle, S. 1997. Green marketing, renewables, and free riders: increasing customer demand for a public good. LBNL-40632. Berkeley, CA: Lawrence Berkeley National Laboratory.

9. The position and role of research infrastructures in the materials science network

Mickael Pero

INTRODUCTION

Since the launch of the European Commission's 6th Framework Programme, the importance of instrumental institutions called 'research infrastructures' (RIs) has played an increasing role in the European Research Area, becoming of its one key pillars.[1] These actors can be defined in short as 'essential or unique facilities used by the (international) scientific community to conduct research in any field of science or technology'.[2] In other terms, an RI can be understood as an entity that provides access to cutting-edge scientific equipment and research services to the scientific community, thus promoting the exchange and diffusion of both knowledge and know-how towards all types of research organizations.

The present chapter focuses on one research area where RIs are particularly important, namely materials sciences. This field aims at exploring the fundamental properties and characteristics of materials, a type of research which has a very wide application scope, ranging from metallurgy to nanotechnology. The role of RIs within this field is mainly to support the research of scientific teams coming from a large number of organizations (national and international) to conduct experiments on top-class scientific instruments, managed by highly trained engineers and technicians.

To investigate the role and position of RIs in the materials science domain, this chapter combines a social network analysis (SNA) as well as a statistical one, a mix which happens to be suited to the evaluation of an actor's position and role within a scientific system. On the one hand, the SNA approach has the advantage of capturing the complexity of interorganizational research, and delivers a static as well as dynamic view of the system. In addition, the reason for experimenting with such a tool in this context is that evaluating the returns from science in monetary terms would need the pricing of scientific output, which in the case of fundamental science is almost impossible to achieve. On the other hand, a statistical method is used to test the effects of conducting research with an RI, in particular concerning international collaborations.

LITERATURE BACKGROUND

The topic of this study relates to the economics of science, where the general aim is to investigate the role and effect of science on society, therefore acknowledging the strong evidence indicating a relationship between scientific performance and economic growth (Stephan, 1996; Dasgupta and David, 1994; Salter and Martin, 2001). Two branches of literature are briefly mentioned here: one that measures the increase in scientific output resulting from a change in the technological and scientific environment, and one evaluating the scientific performance of a research actor.

The first branch of literature focuses on the effect of technology on scientific performance. One approach explores the scientific effects of the introduction of a new technology on scientific outcomes (Goldfarb and Agrawal, 2008; Ding et al., 2010). Another approach, which is closely related to the RIs, is to measure the importance of scientific instruments developed in fundamental scientific environments for both industrial innovations and scientific progress (Von Hippel, 1988; Rosenberg, 1992; Price, 1984). One last approach that could be mentioned is the effect of public scientific investment on scientific performance (Lane and Bertuzzi, 2011).

The second branch of literature seen here focuses on the effect of a research institution on the scientific, technological and economic development. Well-known studies of this type have been taken into consideration, such as the economic effect of an RI from SandT procurement contracts (Autio et al., 2003; Autio et al., 2004). Other studies have measured spillovers by linking patented inventions to basic science (Meyer, 2000; Tijssen, 2002; Foray and Lissoni, 2009), patent-to-patent citations (Verspagen, 1999), scientific mobility between research institutions (Zellner, 2003), or the efficiency of the research itself (Arora et al., 1998).

With regards to methods, on the one hand this chapter adopts a social network analysis that has already been used for evaluating research-related phenomena such as the effect of a research programme on network structures (Breschi et al., 2009), or the position of academic inventors (Lissoni et al., 2011) as well as non-academic inventors (Balconi et al., 2004). On the other hand, the statistical method is based on a modified gravity model. It has been used to identify determinants of collaborations mostly by testing the relevance of geographic proximity but also cognitive, institutional or social proximity (Boschma, 2005), as well as technological proximity (Montobbio and Sterzi, 2011). In this chapter, an RI proximity factor is introduced to test whether Italian research institutions having been supported by Italian RIs affects the probability to collaborate internationally.

Overall, this chapter focuses on RIs' scientific involvement and the support of the Italian materials science network. The added value of this chapter is to shed light on RIs which have been subject to little research. For instance, most of the literature focuses on other types of research organizations such as universities.

HYPOTHESES

The first hypothesis concerns the position and role of the RI within its scientific network. The nature of the facility being to generate new data and technical knowledge for research, an RI in materials science is expected to attract leading scientists within its field. In network terms, we would expect a central position of RI research. In addition, RIs are mostly located within scientific parks; thus their role would be expected to connect external actors to their own scientific cluster.

Hypothesis 1: RIs hold a central brokerage position in the Italian materials science network.

The second hypothesis that is tested concerns the expected effect on international visibility of local research institutions that have collaborated with RIs. The idea is to see whether the presence of RIs in Italy enhances the number of international collaborations for Italian research organizations through common research projects. If that were true, RIs could be considered as 'bridging' international scientists to local ones, and facilitating an increased knowledge and know-how exchange between partners (Bozeman and Corley, 2004).

Hypothesis 2: Italian cities which perform research together with RIs increase the probability of international collaboration in addition to previous bilateral collaborations.

As a preliminary comment, note that for both datasets a group of Italian RIs is identified among the set of research organizations. It is composed of Elettra Sincrotrone Trieste (Elettra ST) as well as the National Institute for Nuclear Physics (INFN), which hosts the two large RIs of Legnaro and Frascatti. These RIs are defined as such by European Strategic Forum for Research Infrastructures (ESFRI) and the European Commission.[3]

HYPOTHESIS 1: TESTING THE CENTRALITY AND BROKERAGE ROLE OF RIs ON THE ITALIAN MATERIALS SCIENCE NETWORK

Method: Social Network Analysis

From the dataset, the available affiliation network (that is, article * affiliation matrix) is converted to a one-node network (that is, article * article matrix): the dataset now has identical rows and columns comprising the nodes of the network (that is, articles) and the ties between nodes (that is, common authorship). The reason why articles are used as nodes instead of institutions is to permit institutional activities to overlap on several network subsets, what I call 'clusters'[4] in accordance with network theory. In addition, the advantage of this configuration is the focus on research, where the position and role of articles in the network will indirectly assess the related research institutions (and in particular RIs). Also, by taking an institutional level of analysis instead of an article level of analysis, structural properties such as centrality of the RI would appear isolated due to the difference in size with institutions whose central objective is to publish. Therefore, little information would be given on the actual role of RI-related research in the network.

The conversion from affiliation to one-node network only makes sense if the set of research institutions affiliated in articles are known to each other, giving an informative substance for each relation. Indeed, criticism arose from the observation of co-authorship one-node networks in scientific collaboration where articles can reach very large number of authors, therefore suggesting that there was no real contact between them, and therefore providing misleading structural properties of networks (Newman, 2001a, 2001b). Indeed, some tests using the dataset of this chapter as a one-node network of scientists reveal features that are subject to the criticism noted above. For instance, using the average number of authors per publication should provide evidence of the proximity between authors; it is found that scientific publications tend to be characterized by numerous authors, with an average of five authors per article and 5 per cent of articles having a team size above ten members. These numbers question the accuracy of collaborations between scientists, and are largely above the number usually accepted for inventor collaborations (Lissoni et al., 2011).

However, the present chapter focuses on networks of organizations and investigates network measures that are only relevant to institutions, instead of scientists. Using this level of analysis, the network depicts an average of three organizations per publication, with a standard deviation of one organization, and only 1 per cent of articles having more than six organizations as author affiliations. With this number, it is reasonable to assume that close institutional contact exists between research organizations. One can observe that using this level of analysis enables avoiding the 'social co-authorship' internal to organizations which might be the cause of such a large number in the case of a one-node scientists' network.

Coming back to the applied methodology, one goal is the categorization of articles into clusters. In order to achieve this, I apply to the article * article dataset an algorithm called the 'Louvain method' which is a community detection approach that maximizes the 'modularity'[5] of the partition (Blondel et al., 2008). In simple terms, this (iterative) two-step method identifies, firstly, small communities that are well connected, and secondly, communities as nodes to build up the network. As a result, for each year under study, a certain number of clusters with their related modularity are generated.[6]

From Table 9.1, one can notice the relatively limited number of clusters and the increasing level of modularity over the years. These elements show that the diversity of the Italian network of materials sciences appears to be quite limited, but with clusters which tend to concentrate relatively more over time. Each article of the network data can now be associated to a cluster. In that way, all the present research institutions can be tracked with respect to their research involvement.

An additional feature of the network that is investigated concerns the 'brokerage' roles of institutions, and in particular of research infrastructures. The information delivered enables understanding the types of relations an RI plays between as well as within clusters. Based on the seminal paper by Gould and Fernandez (1989) and the application of the concept in Pajek by de Nooy et al. (2005), the brokerage roles and structural positions are investigated to uncover the flow or exchange of scientific information.

Table 9.1 Clusters and modularity per year

	1984	1989	1994	1999	2004
Clusters	6	4	6	5	4
Modularity	0.074	0.144	0.173	0.281	0.213

Data

The data originate from the Scopus[7] database and consist of a list of scientific articles from 30 peer-reviewed journals[8] in materials sciences over five years (1984, 1989, 1994, 1999 and 2004). In order to limit the variations of articles in the selected sample and to avoid the general increase in scientific production, the journals are filtered with respect to two main conditions: the presence of articles in each of the five years under scrutiny (1984, 1989, 1994, 1999 and 2004), and a maximum standard deviation of 500 articles per journal in the five one-year periods. An additional condition is to retain articles which have had at least one Italian author, since the focus of this chapter is on Italian related networks. Note that 1984 has been selected as the first year under scrutiny because it is the first year with complete information for the 30 selected journals. This raw dataset consisting of a list of articles with additional characteristics – such as a list of the author's affiliations as well as the number of citations – is then converted into an article * affiliations matrix. From this new format, articles with incomplete information or a single author are dropped, since the interest lies in the network. Also, the latter operation should not affect the study of the network since connections are represented by common authors, and isolates would be disregarded in any case. In addition, in the field of materials sciences, single authors appear to be scarce (Wuchty et al., 2008). As a result of the process, a dataset composed of 702 articles from 72 Italian research institutions is produced (mostly universities but also private and public research organizations or PROs, and of course RIs). The dataset is then broken down by years, and additional cleaning is performed to remove isolates from every period.

Out of these operations, variation of data over all five one-year periods remains small: affiliations vary little (standard deviation of 3), as do the number of publications (standard deviation of 12).

Results

The evolution of the article * article network can be illustrated by Table 9.2. The number of articles over the years is relatively constant, therefore

Table 9.2 Sample nodes and edges (i.e. ties) per year

	Year				
	1984	1989	1994	1999	2004
Nodes (articles)	140	147	154	134	123
Edges (authorship link)	5320	7440	7302	5906	6194

Table 9.3 Network indicators

	Year				
	1984	1989	1994	1999	2004
Network density	0.55	0.34	0.31	0.33	0.41
Network closeness centralization	0.3	0.18	0.37	0.47	0.39
Network betweenness centralization	0.046	0.048	0.036	0.04	0.032

limiting the number of journals controls for the increasing trend in article production. Additional indicators from Table 9.3 can be looked at to analyse this evolution.

The density[9] of the network follows a U-curve, showing that in the case of the field of materials science, the number of co-authorships goes down until 1994 and then goes back up. Over the period, about 40 per cent of all possible connections are present, depicting a well-connected network.

Closeness centralization[10] measures the variation between the closeness centrality of the nodes. It is defined by its closeness to all other nodes. In this case, the indicator increases until reaching a peak in 1999, and stays at a relatively high level in 2004. This means that some key articles – as is usually found in science – are very central in this network, meaning a clear core–periphery structure.

On the other hand, the betweenness centralization reaches very low levels. This indicator measures the variation between the betweenness centrality of nodes. It is defined by how present the node is between all other nodes. Overall, the meaning of this low figure is that all nodes happen to be between others, and no isolated articles are present (that is, due to data treatment).

Research infrastructures and research centrality

The first indication of the relative position of RIs is their overall research output compared to other types of institutions. Summing up the five years, the RI-related group publishes 131 articles which are cited 1207 times, while the non-RI group publishes 567 articles which get 6671 citations. When comparing the two groups statistically, a significant difference can be noticed, meaning that the group involving RIs seems to have a lower performance than the non-RI group. This is mostly due to the difference in numbers of members comprising the two groups (20 fewer actors for the RI group). However, no significant differences can be found when comparing both groups' citations per publication figures (t-test at the 0.05 level).

Refining the data, it is interesting to see that despite the absolute difference in overall performance, RI research happens to be central in the network. Once again, the two centrality measures are used, but this time at the article level. The comparison of the RI group and non-RI group leads to two main results. First, research involving RIs has a similar closeness centrality than research not involving RIs, showing a similar level of connectedness. Second, research with RI involvement depicts a higher betweenness centrality than when RIs are not involved, a higher figure which shows on average a stronger influence of their type of research. This evidence would therefore support the first hypothesis of the chapter concerning the central position of the RIs in the network.

Research infrastructures and clusters

The role of the RIs within the materials science network can be seen by applying a cluster analysis to the network. This would give evidence for hypothesis 1 about the role of RIs towards other organizations. Figure 9.1 summarizes the evolution of the clusters.

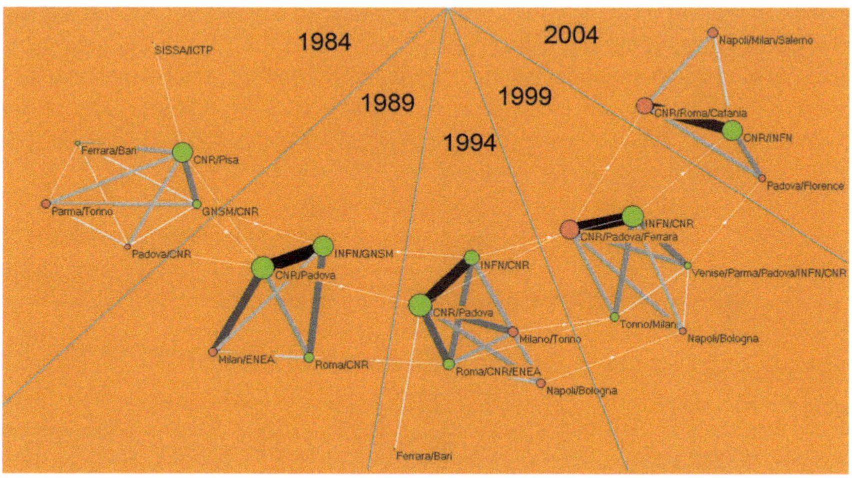

Note: The figure was created from a supermatrix containing both intra-period and inter-period cluster information. Green dots represent clusters with at least one RI involvement and red dots represent clusters without any RI involvement, and their size represents the number of members composing them. The darkness of the ties between clusters indicates the intensity of co-authorship. The name of the clusters relates to the two largest research contributors and finally the thin white arrows show the inter-period cluster similarity as measured by the Pearson correlation indicator. Concerning the latter, a score of at least 0.5 was retained for a link to exist between periods.

Figure 9.1 Cluster network evolution

Thanks to this new configuration it is possible to investigate the cluster position of RIs, and for several cluster-based indicators to be developed. The cluster diversity indicator measures the proportion of clusters where RIs are directly involved.[11] The indicator appears constant through the five periods with on average a presence of RIs in 26 per cent of clusters, whereas on average non-RI research institutions are present in 40 per cent of clusters. RI research appears to focus on specific clusters. The cluster publication indicator measures the average number of cluster publications for RI and non-RI research. The RI-related figure is smaller every year with an average difference of about 25 articles, indicating a lower scientific production in those clusters.

In a similar fashion, a cluster citation indicator measures the average number of cluster citations for RI and non-RI research. In this case the difference is important, with non-RI exceeding RI-related research every year, with a difference on average over the whole period of about 400 citations. Thus, the figure tends to show an involvement of RI research in lower knowledge diffusion clusters. In addition, the cluster members' diversity indicator was measured and captures the average number of cluster members that RI and non-RI groups interacted with. This indicator shows a systematically smaller figure for RI-related research, meaning that RIs tend to be concentrated in selected clusters. From the evidence above, RIs appear to have their own particular position which is not similar to non-RIs actors, and more involved towards specialized clusters.

Research infrastructures and research brokerage role

From the previous sections, the position of the RI research within the materials science network has been investigated, but it would be useful to know, within or between clusters, what its role is. To identify this, four categories of brokerage roles from research collaborations between institutions are proposed (Table 9.4), as inspired and adapted from Gould and Fernandez (1989).

The position of articles with at least one author from an RI can be translated into one of the roles defined in Table 9.4. Out of this operation, the total number of articles which satisfy a role's condition is computed (Figure 9.2).

From the result, the following statements can be made: the research performed by RIs will, on very few occasions, be connected with two other research projects neither belonging to the same cluster nor connecting two projects from another cluster. In decreasing cases over time, the RI research will link two different research projects from two different groups. On the contrary, in most cases, RI research's main role, from 1984 to 2004, has been to connect its cluster to other research organizations

Table 9.4 Categories of brokerage roles

Brokerage role	Brokerage role from Gould and Fernandez (1989)
Coordinating research: research which takes place within a cluster	Coordinator
Itinerant research: research involving the support of two actors from a foreign cluster	Itinerant broker
Representative or gatekeeping research: links between its own cluster to a foreign cluster	Representative broker or gatekeeper. Different roles when directed network
Liaison research: connect two foreign clusters	Coordinator

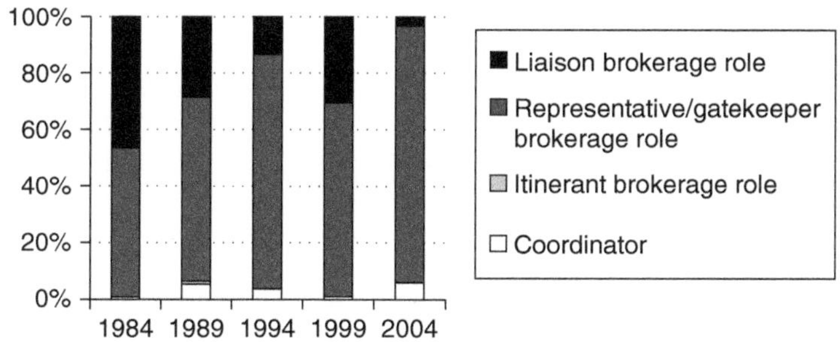

Figure 9.2 RI brokerage roles

from a different cluster. This means that RI research has been embedded during that period into familiar clusters, aimed at representing them towards other clusters. However, when looking at the brokerage role of non-RI research, we find a similar pattern, which means that overall the distribution of brokerage roles between RI and non-RI research tends to be similar. This would mean that brokerage roles at the article level tend to be governed by the same reasons mentioned above, and are not structures specific to RI-related research. Therefore, relating the absolute brokerage roles of RI research to the total of brokerage roles of all institutions might provide additional insight (Figure 9.3).

Indeed, what we find is that within the Italian materials science network, the role of coordinator of RI research is growing over time, controlling for total brokerage roles, until reaching about 50 per cent of the total brokerage role for RI research role and more than 90 per cent of coordinator role across institutional types (RI and non-RI). This result suggests that

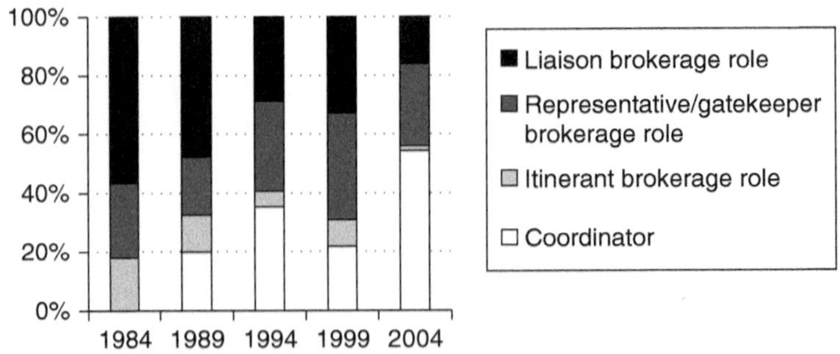

Figure 9.3 *RI brokerage (proportion)*

RIs tend to have a strong coordinator role, meaning extensive interaction within their cluster, hence supporting the idea of RIs embedded in long-standing relations with a given set of actors. Logically, the results also show the marginal existence of itinerant research bridging two 'foreign' clusters. One explanation is that fewer clusters emerge over time due to an increasing integration of research communities, thus enhancing the coordinator role of RIs in Italy.

HYPOTHESIS 2: THE EFFECT OF RI SUPPORT ON INTERNATIONAL COLLABORATIONS

Empirical Model

In order to test this second hypothesis, the statistical empirical model used here is a modified gravity model. Gravity models applied to economics have become increasingly popular in various topics of study. Although subject to criticism for the lack of micro-foundations, the validity of this family of models was re-established by Anderson (1979) and Anderson and Van Wincoop (2003) in studies concerning international trade. In this section this framework will be applied to international research collaborations between Italian and non-Italian research institutions. Related topics have been applied to identify determinants of collaborations, testing geographic proximity but also cognitive, institutional and social proximity (Boschma, 2005), and also technological proximity (Montobbio and Sterzi, 2011).

This method is used here to test the second hypothesis of the chapter: whether an 'instrumental' factor (that is, RI proximity) could explain part of the international collaborations between Italian and foreign institutions. In other words, whether having collaborated with an (Italian) RI leads to a higher chance of common research projects between domestic and foreign research institutions. Other 'classic' proximity measures are being tested in the model, such as geographic, institutional and social measures.[12]

The gravity model is built at the city level. Drawing on the literature applying this model to international collaboration (Montobbio and Sterzi, 2011; Plotnikova and Rake, 2011), I enounce the basic gravity model form:

$$C_{ij} = C_0 * \frac{M_i * M_j}{d_{ij}}$$

where C_{ij} represents the collaboration intensity between Italian city i and non-Italian city j; C_0 is a proportional constant; $M_{i(j)}$ is the mass of city $i(j)$ measured in terms of publication count; and d_{ij} is the proximity between the two cities. Note that d_{ij} comprises the several measures of proximity mentioned above. The 'unconstrained' gravity model is:

$$\ln C_{ij} = \ln C_0 + \beta_1 \ln M_i + \beta_2 \ln M_j - \beta_3 \ln d_{ij} + \epsilon_{ij}$$

where ϵ_{ij} is assumed to be independent and identically distributed.

In order to test the second hypothesis, the above equation needs to be adequately estimated. However, the type of regression used for this equation depends on the nature of the dependent variable collaboration. Indeed, the data behind the variable collaboration has several characteristics that have to be taken into account.

First, it is a count model and as such depicts highly non-normal distribution which makes ordinary least squares (OLS) an improper estimation due to the failure of the homoscedasticity assumption. A usual way to deal with this problem is the use of the Poisson regression (Santos Silva and Tenreyro, 2011) that can be stated as the following:

$$PR[C_{ij}] = \frac{\exp(-\mu_{ij})\mu_{ij}^{C_{ij}}}{C_{ij}!}, \; (C_{ij} = 0,1,\ldots)$$

where the conditional mean is linked to an exponential function of the set of independent variables (Burger et al., 2009; Plotnikova and Rake, 2011; Flowerdew and Aitkin, 1982):

$$\mu_{ij} = E(C_{ij}) = e^{\beta_0 + \beta_1 \ln M_i + \beta_2 \ln M_j - \beta_3 \ln d_{ij}}$$

Last but not least, the dependent variable is overdispersed with a variance significantly larger than the mean, around three times larger. In order to overcome this problem a suitable candidate is the negative binomial as proposed by Burger et al. (2009). The use of this regression instead of the Poisson regression is confirmed by obtaining a dispersion parameter (that is, alpha) significantly different from zero for all tested regressions. Integrating this above parameter in the negative binomial distribution as well as the dispersion parameter alpha delivers (Greene, 2002):

$$P[C_{ij}] = \frac{\Gamma(\alpha^{-1} + C_{ij})}{\Gamma(C_{ij} + 1)\Gamma(\alpha^{-1})} \left(\frac{\mu_{ij}}{\mu_{ij} + \alpha^{-1}}\right)^{C_{ij}} \left(\frac{\alpha^{-1}}{\mu_{ij} + \alpha^{-1}}\right)^{\alpha^{-1}}$$

One last problem faced by the dependent variable is the excess number of zeros which discards the use of ordinary count models such as the negative binomial regressions (Greene, 2002; Santos Silva and Tenreyro, 2011). Indeed, data show that over a total of 12 848 potential collaborations, only 3532 are present (about 27 per cent). Also note that significant Vuong tests favour the use of the zero-inflated negative binomial (ZINB) over the standard negative binomial. Thus the inspection of the data would suggest the use of the zero-inflated negative binomial models (with robust errors).

Coming back to the 'inflated' part of the model, an observed zero could be due either to no contact between organizations (true zero), or contact but no outcome (uncertain zero):

$$P[C_{ij}] = \begin{cases} \varphi_{ij} + (1 - \varphi_{ij})\left(\frac{\alpha^{-1}}{\mu_{ij} + \alpha^{-1}}\right)^{\alpha^{-1}} & \text{if } C_{ij} = 0 \\ (1 - \varphi_{ij})\frac{\Gamma(\alpha^{-1} + C_{ij})}{\Gamma(C_{ij} + 1)\Gamma(\alpha^{-1})}\left(\frac{\mu_{ij}}{\mu_{ij} + \alpha^{-1}}\right)^{C_{ij}}\left(\frac{\alpha^{-1}}{\mu_{ij} + \alpha^{-1}}\right)^{\alpha^{-1}} & \text{if } C_{ij} > 0 \end{cases}$$

The zero-inflated negative binomial aims at differentiating between certain and uncertain zeros. Certain zeros are defined by a situation where two institutions have never been in contact due to their lack of visibility, resources or scientists not being connected to the scientific community. Uncertain zeros are defined by the situation where two institutions are in contact and visible but no common article (but possibly another outcome) was produced. In this situation, inflated zeros could be explained, among other explanatory variables, by the institutional mass because of the importance of the visibility provided by the size of a research institution.

Data

The publications dataset is extracted from the Scopus database and concerns more than 2000 publications in the materials science subject category. The data are used at the city level and comprise all collaboration pairs as measured by scientific co-authorship. The period under scrutiny spans from 2003 to 2005. Setting the level of analysis at the city level appears most adequate for the topic since it reveals more fined-grained information, a clear advantage compared to a regional level of analysis.

The list of dyadic collaborations (that is, *collab*) between Italian and non-Italian cities is then generated (that is, 12 848 undirected pairs) along with gravity model-inspired variables capturing masses and commonly integrated proximity measures (Boschma, 2005), and controls.

The first 'classic' determinant of the attraction between two cities is their mass (*lnMassI* for Italian cities and *lnMassNI* for non-Italian cities). This is measured by the number of publications originating in that city in 2004 in the materials sciences subject category. This information is retrieved for each Italian and non-Italian partner city.

The first proximity measure is the distance (that is, *lnDist*), which is the 'as the crow flies' geographic distance between collaborating Italian and non-Italian cities. It is computed with the Haversine formula (Sinnot, 1984) with coordinates from the Getty Research Institute. For this factor a negative sign is expected (Audretsch and Feldman, 1996). From the evidence in the literature, it is tested whether this factor holds as affecting international collaboration in materials research.

The institutional proximity representing the similarity between the types of institutions involved, namely universities, firms and PROs, is calculated by a similarity indicator.[13] This measure (that is, *lnInstProx*) captures the types of institutions present for each city and tests their importance for collaboration. Therefore if one city hosts two different research organizations, the city is given the aggregated mass and given multiple institutional attributes. A positive effect on collaborations is expected here due to institutional proximity. For instance, an institution in Italy would relate more easily to an institution outside of Italy if they belong to the same type (excluding universities belonging to the same network or community).

The fourth measure, namely social proximity (that is, *lnSocProx*), is also built to identify the importance of past collaborations on current collaborations. This variable is measured by collaborations that took place in 2002 (one year before our sample, to avoid endogeneity issues) for each city pair. Here, a positive sign is expected. The argument behind this is that past collaborations increase trust and stronger commitments between partners, and therefore future collaborations (Boschma, 2005). Since we are measuring the log of previous collaboration and in many cases there are no previous collaborations, one unit is added to the social proximity measure before taking the log.

Note that the RI category is not included in these indicators since they act as support for other research organizations. However, the implication of RIs in scientific collaborations is introduced via an RI proximity variable that accounts for the number of common projects two cities had with an RI during the year just before the period under scrutiny, in 2002. In networking terms, it accounts for a triad between a dyadic collaboration and an RI. Note that the social proximity variable *lnSocProx* previously mentioned contains both dyads as well as triads, since all triangular collaborations are present as dual collaborations. Therefore to disentangle the two types of information and reveal the contribution of triads for the probability to collaborate, the two non-overlapping variables *lnDyad* and *lnTriad* are generated from *lnSocProx*. For computational purposes, in the case of no previous triad, one is added to the measure before taking the log. A positive sign of the coefficient capturing the role of RIs on Italian international collaborations is expected due to the 'instrumental' attraction of RIs towards foreign institutions.

Finally, controls are added. Geographic ones at the continental level are added to cover possible scientific agreements factors (Europe, the Americas, Asia and Australia, and Africa). Dummies for the three years are included to control for period fixed effects. The table of summary statistics (Table 9.5) gives us a first impression of those variables.

As mentioned above, only 27 per cent of city-pair collaborations have occurred and depict a high standard deviation as compared to the

Table 9.5 Summary statistics (cross-section)

	Mean	SD	Min.	Max.
Collaboration	0.2735834	0.9340915	0	15
Mass Italian city	38.93182	35.31259	3	149
Mass non-Italian city	76.70548	95.41336	1	956
Distance	3568.045	3708.872	116.1532	16697.76
Institutional proximity	0.798101	0.5001964	1	3
Social proximity	0.069972	0.3635634	1	10
Dyad	0.042591	0.24948	1	6
Triad	0.023817	0.2047753	1	9
Europe	0.6438356	0.4788832	0	1
Americas	0.1541096	0.3610678	0	1
Asia	0.1712329	0.376727	0	1
Africa	0.0308219	0.1728417	0	1

mean, mostly due to the one/zero pattern of the dependent variable. This confirms the use of the negative binomial instead of the Poisson for the regression. Italian city masses are about half of non-Italian city masses, but show a lower dispersion than foreign masses. On average, about 3500 km separates each dyad, the low end being due to European collaborations, close to the borders, whereas the high end is due to very distant intercontinental collaborations (Cagliari with Menai in New South Wales, Australia with more than 16000 km). Concerning institutional proximity, 80 per cent of city-pairs have a similar institutional set-up (mostly based on a single university). Social proximity tells that only 7 per cent of dyads collaborated in 2002. This number is then further differentiated between dyads (collaboration without RIs) and triads (collaboration with an RI). The first is naturally higher, with 4.3 per cent of existing previous collaboration for dyads as compared to 2.4 per cent for triads. These figures have to be carefully interpreted due to relatively high standard deviation. The same care has to be taken for the RI proximity variable: both variables depict skewed patterns. Finally, most international collaborations of Italian research institutions take place with European partners (64 per cent), the lower partner continent being Africa with only 3 per cent of collaborations.

Results

The first results of the tested regressions start by accounting for the count nature of the dependent variable *collaboration*. Based on the above discussion, Table 9.6 depicts Poisson and negative binomial regressions.

Regressions (1) to (3) depict Poisson estimates. The first regression concerns the 'core' variables of the gravity model. As expected, the mass of Italian as well as international partners are positive and significant. Similarly, the distance coefficient is negative and significant, in line with previous findings. Note that the high negative number for the intercept is due to the evaluation of the three other variables at 0 which is out of the range of possible observations. In the second regression the social proximity measured by previous collaboration between cities (that is, in 2002) is introduced. The coefficient is positive and significant: for each additional unit of previous collaboration, the probability to collaborate increases by a factor of three.[14] This strongly suggests the existence of a trust factor between dyads which increases the chance for future collaborations.

In regression (3) the social proximity variable is divided in two independent groups.[15] Indeed, this variable contains two pieces of information: previous collaborations which involved a research infrastructure (*lntriad*) and those that did not (*lndyad*). By replacing these two variables in the

Table 9.6 Poisson and negative binomial regressions

Equation	Variables	(1)	(2)	(3)	(4)	(5)	(6)
collab	*lnmassi*	1.038***	0.829***	0.805***	1.022***	0.845***	0.830***
		(0.0290)	(0.0301)	(0.0309)	(0.0258)	(0.0256)	(0.0265)
	lnmassni	0.489***	0.328***	0.322***	0.412***	0.313***	0.306***
		(0.0353)	(0.0319)	(0.0322)	(0.0226)	(0.0220)	(0.0227)
	lndist	−0.539***	−0.350***	−0.358***	−0.527***	−0.438***	−0.440***
		(0.0633)	(0.0639)	(0.0634)	(0.0680)	(0.0678)	(0.0684)
	lndyad			1.498***			1.810***
				(0.0737)			(0.0749)
	lntriad			1.231***			1.554***
				(0.0722)			(0.0958)
	lninstprox			0.210***			0.154**
				(0.0804)			(0.0771)
	europe			−0.00135			0.155
				(0.598)			(0.637)
	americas			0.688			0.878
				(0.814)			(0.798)
	asia			0.102			0.302
				(0.657)			(0.733)

	lnsocprox		1.375***			1.715***	
			(0.0564)			(0.0599)	
	Constant	−4.194***	−4.433***	−4.425***	−4.047***	−4.104***	−4.138***
		(0.652)	(0.623)	(0.617)	(0.649)	(0.650)	(0.654)
lnalpha	Constant				0.968***	0.579***	0.573***
					(0.0575)	(0.0640)	(0.0643)
	Time FE	yes	yes	yes	yes	yes	yes
	Country FE	yes	yes	yes	yes	yes	yes
	Observations	38 544	38 544	38 544	38 544	38 544	38 544
	stataaic	21 345	20 206	20 183	20 278	19436	19432
	statabic	21 909	20 779	20 791	20 852	20018	20048
	ll	−10 606	−10 036	−10 020	−10 072	−9650	−9644
	LR test				1068.29***	771.65***	752.2***

Note: Robust standard errors in parentheses (*** $p < 0.01$, ** $p < 0.05$, * $p < 0.1$).

regression, I can test the relevance of triad collaboration while taking into account pre-existing dyadic links between partners. What is found is that both explanatory variables are positive and significant and both increase the probability to collaborate; by a factor of 3.5 for previous dyadic collaborations and by a factor of 2.4 for previous triadic collaboration involving an RI. This is the first empirical evidence of the importance of RIs in affecting collaborations between Italian research organizations and international partners.

However, as explained previously, a negative binomial regression is more adequate than a Poisson based on the overdispersion of the dependent variable. This can be deduced from a likelihood test which describes alpha as significantly different from 0. Regressions from (4) to (6) provide the related estimates. The coefficients show similar results to the Poisson regression. However, it can be noticed that the variables of interest, dyad and triad, depict higher positive coefficients, raising the probability to collaborate from a unit of previous dyads by a factor of 5.1 and triads by a factor of 3.7.

One limitation of these two models is the excessive amount of zeros which are not taken into account. Differentiating between certain and uncertain zeros as explained before can be introduced in the models by proposing both zero-inflated Poisson and negative binomial regressions. Starting with the zero-inflated Poisson regression (ZIP) used in columns (1) to (3) of Table 9.7, what is found are lower coefficients due to the introduction of a logit process to explain excessive zeros. Concerning the excessive zeros, the city masses as well as the distance are significant in explaining the certain zero group. It can be recalled that the certain zero group is defined by a situation where two research organizations have no collaboration due to impossibility (that is, lack of visibility, budget or absence from key scientific communities), whereas the uncertain zero group is for example two organizations being in contact but not having written a common article. Concerning the certain zero group, the inflated part of the estimation shows that the higher the city mass, the lower the odds of belonging to the certain zero group. Indeed, the larger the city the lower the chance of zero collaborations due to a lack of visibility, resources or network. Also, the higher the distance between institutions, the higher the odds of belonging to the certain zero group.

Concerning the estimates, consistent results are found with the previous regressions. For instance, the social measures increase the probability to collaborate by a factor of 1.1 for dyads and 0.9 for triads (in column 3) although to a much lower extent than previously. The introduction of dyads and triads in explaining inflated zeros being highly significant is sensible, since dyads having had past collaborations cannot explain zero

Table 9.7 Zero-inflated Poisson and zero-inflated negative binomial regressions

Equation	Variables	(1)	(2)	(3)	(4)	(5)	(6)
collab	*lnmassi*	0.430***	0.277***	0.246***	0.536***	0.373***	0.347***
		(0.0431)	(0.0408)	(0.0422)	(0.0531)	(0.0461)	(0.0482)
	lnmassni	0.233***	0.154***	0.141***	0.276***	0.164***	0.147***
		(0.0324)	(0.0297)	(0.0308)	(0.0352)	(0.0296)	(0.0310)
	lndist	−0.391***	−0.298***	−0.317***	−0.439***	−0.369***	−0.379***
		(0.0658)	(0.0633)	(0.0640)	(0.0719)	(0.0696)	(0.0702)
	lndyad			0.740***			0.951***
				(0.0676)			(0.0815)
	lntriad			0.643***			0.904***
				(0.0721)			(0.0962)
	lninstprox			0.252**			0.206*
				(0.105)			(0.110)
	americas			0.736			0.831
				(0.510)			(0.515)
	asia			0.233			0.259
				(0.397)			(0.404)
	africa			0.0719			0.0320
				(0.646)			(0.662)
	lnsocprox		0.690***			0.920***	
			(0.0519)			(0.0688)	
	Constant	−0.553	−0.567	−0.524	−1.406*	−1.058	−1.008
		(0.683)	(0.656)	(0.693)	(0.743)	(0.703)	(0.732)
inflate	*lnmassi*	−0.804***	−0.781***	−0.801***	−0.852***	−0.826***	−0.848***
		(0.0529)	(0.0536)	(0.0557)	(0.0706)	(0.0662)	(0.0684)

Table 9.7 (continued)

Equation	Variables	(1)	(2)	(3)	(4)	(5)	(6)
inflate	*lnmassni*	−0.276***	−0.242***	−0.252***	−0.268***	−0.284***	−0.304***
		(0.0361)	(0.0362)	(0.0380)	(0.0471)	(0.0433)	(0.0474)
	lndist	0.202***	0.199***	0.184***	0.231***	0.191***	0.175***
		(0.0391)	(0.0421)	(0.0430)	(0.0576)	(0.0520)	(0.0531)
	lninstprox			0.122			0.0647
				(0.152)			(0.186)
	lndyad			−2.195***			−4.241***
				(0.214)			(1.218)
	lntriad			−2.149***			−2.921***
				(0.265)			(0.709)
	lnsocprox		−2.194***			−3.695***	
			(0.161)			(0.648)	
	Constant	3.779***	3.721***	3.864***	2.841***	3.392***	3.617***
		(0.398)	(0.416)	(0.422)	(0.590)	(0.499)	(0.510)
lnalpha	Constant				−0.0967	−0.384***	−0.381***
					(0.158)	(0.139)	(0.141)
	Time FE	yes	yes	yes	yes	yes	yes
	Country FE	yes	yes	yes	yes	yes	yes
	Observations	38 544	38 544	38 544	38 544	38 544	38 544
	stataaic	20 323	19 303	19 303	20 192	19 187	19 187
	statabic	20 922	19 920	19 970	20 800	19 812	19 863
	ll	−10 092	−9580	−9573	−10 025	−9521	−9515
	LR test				133.17***	118.10***	117.42***
	Vuong test	9.51***	10.13***	10.09***	3.58***	7.24***	7.22***

Note: Standard errors in parentheses (*** $p < 0.01$, ** $p < 0.05$, * $p < 0.1$).

ones due to a lack of visibility. A new result that can be noted is the significance of the institutional proximity which increases the probability to collaborate by 28 per cent.

However, as mentioned before the ZIP regressions suffer from the overdispersion of the dependent variable. This can be seen from the likelihood test giving alpha significantly different from 0 for regressions (4) to (6), suggesting that the ZINB model is preferred to the ZIP model. In addition, the relative goodness of fit of the ZINB statistical model as compared to the ZIP model can be measured by the Akaike information criterion (AIC) which constitutes a trade-off between the number of included parameters and the maximized value of the likelihood function of the estimated model. In simple terms, the model with the lowest AIC figures the better. To accompany that criterion, another one called the Bayesian information criterion (BIC) is computed which is very similar to the AIC criterion but penalizes more the inclusion of additional parameters. Again, the model with the lowest figure shows a better 'efficiency' between the quantity and quality of included variables.

Looking at Table 9.7, the ZINB models appear as fitting better the ZIP for testing the hypothesis. Coefficients are relatively similar to the zero-inflated Poisson regression with some exceptions, and they tend to depict higher figures by taking care of the overdispersion problem. Results are as follows. The original variables of masses of research organizations and distance are significant and behave in the same way as previous regressions. The mass of Italian cities appears to be more important than that of the non-Italian partner cities. As could be expected from the previous findings, the sign of the distance coefficient is negative and significant: the higher the distance, the lower the probability of having international collaborations. The variable of interest measuring the impact of triads on the probability to collaborate mimics the dyad variable with a factor of 0.9 against 1.1. This result confirms the hypothesis that Italian cities which perform research together with RIs increase the probability of international collaboration in addition to previous bilateral collaborations. Previous dyads affecting collaboration probabilities can be explained by the importance for partnerships to build trust over time, and when trust is established it will positively affect future collaborations. As explained before, triads comprise an additional actor: research infrastructures. There are some clear indications of their positive influence on the probability to collaborate alongside dyads. A relevant interpretation is that the objective of triad research is not the same as that of dyad research. More precisely, the presence of an RI denotes a common need from two research institutions to use specialized infrastructure necessary for the advancement of research that is not available in-house. Dyads on the other hand denote collaborations which are more related to data treatment for a given topic than data generation.

Institutional proximity is slightly positive and significant. Therefore similarities between cities hosting institutions of the same type (mostly universities) has a positive effect on the expected number of collaborations, bearing in mind the low value of the coefficient. Lastly, it can be noted from the continental geographic dummies that a city in the Americas has significantly more bilateral international collaborations than a non-Italian European city. This result, which appears surprising at first, can be explained by the concentration of American cities collaborating with Italian ones.

Finally, note that the regressions accounted for time and country fixed effects. Dyad fixed effects could not be computed due to the large size of cities in the sample. By introducing country fixed effects, part of the city variation could be captured, but only concerning non-Italian cities. To partly account for country variation over time, time-varying country fixed effects are controlled for in the zero-inflated negative binomial regression. What can be noticed is that only marginal change occurs with respect to the previous regression, suggesting that accounting for country variations over the period does not affect the outcome. Also, the short period under scrutiny (that is, three years) may be too short to experience significant change within countries.

Robustness Check

The section above has looked at the city-pairs collaborations over time; this section looks at the measures affecting international collaborations over the entire 2003–2005 period. Only the zero-inflated Poisson and zero-inflated negative binomial results will be shown here (Table 9.8). Some coefficients decrease, such as the institutional proximity showing lower importance over the whole period. However, in general the cross-sectional regressions show higher estimates. This is due to the fact the numbers show aggregates for the three periods under scrutiny. For example, the probability to collaborate internationally as estimated by the ZINB (column 6) is the increase by a factor of 2.6 instead of 1.1 for previous dyads, and from 2.1 to 0.9 for previous triads. Overall, the cross-sectional results show consistency with the panel results.

Table 9.8 Zero-inflated Poisson and zero-inflated negative binomial regressions

Equation	Variables	(1)	(2)	(3)	(4)	(5)	(6)
collab2	*lnmassi*	0.517***	0.361***	0.337***	0.648***	0.483***	0.458***
		(0.0329)	(0.0329)	(0.0338)	(0.0490)	(0.0428)	(0.0445)
	lnmassni	0.222***	0.151***	0.142***	0.246***	0.155***	0.139***
		(0.0256)	(0.0252)	(0.0259)	(0.0337)	(0.0306)	(0.0322)
	lndist	−0.311***	−0.225***	−0.247***	−0.443***	−0.380***	−0.386***
		(0.0673)	(0.0639)	(0.0647)	(0.0870)	(0.0835)	(0.0842)
	lndyad			0.960***			1.283***
				(0.0538)			(0.0957)
	lntriad			0.800***			1.156***
				(0.0609)			(0.117)
	lninstprox			0.228**			0.192
				(0.0890)			(0.119)
	americas			0.707			0.904
				(0.586)			(0.601)
	asia			0.258			0.314
				(0.470)			(0.480)
	africa			0.174			−0.112
				(0.654)			(0.758)
	lnsocprox		0.885***			1.229***	
			(0.0425)			(0.0798)	
	Constant	−0.516	−0.577	−0.653	−0.930	−0.714	−0.540
		(0.684)	(0.649)	(0.676)	(0.844)	(0.811)	(0.827)
inflate	*lnmassi*	−0.754***	−0.742***	−0.755***	−0.872***	−0.844***	−0.856***
		(0.0462)	(0.0490)	(0.0503)	(0.0802)	(0.0759)	(0.0762)

Table 9.8 (continued)

Equation	Variables	(1)	(2)	(3)	(4)	(5)	(6)
inflate	*lnmassni*	−0.303***	−0.276***	−0.281***	−0.346***	−0.372***	−0.390***
		(0.0309)	(0.0327)	(0.0338)	(0.0509)	(0.0488)	(0.0512)
	lndist	0.246***	0.254***	0.244***	0.310***	0.274***	0.252***
		(0.0357)	(0.0391)	(0.0397)	(0.0691)	(0.0622)	(0.0634)
	lninstprox			0.112			0.109
				(0.143)			(0.226)
	lndyad			−2.061***			−58.24
				(0.201)			(20.728)
	lntriad			−2.174***			−3.640*
				(0.301)			(1.890)
	lnsocprox		−2.114***			−21.05	
			(0.165)			(1,036)	
	Constant	3.018***	2.857***	2.927***	1.904***	2.476***	2.710***
		(0.338)	(0.367)	(0.371)	(0.658)	(0.576)	(0.584)
lnalpha	Constant				0.322***	−0.0837	−0.112
					(0.110)	(0.0913)	(0.0966)
	Country FE	yes	yes	yes	yes	yes	yes
	Observations	12848	12848	12848	12848	12848	12848
	stataaic	14130	13323	13321	13648	12997	13000
	statabic	14638	13845	13888	14162	13527	13575
	ll	−6997	−6591	−6584	−6755	−6428	−6423
	LR test				484.94***	327.88***	322.56***
	Vuong test	10.33***	10.18***	10.23***	3.51***	5.70***	5.68***

Notes: Standard errors in parentheses. *** $p < 0.01$, ** $p < 0.05$, * $p < 0.1$.

CONCLUSION

The aim of this chapter has been to test two hypotheses: on the one hand that research infrastructures hold a central position in the materials science network, and on the other that being connected to an RI for conducting international collaborations has a positive effect.

Concerning the first question, a social network methodology was proposed and evidence shows that RI research in materials science is central, which means that it is involved in key research, and is therefore influential over the network. In addition, within the group of articles involving RIs, several articles show high centrality, meaning an important role of few but key articles. From these general centrality aspects, a complementary cluster analysis is conducted. Results show that RI research is on average involved in a lower number of clusters compared to the non-RI group. However, if we take RI research independently from the cluster division, we find that RI-related research has a similar performance in terms of citations per publication performance with respect to the non-RI group (t-test at the 0.05 level). These two observations suggest a beneficial role of RI research towards specific clusters, which is in line with the support and 'cohesion' role of the initial objectives of the RIs. The way in which RIs conduct their 'cohesive' role can be understood by analysing the brokerage position of their research. The results obtained indicate that RIs increasingly coordinate research projects within their own cluster as opposed to research projects outside their cluster.

In the second question addressed by this chapter, the effect of RIs on Italian international scientific collaborations has been investigated. In order to test the hypothesis of a positive effect, a modified gravity model was proposed. Setting collaborations as the dependent variable, a series of variables and proximity measures were introduced in the model. The tested variable of interest concerned an RI proximity variable which described a triadic relation between Italian and non-Italian research organizations, and a research infrastructure. This variable was interpreted as how the presence of a research infrastructure in city-pairs collaborations (and their respective institutions) affects the probability to collaborate. The tested regressions point to a positive and significant role of triads on the probability to collaborate at the international level. Note that alongside the triad factor, the dyad factor defined by previous bilateral collaboration also affects – to a slightly higher degree – the international collaboration. The two variables can be interpreted as positively affecting the probability to collaborate, both from trust being built between partners as well as a common need in using cutting-edge instruments.

To conclude, it appears that RI research generally fulfils its inherent objectives: RIs appear central and support a large number of research organizations in the network. Also, they are involved in the creation of central knowledge and know-how towards more specialized and not necessarily central clusters. They affect the scientific network of Italian research organizations by increasing the probability of international collaborations to connected cities. This result suggests the role of RIs in improving visibility of local research organizations in the global materials science field.

NOTES

1. http://ec.europa.eu/research/era/pdf/era_gp_final_en.pdf. World-class research infrastructures, integrated, networked and accessible to research teams from across Europe and the world, notably thanks to new generations of electronic communication infrastructures.
2. Definition retained by the European Commission-funded Research Infrastructures: Foresight and Impact (RIFI) project, 2010.
3. http://www.RIportal.eu.
4. 'Clusters' are to be understood in the network theory jargon and not in the economic sense: clusters consist of groups in which articles are assigned.
5. The modularity of a partition is a scalar value between –1 and 1 that measures the density of links inside communities as compared to links between communities.
6. Use of the open source application GEPHI (http://wiki.gephi.org).
7. http://www.scopus.com.
8. *Inorganica Chimica Acta*, *Journal of Organometallic Chemistry*, *Solid State Communications*, *Macromolecules*, *Thin Solid Films*, *Journal of Non-Crystalline Solids*, *Journal of Materials Science*, *Applied Surface Science*, *Polyhedron*, *Journal of Nuclear Materials*, *Solid State Ionics*, *Cryogenics*, *Journal of Fluorine Chemistry*, *Polymer Degradation and Stability*, *Corrosion Science*, *European Polymer Journal*, *Journal of Power Sources*, *Journal of Membrane Science*, *Physics and Chemistry of Minerals*, *Journal of Elasticity*, *Materials Chemistry and Physics*, *Materials Letters*, *Oxidation of Metals*, *Polymer Bulletin*, *Cement and Concrete Research*, *Solar Energy*, *Journal of Physics and Chemistry of Solids*, *Materials Research Bulletin*, *Journal of Electrostatics and Ultramicroscopy*.
9. The density of a network is the proportion of edges that are present out of all the edges that could be there.
10. Closeness centralization is the variation in the closeness centrality of vertices divided by the maximum variation in closeness centrality scores possible in a network of the same size.
11. Many scientists, unfortunately, do not acknowledge the use of the RIs in their publications. The current study, therefore, analyses publications where RI-based scientists are directly involved.
12. One of them, namely cognitive proximity – usually measured by scientific field – is not present as a factor since the research focuses on one homogeneous field of physics: materials science.
13. As measured by a modified Hamming distance which sums the institutional similarity between pairs of cities.
14. $Exp(1.375) - 1 = 2.955$
15. Correlation between the two variables ~ 0.

REFERENCES

Anderson, J. (1979). A theoretical foundation for the gravity equation. *American Economic Review*, 69, 106–116.

Anderson, J.E. and Van Wincoop, E. (2003). Gravity with gravitas: a solution to the border puzzle. *American Economic Review*, 93 (1), 170–192.

Arora, A., David, A. and Gambardella, A. (1998). Reputation and competence in publicly funded science: estimating the effects on research group productivity. *Annales d'economie et de statistiques*, 49–50, 163–198.

Audretsch, D. and Feldman, M. (1996). R&D spillovers and the geography of innovation and production. *American Economic Review*, 86 (3), 630–640.

Autio, E., Bianchi-Streit, M. and Hameri, A.-P. (2003). Technology transfer and technological learning through CERN's procurement activity. CERN Education and Technology Transfer Division, working paper 2003-5.

Autio, E., Hameri, A.-P. and Vuola, O. (2004). A framework of industrial knowledge spillovers in big-science centers. *Research Policy*, 33, 107–126.

Balconi, M., Breschi, S. and Lissoni, F. (2004). Networks of inventors and the role of academia: an exploration of Italian patent data. *Research Policy*, 33, 127–145.

Blondel, V.D., Guillaume, J.-L., Lambiotte, R. and Lefebvre, E. (2008). Fast unfolding of communities in large networks. *Journal of Statistical Mechanics: Theory and Experiment*, 10, P10008 (12pp).

Boschma, R. (2005). Proximity and innovation: a critical assessment. *Regional Studies*, 39 (1), 61–74.

Bozeman, B. and Corley, E. (2004). Scientists' collaboration strategies: implication for scientific and technical human capital. *Research Policy*, 33 (4), 599–616.

Breschi, S., Cassi, L., Malerba, F. and Vonortas, N. (2009). Networked research: European policy intervention in ICTs. *Technology Analysis and Strategic Management*, 21 (7), 833–857.

Burger, M.J., Van Oort, F.G. and Linders, G.-J.M. (2009). On the specification of the gravity model of trade: zeros, excess zeros and zero-inflated estimation. *Spatial Economic Analysis*, 4 (2), 167–190.

Dasgupta, P. and David, P. (1994). Towards a new economics of science. *Research Policy*, 23, 487–521.

Ding, W., Levin, S., Stephan, P. and Winkler, A. (2010). The impact of information technology on academic scientists' productivity and collaboration patterns. *Management Science, Articles in Advance*, 56 (9), 1–23.

Flowerdew, R. and Aitkin, M. (1982). A method of fitting the gravity model based on the Poisson distribution. *Journal of Regional Science*, 22 (2), 191–202.

Foray, D. and Lissoni, F. (2009). University research and public–private interaction. In Hall, B. and Rosenberg, N. (eds), *Handbook of Economics of Technical Change*, Amsterdam: North Holland/Elsevier.

Goldfarb, A. and Agrawal, A. (2008). Restructuring research: communication costs and the democratization of university innovation. *American Economic Review*, 98 (4), 1578–1590.

Gould, R. and Fernandez, R. (1989). Structures of mediation: a formal approach to brokerage in transaction networks. *Sociological Methodology*, 19, 89–126.

Greene, W.H. (2002). *Econometric Analysis*. Upper Saddle River, NJ: Prentice Hall.

Lane, J. and Bertuzzi, S. (2011). Measuring the results of science investments. *Science*, 331, 678–680.
Lissoni, F., Llerena, P. and Sanditov, B. (2011). Network of inventors and academics in France. Preliminary draft.
Meyer, M. (2000). Does science push technology? Patents citing scientific literature. *Research Policy*, 29, 409–434.
Montobbio, F. and Sterzi, V. (2011). The globalization of technology in the emerging markets: a gravity model on the determinants of international technological collaborations. Unpublished draft.
Newman, M.E. (2001a). Scientific collaboration networks. I. Network construction and fundamental results. *Physical Review E*, 64 (1), http://link.aps.org/doi/10.1103/PhysRevE.64.016131.
Newman, M.E. (2001b). Scientific collaboration networks. II. Shortest paths, weighted networks, and centrality. *Physical Review E*, 64 (1), http://link.aps.org/doi/10.1103/PhysRevE.64.016132.
Nooy (De), W., Mrvar, A. and Batagelj, V. (2005). *Exploratory Social Network Analysis with Pajek*. New York: Cambridge University Press.
Plotnikova, T. and Rake, B. (2011). Collaboration in pharmaceutical research: exploration of country-level determinants. Draft, DIME Conference.
Price, D.J. (1984). The science/technology relationship, the craft of experimental science, and policy for the improvement of high technology innovation. *Research Policy*, 13, 3–20.
Rosenberg, N. (1992). Scientific instrumentation and university research. *Research Policy*, 21, 381–390.
Salter, A. and Martin, B. (2001). The economic benefits of publicly funded basic research: a critical review. *Research Policy*, 30 (3), 509–532.
Santos Silva, J. and Tenreyro, S. (2011). Further simulation evidence on the performance of the Poisson pseudo-maximum likelihood estimator. *Economics Letters*, 112 (2), 220–222.
Sinnot, Roger W. (1984). Virtues of the Haversine. *Sky Telescope*, 68, 159.
Stephan, P. (1996). The economics of science. *Journal of Economic Literature*, 34 (3), 1199–1235.
Tijssen, R. (2002). Science dependence of technologies: evidence from inventions and their inventors. *Research Policy*, 31, 509–526.
Verspagen, B. (1999). Large firms and knowledge flows in the Dutch R&D system: a case study of Philips Electronics. *Technology Analysis and Strategic Management*, 11 (2), 211–233.
Von Hippel, E. (1988). *The Sources of Innovation.* New York and Oxford: Oxford University Press.
Wuchty, S., Jones, B.F. and Uzzy, B. (2008). Multi-university research teams: shifting impact, geography, and stratification of science. *Science*, 322, 1259.
Zellner, C. (2003). The economic effects of basic research: evidence from embodied knowledge transfer via scientists' migration. *Research Policy*, 32 (10), 1881–1895.

PART III

Strategic challenges for policy makers in shaping technology development

INTRODUCTION

The nature of technological innovation has changed significantly. The everlasting debate of what comes first – technological invention that creates new markets, or identifiable market demands that direct technological searches – is replaced by aspirations to resolve grand societal challenges such as sustainability, energy, health and ageing with technological advances. To achieve impact, emergent technologies have to go well beyond successful commercialization and demonstrate potential for increasing societal well-being.

Focus on societal impact creates new challenges for policy makers in shaping trajectories of emerging technology development. Firstly, policy makers are not only providers of public funding for scientific research in universities and for-profit organizations (Leydesdorff et al., 2013); they are increasingly actors in wider social coordination that brings together multiple constituencies with diverse interests in the development of emergent technology (Guston, 1999). More importantly policy makers are often in a position to orchestrate pluralistic processes that shape policy towards a particular emergent technology. Secondly, policy makers face challenges of how to make sense of the impact that is achieved by emergent technology. The conventional economic and scientometric measures do not necessarily reflect the societal impact of investments into emerging technologies. Neither is it conclusive how different fiscal measures in the domain of government policy influence the adaptation and diffusion of emergent technologies. Although most countries share similar environmental goals and deploy a wide set of fiscal measures to promote the adaptation and diffusion of emergent technologies for achieving goals such as greenhouse gas emissions reductions and energy savings, their choice of financial and fiscal instruments varies, depending mainly on the national priorities.

Empirical evidence of clear relations between policy action on and outcomes of particular innovation systems is difficult to obtain and is rather scarce. Policy makers usually apply a bottleneck analysis (Arnold, 2004), identifying market and structural imperfections and deciding where and how to intervene. One of the few examples of research aiming at analysing the effectiveness of policies addressing emergent technologies concerns biotechnology. Enzing and Reiss (2008) investigate the effectiveness of biotechnology policies in European countries since the 1990s. Using an input–output approach, the mix of policy instruments applied by various European countries to support biotechnology was compared to the biotechnology performance of these countries, considering the whole innovation system. Enzing and Reiss (2008) conclude that a well-balanced mix of generic and biotech specific measures is best suited for supporting the development of biotechnology.

Chapters in this part address both types of challenges faced by policy makers when they engage with dynamics of technology development. Two chapters, from Wolf and Wilkins (Chapter 13) and van Doren and Reiss (Chapter 10), explicitly address the social complexity of the policy-making processes where knowledgeable actors utilize cultural resources such as discourse, institutional logics and professional identities in order to negotiate meaning around emergent technologies. Although not directly building on process tradition in management research (Langley et al., 2013), both chapters are essentially concerned with the emergence, development and growth of technology in general and innovation policy in particular. They illuminate the role of tensions and contestations that characterize the political process of meaning creation around an emergent technology. Most importantly both chapters recognize that multilevel interactions among actors contribute to the emergence and development of technology and associated innovation policy.

Wolf and Wilkins build on institutional theory when describing the process of meaning creation in transorganizational form such as the European Technology Platforms. These hybrid organizations provide a permanent organizational structure for members of an emerging organizational field (Meyer et al., 2005). In these relational spaces members constantly negotiate shared meaning by increasing interaction among a disparate set of constituencies. European Technology Platforms provide participants with the sense of being involved in a common socio-technical issue and equip them with the power to take institutional action, with the impact on innovation policy development. Wolf and Wilkins recognize the process of innovation policy development in institutionally complex settings as deliberative, and draw heavily on political science literature. Dispersed power structures prevent individual members from coercing a particular meaning, and innovation policy processes unfold in the absence of both strict decision rules and dominant actors capable of imposing unity and coherence around the meaning of technology and societal issues (Hale and Held, 2011). Political deliberation is essentially needed, because economical, technical and social issues are increasingly changing autonomous organizations into interdependent actors. Even governments, as powerful actors, are increasingly unable to mobilize resources within their own realm, and so they become dependent on collaborations with actors outside their direct control.

Van Doren and Reiss recognize communication as a main activity and language as a main resource to negotiate definitions of emerging technology such as synthetic biology. They share the belief with the authors of Chapter 13 that social coordination and mutual understanding among dispersed actors is achieved by communicative actions. The challenge they identify is the availability of multiple discursive resources that are potentially contradictive and incompatible. They build on the notion of trading zones (Galison, 1997) that focuses on coordination between members of different scientific disciplines when they confront interdisciplinary challenges. Van Doren and Reiss suggest that synthetic biology is better approached and understood as a boundary object (Gorman, 2002). This enables actors to integrate independent tasks without the need for shared meaning. More intriguingly, however, the authors indicate that synthetic biology should move from being only a boundary object, and become a category determined with essential properties, which will help with the legitimization and regulation of this emerging technology. They recognize that such a categorization is a complex process determined by both historical development of the discipline and the influence of social groups on the qualification of particular categories.

Chapter 12 from Dombrovski and Chapter 11 from Drenkovska address a different type of challenge faced by policy makers. They do not investigate the complexity of the policy-making process, but focus on measuring the economic and social impact of emergent technology. Emergent technology-driven innovation often leads to outcomes not well captured in conventional national accounting. New quantifiable measures and valuation techniques that synthesize different levels of the complex process of emergent technology innovation benefit policy makers and provide a more balanced view of the efficiency of the research policy and governmental interventions. The chapters address these policy challenges from different perspectives. Dombrovski is concerned with investigating aggregate effects of fiscal policy on the diffusion of technologies that enable sustainable energy generation. Drenkovska tackles the problem of unobservables in innovation management research. This is timely, because emergent and developing countries are increasing their innovation output despite their appropriability regimes, and research and development (R&D) intensity suggest they should lag behind their developed peers.

Dombrovski argues that a move towards an energy system based on renewables comprises large-scale changes in current energy systems. Such a system transition requires more than the accumulation of a number of individual technological or organizational innovations. Rather, the whole system needs to change radically, including, for example, agents, their roles and interactions; markets, and their international configuration and access conditions; the division of labour along value chains; supply–demand relations; and not least, the rules of the game. Usually such system transitions include transboundary movements of technical systems, political fields, agents and models. It is obvious that political governance of system transitions requires new concepts, approaches and instruments (Flanagan et al., 2011). In this complex context Dombrovski analyses selected policy instruments with particular emphasis on their impact and effectiveness and finds that there is no single preferred policy instrument. Instead, she pleads for an optimal mix of the most efficient energy policies to encourage emergent technologies.

Drenkovska's chapter builds on increasing interest by economists involved with development topics in understanding the role of innovation for developing countries, and in particular whether the growth stemming from innovative activities is inclusive (Chataway et al., 2013). This requires an understanding of the nature of innovative activities taking place in low-income countries. What is observed is the importance of local, bottom-up and grassroots innovations, where the innovation taking place does not only represent an innovative activity, but is also a factor of social transformation (Smith et al., forthcoming). Yet, this stream of research faces difficulties regarding both the collection and the quality of data. Indeed, official databases as well as existing questionnaires tend to be incomplete or unreliable. Drenkovska's chapter suggests improvements to most of the existing questionnaires. In particular, based on the Community Innovation Survey (CIS), she tries to overcome existing limitations by the 'introduction of new questions and putting more weight on others', and to capture in a better way innovative activities in developing countries.

It is indicative that all four chapters build on different disciplinary traditions within the social sciences, despite addressing the same phenomenon. The economics perspective meets innovation management traditions, and both are complemented by perspectives building on institutional theory and the sociology of technology. Advances in science and engineering are often attributed to interdisciplinary research. This part of the book demonstrates that interdisciplinary research among social science disciplines could produce relevant knowledge if a phenomenon is investigated with rigorous methodological apparatus without ignoring the intrinsic social complexities.

Krsto Pandza, Thomas Reiss and Joze P. Damijan

REFERENCES

Arnold, E. (2004). 'Evaluating research and innovation policy: a systems world needs systems evaluations'. *Research Evaluation*, 13(1), 3–17.
Chataway, J., Hanlin, R. and Kaplinsky, R. (2013). 'Inclusive innovation: an architecture for policy development'. IKD Working Paper No. 65.
Enzing, C. and Reiss, T. (2008). 'The effectiveness of biotechnology policies in Europe'. *International Journal of Biotechnology*, 10(4), 327–340.
Flanagan, K., Uyarra, E. and Laranja, M. (2011). 'Reconceptualising the "policy mix" for innovation'. *Research Policy*, 40(5), 702–713.
Galison, P. (1997). *Image and Logic: a Material Culture of Microphysics*. Chicago, IL: University of Chicago Press.
Gorman, M.E. (2002). 'Levels of expertise and trading zones: a framework for multidisciplinary collaboration'. *Social Studies of Science*, 32, 933–938.
Guston, D.H. (1999). 'Evaluating the first US consensus conference: the impact of the "citizens" panel on telecommunications and the future of democracy'. *Science, Technology and Human Values*, 24, 451–482.
Guston, D.H. (2001). 'Boundary organizations in environmental policy and science: an introduction'. *Science, Technology, and Human Values*, 26, 399–408.
Hale, T. and Held, D. (2011). *The Handbook of Transnational Governance: Institutions and Innovations*. Cambridge: Polity Press.
Langley, A., Smallman, C., Tsoukas, H. and Van de Ven, A. (2013). 'Process studies of change in organization and management: unveiling temporality, activity and flow'. *Academy of Management Journal*, 56, 1–13.
Leydesdorff, L., Park, H.W. and Lengyel, B. (2013). 'A routine for measuring synergy in university–industry–government relations: mutual information as a Triple-Helix and Quadruple-Helix indicator'. *Scientometric*, 99(1), 27–35.
Meyer, A.D., Gaba, V. and Colwell, K.A. (2005). 'Organizing far from equilibrium: nonlinear change in organizational fields'. *Organization Science*, 16, 456–473.
Smith, A., Fressoli, M. and Thomas, H. (forthcoming). 'Grassroots innovation movements: challenges and contributions'. *Journal of Cleaner Production*.

10. Communicating synthetic biology: a conceptual framework to position developments in biotechnology

Davy van Doren and Thomas Reiss

INTRODUCTION

Communication has proven to be important in managing processes of innovation related to firm behaviour, industrial network development, public understanding and interdisciplinary research. As a result, the quality of communication is likely to influence the social and economic impact of emerging technologies. Synthetic biology is an emerging field being advocated as a facilitator of future biotechnology development. Although synthetic biology has become familiar to a growing audience over the past decade, its meaning has been debated for over a century. In the past 100 years, synthetic biology emerged within both conceptual and practice-orientated discussions, by describing in various degrees to what extent biological systems could be exploited. In 1912, both Stéphane Leduc (1912) and Jacques Loeb (1912) mentioned synthetic biology in speculations over possibilities to create artificial living systems. Later on, in 1974, Waclaw Szybalski mentioned synthetic biology to describe the application of recombinant DNA technology to generate organisms with new genetic properties (Benner, 2010). Around a decade ago, the concept started to be used to refer to the synthesis of unnatural molecules (Rawls, 2000), unnatural chemical systems (Benner and Sismour, 2005), biology-inspired systems (Pleiss, 2006) and functions not existing in nature (Serrano, 2007).

Synthetic biology's rich history, in combination with its interdisciplinary nature, seems to have caused confusion concerning what synthetic biology actually is (ibid.). There seems no consensus about to what extent synthetic biology's nature is epistemological, technological, procedural, or a balanced combination of these different representations. As a result, it is difficult to judge the current developmental state of synthetic biology. Such uncertainty could lead to both miscommunication within the field and flawed regulation of it. Furthermore, it is unknown to what extent the apparent confusion could impact upon synthetic biology's credibility among stakeholders and the trust of potential end-users in synthetic biology-based products.

Difficult judgement regarding synthetic biology's essence might prove a barrier concerning synthetic biology-related communication efforts within and across the scientific domain. Visual frameworks might be able to improve biotechnology's interdisciplinary communication through considering biotechnology as a boundary object. Although some synthetic biology visual frameworks have been developed, their ability to capture relevant elements or position comparable concepts seems limited. This chapter builds upon previous research by developing a visual framework that captures these essential elements and enables synthetic biology's positioning among past and potential future concepts within biotechnology.

SCIENCE COMMUNICATION: ITS RELEVANCE FOR INNOVATION AND EMERGING TECHNOLOGIES

The Importance of Communication at the Science–Industry–Public Interface

Communication has proven to be essential within and between the domains of state, industry and science. At the level of industry, obstructions to knowledge flows within and between firms are believed to reduce the generation, development and implementation of technologies. It has been claimed that an economic environment in which needs are clearly defined, as opposed to being heterogeneous or diffuse, will tend to stimulate technical innovation in firms (Utterback, 1971). In addition, the relevance of knowledge flows relates to the impact of standardization on innovation, productivity and market structure (Tassey, 2000). Regarding the importance of individuals as agents of research and development (R&D) and technology transfer (Rothwell and Robertson, 1973), standards are communicated artefacts that enable coordination between different actors involved in technology development with regard to specified products, processes, formats or procedures (Wegberg, 2004).

In addition to its importance at the level of industry, communication and linguistic issues are also crucial at the science–public interface. There are several factors that make public understanding of science important, including the need to act intelligently in relation to evaluating science policy issues (Hartz and Chappell, 1997), to make rational personal choices (Nelkin, 1995) and to distinguish between responsible and irresponsible scientific research (Shortland and Gregory, 1991). However, as the explanation of science involves translation from one context to another, overcoming large quantities of evolved jargon has become challenging (ibid.).

The converging and diverging nature of scientific language also proves to be challenging within the scientific domain itself. Although the birth of new scientific fields and new vocabularies has been a defining aspect of scientific progress, growing specialization has also generated an increasing range of opportunities for the sharing of language through interdisciplinary research (Montgomery, 2004).

The Meaning of Biotechnology

The dynamic nature of scientific language also seems relevant for the emerging synthetic biology field. However, the existence of heterogeneous definitions is not a new phenomenon for the biotechnology domain. The need for a biotechnology definition became apparent in the early 1980s, when Japan, the USA and several European Economic Community (EEC) countries wanted to respond to reports hyping biotechnology as the future foundation of a new society. Biotechnology's rapid evolution and social integration has led to multiple definitions being developed and discussed within and between several interest groups (Bud, 1991). In 1982, the Organisation for Economic Co-operation and Development (OECD) published a report including 11 definitions of biotechnology (Bull et al., 1982). Possible drivers for the rapid creation of such diversity have been suggested: a multitude of stakeholders exists, biotechnology is an emerging developing technology, and there are different perspectives on biotechnology – it being regarded as a new invention (revolution) versus a development of the existing technology (evolution) (Bud, 1991).

This heterogeneous set of definitions had important ramifications in biotechnology history (US Congress, Office of Technology Assessment, 1988). The definition applied for describing biotechnology could affect how and to whom available research funds were provided, as well as the scope of existing regulations related to both biotechnology-based products and processes (ibid.). In addition to potentially influencing the development and dispersion of biotechnology products and processes, different definitions might have hampered both the assessment of markets and the comparison of results between nations or sectors using similar technologies. Finally, the existence of a heterogeneous set of biotechnology definitions decreased transparent communication of progress and future intentions.

Biotechnology as a Boundary Object: The Role of Visual Frameworks in the Interdisciplinary Synthetic Biology Field

It could be argued that biotechnology's multidisciplinary relevance and absent definitional consensus are indicators for approaching biotechnology as a boundary object. A boundary object is a shared space that allows different communities of practice, having different information and work requirements, to work together without consensus over its exact definition (Star, 2010). It is adaptable to local needs and the constraints of the parties involved, yet robust enough to maintain a common identity across sites (Star and Griesemer, 1989). As for biotechnology, a similar argument was made with respect to resilience, originally defined by Holling (1973) as a descriptive ecological concept, but over time taken up by diverse domains leading to a dilution of its original specific meaning (Brand and Jax, 2007).

Boundary objects, also related to trading zones denoting interdisciplinary partnerships in which combined perspectives results in shared language (Galison, 1997), could enable collaborative work and promote intercommunal negotiation (Brown and Duguid, 2001). As a consequence, the use of boundary objects has been extended to several management and organization fields, with respect to change processes (Briers and Chua, 2001), organizational memory (Cacciatori, 2003), new product development (Carlile, 2002; D'Adderio, 2004), project management (Yakura, 2002) and shared documents and tools (Brown and Duguid, 2001). With respect to tools, a multidimensional visual framework could facilitate approaching biotechnology as a boundary object. Through determining the boundaries of biotechnology, such a framework could then assist in developing a general understanding of how synthetic biology could be interpreted.

There have been some previous attempts to develop visual frameworks for synthetic biology. For example, Cambray et al. (2011) visually mapped different research fields and technologies based on a number of drivers shaping the space of synthetic biology. However, this framework did not clarify the explicit properties that make synthetic biology potentially different from other biotechnology concepts, hampering the positioning of synthetic biology within biotechnology. Furthermore, this framework's use of the concepts of design and discovery seems to ignore past and future joint scientific and engineering interdisciplinary efforts. Another attempt aimed to visualize different degrees of biological system modification (De Lorenzo, 2010). Focusing on the engineering nature of synthetic biology developments, this framework illustrates to what extent engineering principles are required to increasingly modify biological systems. However, this framework does not make clear how, and to what extent, other factors than engineering could drive developments in synthetic biology.

Towards a New Visual Framework for Biotechnology

Considering limitations of previous visual frameworks, this chapter develops a visual framework by means of analysing biotechnology's historical development. By enabling the identification of inherent characteristics and developments of biotechnology, the presented visual framework could facilitate both socialization and fact absorption, improving the transfer of knowledge within a community of practice (Grabher, 2002). In addition, this framework could harmonize the different perceptions that are currently hampering transparent and comprehensive communication concerning biotechnology between different communities of practice. Finally, it could coordinate policy makers in developing specific regulations related to the unique and inherent properties of synthetic biology.

In constructing the visual framework, two methodological steps are taken. First, a literature-based historical analysis is carried out to identify key characteristics regarding biotechnology development. In order to structure the biotechnology movement, temporal-based phases are defined for categorizing biotechnology developments. And second, the identified key characteristics are used as input for the development of the new framework.

A HISTORICAL ANALYSIS OF BIOTECHNOLOGY DEVELOPMENT

Assessing Key Characteristics of the Biotech Movement: A Historical Assessment

For this assessment, biotechnology is regarded as the use of biological systems to execute a certain function. To assist in the identification of inherent characteristics of biotechnology, it is necessary to create temporal divisions within its history that can then be analysed individually and compared globally. Within this context, such a temporal division has been made based on the nature and degree of modification of biological systems. Arguably, a biological system can be characterized based on: (1) the procedures used for its production; or (2) the extent it is being modified in relation to its natural form. Based on this approach, the historical development of biotechnology has been assessed according to four phases:

- Phase 1: Using non-modified (natural) biological systems.
- Phase 2: Using selected non-modified (natural) biological systems.

- Phase 3: Using selected modified (natural) biological systems.
- Phase 4: Using engineered (unnatural) biological systems.

In Table 10.1, a summary is provided concerning the analysis of these four phases of biotechnology development.

Phase 1: Using non-modified (natural) biological systems: the importance of implementation and enabling technologies

Purposeful use of biological systems is an ancient human behaviour. Most of the initial use of biological systems was not based on knowledge concerning the inherent logic on which biological systems operate. Most

Table 10.1 Summary of the analysis of the phases of biotechnology development

Phase	Description phase
Phase 1 – Initial application and experimentation (What to modify?)	Organisms were used because of known functions they performed. Parallel to such use, several processes were applied in optimizing environmental conditions, leading to improved output of such functions. Due to the absence of tools enabling detailed observation of biological systems, the responsible biological structures and processes responsible for observed biological activity were unknown.
Phase 2 – Understanding is limited by technical resolution (Can we modify?)	In addition to applied processes to improve on environmental conditions, selection procedures were developed to improve the output of biotechnology. Much of the understanding related to such selection procedures, focused around motors of evolutionary change, was fuelled by the invention of several technical aids enabling detailed observation of biological systems, thereby initializing the development of theories explaining biological operation.
Phase 3 – Modification is limited by complexity (How to modify?)	After having identified the forces behind the adaptation of species, the discovery of DNA as responsible molecules for genetic inheritance set off industrial biotechnology. However, the complexity of biological operation limited the extent to which natural biological systems could be modified.
Phase 4 – From modification to creation (How to design?)	In order to bypass limits of modification related to biological complexity, engineering principles have been introduced in biotechnology to modify or design systems that are less complex, behave predictably in specific settings and are tailored for delivering specific functions.

likely, the initial knowledge regarding characteristics and properties of living systems was acquired as a consequence of application, experimentation and experience. Based on such incremental build-up of biology-specific knowledge, species were eventually identified that displayed useful properties or functions. For example, the earliest chemical evidence of beer brewing dates back around 3500–3100 BC in the Zagros Mountains of Western Iran, in times when beer was seen as a prayer to the goddess Ninkasi (McGovern, 2009; Wilford, 1992). However, almost 5000 years passed until Pasteur in 1857 laid the foundation for how the fermentation conversion process could be explained (Wrotnowska, 1981). Many of the initial uses of biological systems can therefore be assessed as a black box approach; the lack of systematic biological understanding did not hamper their use as such.

Nevertheless, the initial application of natural biological systems was supported by several enabling technologies, which facilitated increased qualitative and quantitative returns. Comparable to biological systems, the understanding of these enabling technologies lagged far behind their application. The operation of fertilization, driving the dawn of agriculture during the Neolithic Revolution (Baker, 2008), was not understood until its underlying principles were revealed in the nineteenth century through the combined insight of Sprengel's and Liebig's 'Law of the Minimum'[1] (van der Ploeg et al., 1999) and Justus von Liebig's discovery of nitrogen as an essential plant element (Smith et al., 1999). Similarly, a true understanding of pesticide operations, of which first applications date back to ancient Sumer in Mesopotamia, only started to emerge well into the twentieth century. Only after the effect of the industrial-scale introduced pesticide dichlorodiphenyltrichloroethane, better known as DDT, on reproduction capabilities of fish-eating birds in the 1950s (Allison et al., 1963) became apparent did its exact operation and influence on nerve system functioning (Yeager and Munson, 1945) become a target of research.

Phase 2: Using selected non-modified (natural) biological systems: increased importance of knowledge

As was the case for many technology trajectories, biotechnology first went through a stage of experimental failures and successes before it was attempted to explain and understand these (Carlson, 2010). Millennia of using natural organisms passed in ignorance of actual understanding of what enabled these organisms to provide their services. For such understanding, technological aids were required to observe the basic structural units of life, which did not appear until the seventeenth century, when Galileo developed his compound microscope in 1625 (Gould, 2000). Soon after, Antonie van Leeuwenhoek observed the first cells of protozoa and bacteria using a self-made microscope (Van Zuylen, 1981).

Inspired by the possibilities of studying life, and for the first time since its initial acknowledgement by Hippocrates and Aristotle, biology started to flourish (Magner, 2002). The nineteenth-century scientific discoveries of Charles Darwin (differential individual reproduction of natural selection) and Gregor Mendel (patterned inheritance) laid the basis for a first true understanding regarding the processes influencing or affecting life's operation. Although domestication of agricultural plants has a history dating back many thousands of years, laying the theoretical foundation of heredity formed the basis of more deliberate and better understood selection and breeding processes of superior natural biological systems. In addition, cross-breeding, an optimization technique currently widely used in agriculture, forestry and livestock management, was deliberately introduced to steer the genetic make-up of organisms and populations.

Fuelled with new technological possibilities and greater understanding of microbiology, twentieth-century scientists started to explore ways to deliberately manufacture specific products using biological systems. The initial main promises of biotechnology were the solving of immense social and environmental problems, laying a foundation for an upcoming Green Revolution by providing bio-based solutions related to global food shortage, lack of high-quality fermentation products and shortage of raw materials (Bud, 1991). But biotechnology was more than just green. The need for acetone in explosives production during the First World War induced Weizmann in 1917 to successfully create a culture of *Clostridium acetobutylicum* able to convert starch on an industrial scale (Springham et al., 1999). Another example is the first purification of antibiotics of Fleming's discovered mould *Penicillum* in 1928. In 1940, penicillin became available on a large scale for medicinal use to treat bacterial infections in humans (Thieman and Palladino, 2003).

Phase 3: Using selected modified (natural) biological systems: start of unnatural optimization

Since the first uses of biotechnology, the two once independent concepts of understanding and application had grown closer. The increased interaction of humans with biological systems led to improved understanding of biology. Through this understanding of life, people started to use this knowledge for the manipulation of natural systems, in order to improve on their intrinsic functions.

The trend of increased importance for understanding became apparent when humans tried to dig deeper towards the core of life. By the 1950s, the interdependent development of both life sciences and technological improvements had realized huge steps in penetrating the different layers of biological systems. The identification of DNA, being the motor of change within the earlier revealed evolutionary and hereditary processes, precluded the taking-off of genetics. Once DNA could be observed, it could be targeted. From 1973, when recombinant DNA technologies became available and biology turned into a laboratory science, a number of spectacular breakthroughs followed, including the production of synthetic insulin, human growth hormone and interferon (Yi, 2008). When in 1980 the United States Supreme Court accepted the patenting of genetically modified *Gammaproteobacteria* capable of breaking down crude oil, the commercialization of industrial biotechnology became a fact. The combination of improved intellectual property rights, the introduction of enforced legislation and strengthened demand for medical and pharmaceutical products led to immense growth and development of an industrial sector using biomass conversion carried out by (parts of) living organisms.

In order to compete with the more mature chemical sector for producing similar compounds, it was necessary to improve on both industrial biotechnology's[2] absolute productivity and its relative productivity, including the processes involved therein. Trying to realize such improvements, different levels of biological systems started to become focal points, including the organism level,[3] the sub-organism level[4] and the supra-organism level.[5]

In the optimization process of biological systems, the rational modification of its core components stands central. Although the elements responsible for observed behaviour had been identified, the interaction between individual components and processes, as well as the total effect of cellular behaviour on its environment, are essential features that needed to be accounted for within modification efforts of biological systems. Although improved technological sophistication enabled scientists to observe and alter microscopic biological components, elucidating the specific biological components and control mechanisms inherent to biological system functioning lagged significantly (Ramakrishnan, 2002; Sims III et al., 2004; Moore, 2009). A limited understanding of the multilevel and stochastic operation, complex regulation and emergent properties of biological systems created challenges in altering biological systems and their processes in an efficient and cost-effective manner (Sprinzak and Elowitz, 2005; Veening et al., 2008; Mitarai et al., 2008).

In the past decade, the explanatory power of systems biology – supported by technological tools including *in silico* modelling simulation,[6] DNA sequencing and high-throughput omics analysis – has improved our systemic understanding concerning genetic circuit operation, cell metabolism and phenotype–genotype relations. In combination with improved resolution of tools able to modify biological systems, including DNA synthesis, genetic pathway insertion and codon optimization, the use of biotechnology has become integrated in the fields of medicine, agriculture and bioremediation.

Phase 4: Using engineered (unnatural) biological systems: towards rational design

During the first three described phases of biotechnology's history, there has been a gradual transition from: (1) a black box approach of biological systems; to (2) the increased importance of knowledge in explaining biological behaviour; towards (3) the rise of technological progress required to observe and modify the structures making up final biological phenotypes. However, there seem to be limits to what extent improved understanding and technological sophistication are enough to satisfy the demand for the increasingly complex products and functions that biological systems could potentially deliver.

In order to overcome complexity linked to rational and functional modifications of fundamental biological components, it has been argued that the introduction of engineering principles is required (Endy, 2005). Engineering biological systems follows the same classical engineering principles as applied to any other field, these being abstraction, standardization and decoupling (Andrianantoandro et al., 2006; Haseltine and Arnold, 2007; Fung et al., 2005) (Table 10.2).

There has already been some progress concerning these engineering principles within biotechnology. Abstraction, through targeted gene deletion or foreign gene implementation, deletion and integration, has to some extent resulted in improved production processes (Vinuselvi et al., 2011). Furthermore, several communities have emerged that are building databases of standardized parts.[7] Concerning decoupling, compartmentalization of biological systems has been shown to introduce orthogonality and increase the output of metabolic reactions.[8] However, it could be questioned whether the use of engineering principles in current biotechnology practice has already become integrated at the cellular level. It remains uncertain when, how and to what extent a truly engineered bottom-up approach in creating biological systems will or could appear.

Synthetic biology aims to tackle such issues. From the conventional understanding of life as it exists, synthetic biology tries to explore what could be (Elowitz and Lim, 2010). In the early twentieth century, Tornier mentioned that the mechanical elegance of living beings should inspire engineers to a new perspective on technology (Tornier, 1901). In this context, synthetic biology can be observed within the engineering domain through the use of natural or unnatural parts[9] to create biological systems that do things inside or outside the natural biology regime. On the other

Table 10.2 Engineering principles in a biological context

Engineering principle	General description	Context biotechnology
Abstraction	The process of streamlining a problem to focus only on pertinent issues	The *de novo* design* or integration (Martin et al., 2003; Ro et al., 2006; Levskaya et al., 2005) of metabolic pathways based on desired functions or phenotypic properties the biological system needs to exhibit (Jarboe et al., 2010). Characterized by minimal genome** creation through removal of non-essential parts (rational strain-engineering) for the system's delivery of a specific function in a cost-effective manner.
Standar-dization	Arbitrary solutions or best practice (Markus and Gelinas, 2006), expressing a shared interest of compliance (Tassey, 2000)	Standardization*** can facilitate the abstraction of biological systems. In modifying biological systems, it is required that circuits function reliably and with known performance characteristics. Standardization of biological parts is needed to ensure that components of a system are compatible and exchangeable.
Decoupling	Dissecting a system into less complicated subsystems by means of orthogonality and/or modularity	Orthogonal biological systems: the incorporation of biological components can interact with existing subsystems. Orthogonality aims to modify subsystems without causing significant disturbances in the action of, or interaction with, such existing subsystems (Weiss and Panke, 2009). Modularity explains coupling characteristics between system components, as a function of the degree to which a system architecture and core operation enables the introduction and integration of components (Sauro, 2008).

Notes:

* *De novo* design: a bottom-up approach of biological system engineering, or artificial protocell creation, which uses materials as input that were not part of pre-existing natural systems, including natural or synthetic molecular components (Szostak et al., 2001). Artificial protocells produce the following interlocking chemical properties: (a) spatial localization of components by containment; (b) utilization of energy and raw materials from the environment by metabolism; and (c) control of the containment and metabolism by chemical information that can be replicated and is able to mutate (Rasmussen et al., 2003). Although artificial protocells have not been developed yet, both the utilization and replication in primitive vesicles, able to self-assemble from self-providing lipid molcules and generate proteins when provided with feedstock ingredients, have already been reported (Noireaux and Libchaber, 2004; Luisi et al., 2006).

** Minimal genomes are strains exhibiting minimal complexity.

*** In comparison with the standardization of non-biological systems, the evolution of living cells is an intrinsic but complicating factor. Newly emerging features can be undesirable when genetic circuits, designed and fabricated by rational, synthetic biological approaches, are installed in the cell (Umenhoffer et al., 2010). Furthermore, randomly arising genetic mutations can potentially hamper proper interaction between biological system parts (Kwok, 2010).

hand, synthetic biology can also be referred to as a scientific discipline contributing to expanding our knowledge of natural and systems biology.[10] In addition, a number of key enabling technologies have enabled improved possibilities in relation to biological system research, including their discovery and characterization, understanding, modification and monitoring.[11] This could usher in increased foundational knowledge research and improved engineering optimization efforts, and might stimulate both bottom-up design and top-down optimization.

Constructing a General Framework for Biotechnology Development

Over the past millennia, the accumulative generation of experience, knowledge and technological possibilities has dramatically increased the output potential of biotechnology. Based on the historical assessment of biotechnology, there seem to be three distinct drivers steering the improved use of biological systems, these being:

1. Understanding biological systems: increased knowledge concerning the operation of biological systems. This dimension stresses the need for science in order to develop biotechnology. Although the initial use of biological systems was not based on any explanatory knowledge, increased understanding created opportunities to select, and in a later historical phase to modify, organisms based on emerging global needs. The derivation of knowledge concerns how action and interaction of components and processes within and between biological systems operate, including those in relation to an external environment.
2. Technological resolution: the extent to which technologies are able to facilitate the observation, modification and reconstruction of biological systems. This dimension stresses the importance of technology required to both facilitate scientific progression as well as translate increased scientific understanding into technological progression, and of providing means to modify biological systems with increased resolution.
3. Engineering principles: specific modes of operation, creation or development for rational and deliberate modification and construction of biological systems. This dimension indicates the requirement of engineering principles (standardization, decoupling and abstraction) to rationally modify biological systems. Engineering enables the creation of new functions, or optimizes both the quantitative and the qualitative performance of existing biotechnology processes (Vancompernolle and Ball, 2005; Schwille, 2011; Kitney et al., 2009).

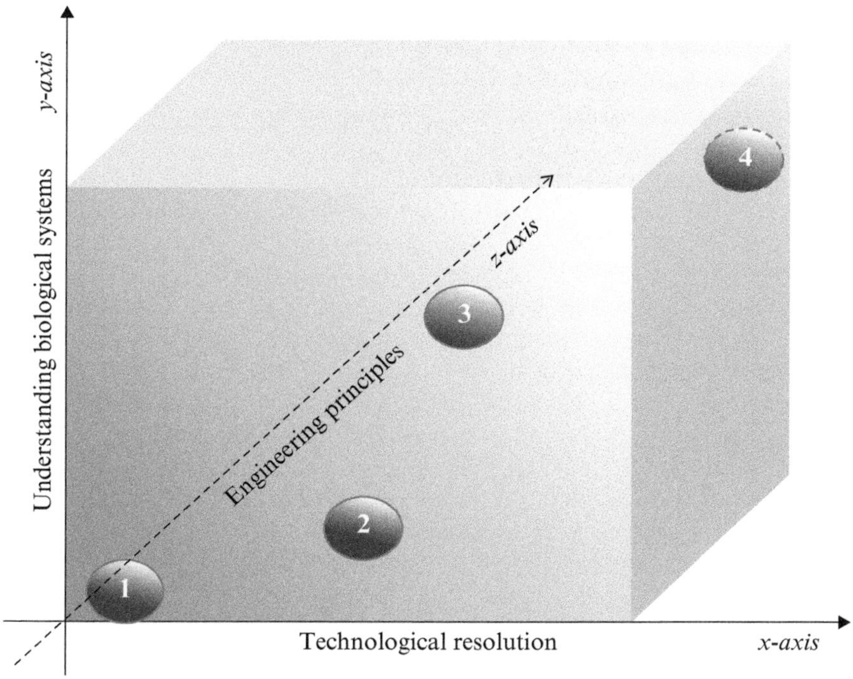

Note: Mapping the four phases of historical biotechnology development, along the three pillars of: (1) increased understanding; (2) increased technological resolution; and (3) increased use of engineering principles.

Figure 10.1 Framework of biotechnology's historical development

Combining insights from the analysed phases with the identified drivers, the above framework has been constructed (Figure 10.1).

DISCUSSION

Positioning Synthetic Biology in the Context of Biotechnology

The different interpretations of synthetic biology can be traced back to the drivers responsible for its emergence in biotechnology's historical development. The two initial distant practices of applying and understanding life have grown closer over the many centuries of biotechnology's existence. Despite the reality of such merging, the individual contributing fields of biotechnology have been blind to fully accept their cross-fertilization, resulting in difficulties on conceptual and definitional harmonization. A similar problem seems to be looming for synthetic biology, where engineers are entering biotechnology's arena.

It could be argued that synthetic biology suffers from a lock-in effect. When it was mentioned for the first time in 1912, its context was of a conceptual and philosophical nature. Now, the creation of unnatural life forms has never been more realistic. Would it be fruitful to introduce a standardized concept that refers to the overall technology that is able, or will be able, to execute rational design of biological systems? Biotechnology's complex combination of technical and cultural roles is characteristic for scientific disciplines in general; but its practical orientation would distinguish it from the typical disciplines (Bud, 1991).

Coming back to the drivers of heterogeneous interpretation of biotechnology, the definitional confusion haunting synthetic biology might be a combination of stakeholder diversity (related to biotechnology's three dimensions) and different perspectives (the relative positioning of synthetic biology according to these three dimensions).

Based on the concept of boundary objects, the developed framework, instead of following a more static and resolute approach of defining synthetic biology, embraces the rich historical fusion of different disciplines throughout the phases of biotechnology development. The framework can be used to map achievements and developments within biotechnology history and to identify its driving forces. The framework's three dimensions indicate the important drivers of knowledge generation, technological sophistication and engineering principles regarding past and future biotechnology development. Based on these drivers, synthetic biology can be characterized as a biotechnology concept aiming to integrate these three dimensions in an interdependent manner.[12]

In this chapter, synthetic biology has been presented within phase 4 of biotechnology's analysed historic development, due to its particular focus on the systematic engineering of biological systems. Nevertheless, the position of synthetic biology in the framework remains arbitrary. In other words, synthetic biology's existence remains subjective. For example, although Craig Venter's creation of *Mycoplasma mycoides* (Lartigue et al., 2009) was spectacular in itself, it remains questionable to what extent the three engineering principles were applied in an interactive manner. As the presence and interaction of these engineering principles can be ranked in a continuum, labelling an occurrence or phenomenon 'synthetic biology' remains subjective. Strictly speaking, it is difficult to reason logically how to set limits of engineering principles regarding synthetic biology. First of all, the measurement of their presence is difficult. And second, where their presence could be assessed, do these engineering principles have ontological limits? Although human abilities might be omniscient, it is unlikely that limits of biological system knowledge will soon be reached. This is different for engineering principles, as these are rooted in human action. Could there be limits for synthetic biology, and will there be a time in which further development of biotechnology will become dependent on other existing or non-existing drivers?

In addition, although it was suggested that the presented framework could facilitate the treatment of the biotechnology domain as a boundary object, some reservations regarding its implementation power need to be mentioned. Boundary objects are 'high-maintenance' items, have no independent potency for alignment, and need legitimate authority or interdependence in order to convey any meaningful knowledge exchange (Sapsed and Salter, 2004). In addition, the role of a broker should not be underestimated with regard to creating political balance between involved actors, making relevant information available and directing the involved community through controlling the flow of information (Kimble et al., 2010). Therefore, although the presented framework could potentially facilitate approaching biotechnology as a boundary object, its validity and legitimacy can only be ascertained through the involvement of all relevant actors.

The Influence of Non-Technological Drivers on Synthetic Biology Development

The aim of this chapter was to provide a framework to delineate ongoing discussions related to biotechnology and synthetic biology. However, it must be noted that the centralization of biotechnology within this framework is arbitrary, meaning that any of the three identified biotechnology drivers are more than just serving as preconditions driving biotechnology development. In addition, biotechnology itself could be seen as a self-enforcing driver, a phenomenon which was already mentioned in describing the initial human use of biotechnology based on application, experimentation and experience (phase 1). Nevertheless, with regard to potential needs for governance support, as well as to the position synthetic biology might occupy in a foreseen future, a biotechnology-centric framework might prove relevant.

The provided framework shows how synthetic biology can be treated as a new subdiscipline within the overall biotechnology concept. As such, it could assist more transparent communication within and between involved research communities, industries, regulators and society as a whole, providing a means to help develop the field.

Development can be regarded from multiple perspectives. One of these perspectives includes the consideration of non-technological drivers affecting the development and diffusion of innovations. According to theories concerning innovation diffusion and innovation systems, factors influencing market access, product regulation and socially constructed technological legitimacy could either enhance or hamper the speed and direction of technological development. Diagnosed ominous links between technological dependency and tangible harm from biotechnology products have caused the European Union to enact precautionary legislation on the intentional release of genetically modified organisms (GMOs) (Levidow and Carr, 1997). Although it could be debated to what extent ethical consideration should be part of regulatory frameworks for biotechnology, affecting the distribution of funding and market access of products, discussions concerning how bioethics and value-embedded science should be considered within regulation (ibid.) is outside the scope of this chapter. Nevertheless, it is clear that discussions on the use of biotechnology in agriculture have affected the shaping of legal restrictions and distribution of research funding in the past decades. Although it has been questioned to what extent such biotechnology has provided benefits or harm (Shiva, 1991) in natural, social and economic dimensions, it can be argued that increased definitional transparency will facilitate the development of a more standardized language to improve communication related to this research field.

It can only be hypothesized to what extent the introduction of engineering principles in biotechnology would increase the potential to facilitate sustainable development. Many of the ethical, social, environmental and economic issues appearing in current debates about the creation and use of genetic modified organisms are also relevant to synthetic biology. Nevertheless, there are indications that the engineering principles might bring some of the required solutions towards more sustainable use of biotechnology, including decoupled engineered production organisms from natural ecosystems through XNA-based orthogonal organisms (Marlière, 2009; European Academies Science Advisory Council, 2010), increased conversion efficiency and output quality through the abstraction of biological systems (OECD et al., 2009), and potential decentralized and distributed biological production based on open source standardization of technological knowledge (Carlson, 2010), stimulating more equal distribution of rewards gained from biotechnology application.

Resolution of the Framework, with Respect to Engineering Principles and Knowledge Development

Another potential issue regarding the presented framework relates to a limited amount of represented resolution. The framework's three dimensions enable the mapping of past and current biotechnology developments. However, if there is also a need for visualization of current and future developments within synthetic biology (for example, biopart library development, XNA integration, minimal cell creation or protocell research), the merging of engineering principles into one dimension limits the resolution of distinction, which limits attempts to map contributions related to the individual engineering principles. One option to solve this is by developing a separate but modular framework dedicated to synthetic biology, splitting up the engineering axis of the biotechnology framework into three distinct axes.

However, a similar argument could be made for the pillar of knowledge generation. Within the presented framework, the development of knowledge has been considered as a straightforward process driven by agent–environment interaction. In reality, this process is obviously much more subtle. The development, but also the interpretation, of knowledge is context-dependent. Local knowledge tends to consist of factual knowledge, skills and capabilities culturally situated as a social product (Antweiler, 1998). The diffusion and translation of local knowledge into a more global knowledge system depends on how such knowledge is being instrumentalized and valued by its examiners, which could lead to different interpretations depending on the level at which knowledge is finally used.

In addition, the way in which knowledge is produced depends on the complexity related to the interactions between involved actors and non-human entities. To know how distributed and heterogeneous cognitive systems work, one has to know about the culture and social organization, as well as about the capabilities of people and artefacts (Giere, 2002). It is being argued that for highly complex research endeavours, expertise is not centralized, but distributed over all involved actors. As a consequence, personal relationships of authority, responsibility and division of final rewards seem to migrate towards a more shared position. Such a phenomenon, being either distributed cognition, collective cognition (Hutchins, 1995) or epistemic cultures (Knorr-Cetina, 1999), depending on how system borders are drawn around the involved agents and technologies, could influence how knowledge production could be governed or managed in a most optimal manner.

Both the inter-actor production of knowledge and the diffusion of such local knowledge towards a global level seem relevant points in the current globalized biotechnology research landscape. However, such reflections do not seem to challenge the idea that knowledge, regardless of how, by whom or in what form it is being produced, has contributed significantly to biotechnology development.

CONCLUSION

Communication is likely to remain an important management process at the science–industry–public interface. The growing interdisciplinary nature of biotechnology will continue to impact upon the quality of communication between its heterogeneous active communities. Also for synthetic biology, both technological and non-technological developments are likely to be affected by the evolution of its multiple languages. The increased presence of engineers, following in general more vendor- and customer-oriented communication approaches compared to literature-focused scientists (Rothwell and Robertson, 1973), is at least one element that is likely to drive more complex communication within the biotechnology domain.

The involvement of heterogeneous communities seems to make it relevant to treat biotechnology as a boundary object. The visual framework presented here aims to facilitate observing biotechnology as a boundary object, through unravelling the confusion rooted in historically heterogeneous biotechnology conceptions, and to enable different communities to position their activities within the overall biotechnology domain.

Although this study appreciates the current nature of heterogeneous languages present within different communities of practice, future research could focus on to what extent the qualities of these specific languages affect communication within biotechnology. It could be argued that the specific jargon used in life sciences does not optimally support communication efforts between all actors involved in developing the evolving biotechnology domain. The emerging field of synthetic biology might have become a victim of this phenomenon. Although biology is more and more associated with being a technology itself (Carlson, 2010), biology in its true sense remains 'the study of life'. Although the presented framework shows that biology has been a crucial driver in biotechnology development, a total mix-up of these different independent concepts might need to be avoided for optimizing communication.

In such a strict sense, synthetic biology would refer to a knowledge field, 'the study of synthetic life forms not existing in nature as such'. As the creation of such life forms calls for technological capabilities to do so, would it be time to introduce the term 'synthetic biotechnology'? Maybe it will emerge once the field is mature enough to deliver on its promises. Evolutionary pressure affecting biotechnology jargon: Darwin might be near, even when synthetic biology tries to avoid him.

ACKNOWLEDGEMENTS

The chapter draws on research carried out by Fraunhofer ISI within a technology assessment project of the Office of Technology Assessment at the German Bundestag. This chapter was also developed in relation to the EU FP7 Marie Curie Research Training Network ManETEI (Management of Emerging Technologies for Economic Impact).

NOTES

1. The 'Law of the Minimum' describes the effect of individual nutrients on crops; growth is not controlled by the total amount of resources, but by the limiting factor.
2. Industrial biotechnology, the application of biotechnology for industry-scale bio-based compound production, is also being labelled 'white biotechnology'. Other large (market) domains for biotechnology include health care ('red biotechnology'), environmental protection and agricultural production ('green biotechnology') (Lorenz and Eck, 2005). Other terms related to biotechnology usage include 'blue biotechnology' (biotechnology use in aquatic environments) (Qin et al., 2011) and 'black biotechnology' (the use of biotechnology in terrorism).
3. Modifications on the organism level include: (a) the optimization, (de)regulation and expression of processes determined by the natural genome; or (b) the integration of foreign processes by means of gene implementation (Parekh, 2004). Advantages are increased productivity and performance characteristics (growth rate, stability, use of input substances) under specific environmental conditions (pH, temperature, oxygen, concentration).
4. Biotechnology at the sub-organism level includes the use of natural catalysts to improve chemical transformations on organic compounds, or biocatalysis (Roberts et al., 1995). Advantages are environmentally acceptable catalysts, being completely degraded in the environment and acting under mild conditions, which minimize problems of undesired side-reactions such as decomposition, isomerization, racemization and rearrangement, which often plague the use of unnatural catalysts.
5. Optimization of industrial biotechnology from a supra-organism-level perspective relates to the biorefinery concept. A biorefinery is the integration of multiple biological conversion processes into one single system, in order to: (a) increase efficiency of total process or produced products; (b) achieve more complete conversion of input material; and (c) decrease by-product production (Clark et al., 2012).
6. *In silico* modelling and the simulation of biological systems is an approach that supports the unravelling of cellular complexity (Zheng and Sriram, 2010). Examples of modelling include decreased complexity of cells (abstraction), unnatural combinations of genetic pathways, design of *de novo* genetic pathways and circuits, and controllability and evolvability of an organism.
7. See for example the Biobricks Foundation, a Registry of Standard Biological Parts, http://partsregistry.org/Main_Page.

8. Based on a presentation made by P. Silver, at the SB 5.0 conference in 2011, Stanford, USA.
9. The use of unnatural parts is a main concern within xenobiology (Schmidt, 2010). Xenobiology, or unusual life forms, could either be found on or beyond Earth, or created in the laboratory.
10. Based on presentations made by both S.A. Benner and D. Endy, at the SB 5.0 conference in 2011, Stanford, USA.
11. Examples of relevant papers are: (a) discovery and characterization – the contribution of systems biology (Ellis and Goodacre, 2011) and the output of omics (Gulati et al., 2009); (b) understanding – the aims of bioinformatics (O'Loughlin et al., 2006) and prediction of biological system dynamics (Zheng and Sriram, 2010); (c) modification – optimization through directed evolution (Kwok, 2010), synthesis of genetic material (Carr and Church, 2009) and combinatorial biosynthesis of genetic material (Jarboe et al., 2010); and (d) monitoring – measuring the interaction between environment and system (Holmes and Gawad, 2010), and biological real-time monitoring and control (Scheper and Reardon, 2008).
12. Generality of the framework: although the framework has been developed for the case of synthetic biology, it could also be argued whether it could be applied for (different aggregation levels of) technology in general. Although this study does not intend to prove this suggestion, it might be interesting to reflect the framework on the historic development of other technologies, or concepts in which these technologies are embedded.

REFERENCES

Allison, D., Kallman, B.J., Cope, O.B. and Van Valin, C.C. (1963) Insecticides: effects on cutthroat trout of repeated exposure to DDT. *Science*, 142(3594), 958–961.

Andrianantoandro, E., Basu, S., Karig, D.K. and Weiss, R. (2006) Synthetic biology: new engineering rules for an emerging discipline. *Molecular Systems Biology*, 2, 1–14.

Antweiler, C. (1998) Local knowledge and local knowing – an anthropological analysis of contested 'cultural products' in the context of development. *Anthropos*, 93(4–6), 469–494.

Baker, M.J. (2008) A structural model of the transition to agriculture. *Journal of Economic Growth*, 13(4), 257–292.

Benner, S.A. (2010) Q&A: Life, synthetic biology and risk. *BMC Biology*, 8(77), 4.

Benner, S.A. and Sismour, A.M. (2005) Synthetic biology. *Nature Reviews Genetics*, 6(7), 533–543.

Brand, F.S. and Jax, K. (2007). Focusing the meaning(s) of resilience: resilience as a descriptive concept and a boundary object. *Ecology and Society*, 12(1), 23.

Briers, M.I. and Chua, W.F. (2001) The role of actor-networks and boundary objects in management accounting change: A field study of an implementation of activity-based costing. *Accounting, Organizations and Society*, 26(3), 237–269.

Brown, J.S. and Duguid, P. (2001) Knowledge and organization: a social-practice perspective. *Organization Science*, 12(2), 198–213.

Bud, R. (1991) Biotechnology in the twentieth century. *Social Studies Of Science*, 21(3), 415–457.

Bull, A.T., Holt, G. and Lilly, M.D. (1982) *Biotechnology: International Trends and Perspectives*. Paris: Organisation for Economic Co–operation and Development (OECD).

Cacciatori, E. (2003) Organisational memory in innovative project-based firms: a boundary object perspective. Paper presented at DRUID Summer Conference, 'Creating, sharing and transferring knowledge: The role of geographical configuration, institutional settings and organizational contexts'. Copenhagen, Denmark, 12–14 June.

Cambray, G., Mutalik, V.K. and Arkin, A.P. (2011) Toward rational design of bacterial genomes. *Current Opinion in Microbiology*, 14(5), 1–7.

Carlile, P.R. (2002) A pragmatic view of knowledge and boundaries: boundary objects in new product development. *Organization Science*, 13(4), 442–455.

Carlson, R.H. (2010) *Biology Is Technology: The Promise, Peril, and New Business of Engineering Life*. Cambridge, MA: Harvard University Press.

Carr, P.A. and Church, G.M. (2009). Genome engineering. *Nature Biotechnology*, 27(12), 1151–1162.

Clark, J.H., Luque, R. and Matharu, A.S. (2012) Green chemistry, biofuels, and biorefinery. *Annual Review of Chemical and Biomolecular Engineering*, 3, 183–207.

D'Adderio, L. (2004) *Inside the Virtual Product: How Organizations Create Knowledge through Software*. Cheltenham, UK and Northampton, MA, USA: Edward Elgar Publishing.

De Lorenzo, V. (2010) Environmental biosafety in the age of synthetic biology: do we really need a radical new approach? *BioEssays*, 32(11), 926–931.

Ellis, D.I. and Goodacre, R. (2011) Metabolomics-assisted synthetic biology. *Current Opinion in Biotechnology*, 23, 1–7.

Elowitz, M. and Lim, W.A. (2010) Build life to understand it. *Nature*, 468(7326), 889–890.

Endy, D. (2005) Foundations for engineering biology. *Nature*, 438(7067), 449–453.

European Academies Science Advisory Council (2010) Realising European potential in synthetic biology: scientific opportunities and good governance. European Academies Science Advisory Council.

Fung, E., Wong, W.W., Suen, J.K., Bulter, T., Lee, S.G. and Liao, J.C. (2005) A synthetic gene-metabolic oscillator. *Nature*, 435(7038), 118–122.

Galison, P. (1997). *Image and Logic: A Material Culture of Microphysics*. Chicago, IL: University of Chicago Press.

Giere, R. (2002) Scientific cognition as distributed cognition. In: Carruthers, P., Stitch, S. and Siegal, M. (eds), *The Cognitive Basis of Science*. Cambridge: Cambridge University Press, pp. 285–299.

Gould, S.J. (2000) *The Lying Stones of Marrakech: Penultimate Reflections in Natural History*. New York: Harmony Books.

Grabher, G. (2002) Cool projects, boring institutions: temporary collaboration in social context. *Regional Studies*, 36(3), 205–214.

Gulati, S., Rouilly, V., Niu, X., Chappell, J., Kitney, R.I., Edel, J.B., Freemont, P.S. and DeMello, A.J. (2009) Opportunities for microfluidic technologies in synthetic biology. *Journal of the Royal Society Interface*, 6(4), 493–506.

Hartz, J. and Chappell, R. (1997) *Worlds Apart: How the Distance between Science and Journalism Threatens America's Future*. Nashville, TN: First Amendment Center.

Haseltine, E.L. and Arnold, F.H. (2007) Synthetic gene circuits: design with directed evolution. *Annual Review of Biophysics and Biomolecular Structure*, 36, 1–19.

Holling, C.S. (1973) Resilience and stability of ecological systems. *Annual Review of Ecology and Systematics*, 4, 1–23.

Holmes, D. and Gawad, S. (2010) Microengineering in biotechnology. *Methods in Molecular Biology*, 583, 55–80.
Hutchins, E. (1995) *Cognition in the Wild.* Cambridge, MA: MIT Press.
Jarboe, L.R., Zhang, X., Wang, X., Moore, J.C., Shanmugam, K.T. and Ingram, L.O. (2010) Metabolic engineering for production of biorenewable fuels and chemicals: contributions of synthetic biology. *Journal of Biomedicine Biotechnology* (online, 18pp).
Kimble, C., Grenier, C. and Goglio-Primard, K. (2010) Innovation and knowledge sharing across professional boundaries: political interplay between boundary objects and brokers. *International Journal of Information Management*, 30(5), 437–444.
Kitney, R. et al. (2009) Synthetic biology: scope, applications and implications. Royal Academy of Engineering.
Knorr-Cetina, K. (1999) *Epistemic Cultures: How the Sciences Make Knowledge.* Cambridge, MA: Harvard University Press.
Kwok, R. (2010) Five hard truths for synthetic biology. *Nature*, 463(7279), 288–290.
Lartigue, C. et al. (2009) Creating bacterial strains from genomes that have been cloned and engineered in yeast. Science, *AAAS*, 325(5948), 1693–1696.
Leduc, S. (1912) *La biologie synthétique.* Paris: A. Poinat.
Levidow, L. and Carr, S. (1997) How biotechnology regulation sets a risk/ethics boundary. *Agriculture and Human Values*, 14(1), 29–43.
Levskaya, A., Chevalier, A.A., Tabor, J.J., Simpson, Z.B., Lavery, L.A., Levy, M., Davidson, E.A., Scouras, A., Ellington, A.D., Marcotte, E.M. and Voigt, C.A. (2005) Engineering *Escherichia coli* to see light. *Nature*, 438(7067), 441–442.
Loeb, J. (1912) The mechanistic conception of life. *Biological Essays*, Chicago, IL: University of Chicago Press.
Lorenz, P. and Eck, J. (2005) Metagenomics and industrial applications. *Nature Reviews Microbiology*, 3(6), 510–516.
Luisi, P.L., Chiarabelli, C. and Stano, P. (2006) From never born proteins to minimal living cells: two projects in synthetic biology. *Origins of Life and Evolution of the Biosphere*, 36(5–6), 605–616.
Magner, L.N. (2002) *A History of the Life Sciences*, rev. and expanded edn. New York: Marcel Dekker.
Markus, M.L. and Gelinas, U.J. (2006) Comparing the standards lens with other perspectives on IS innovations: the case of CPFR. *International Journal of IT Standards and Standardization Research*, 4(1), 24–42.
Marlière, P. (2009) The farther, the safer: a manifesto for securely navigating synthetic species away from the old living world. *Systems and Synthetic Biology*, 3(1–4), 77–84.
Martin, V.J.J., Pitera, D.J., Withers, S.T., Newman, J.D. and Keasling, J.D. (2003) Engineering a mevalonate pathway in Escherichia coli for production of terpenoids. *Nature Biotechnology*, 21(7), 796–802.
McGovern, P.E. (2009) *Uncorking the Past.* Oakland, CA: University of California Press.
Mitarai, N., Sneppen, K. and Pedersen, S. (2008) Ribosome collisions and translation efficiency: optimization by codon usage and mRNA destabilization. *Journal of Molecular Biology*, 382(1), 236–245.
Montgomery, S. (2004) Of towers, walls, and fields: perspectives on language in science. *Science*, 303(5662), 1333–1335.

Moore, P.B. (2009) The ribosome returned. *Journal of Biology*, 8(1), 8.

Nelkin, D. (1995) *Selling Science: How the Press Covers Science and Technology*. New York: W.H. Freeman & Company.

Noireaux, V. and Libchaber, A. (2004) A vesicle bioreactor as a step toward an artificial cell assembly. *Proceedings of the National Academy of Sciences of the United States of America*, 101(51), 17669–17674.

O'Loughlin, T.L., Patrick, W.M. and Matsumura, I. (2006) Natural history as a predictor of protein evolvability. *Protein Engineering Design Selection – PEDS*, 19(10), 439–442.

OECD, Royal Society and US National Academies (2009) Opportunities and challenges in the emerging field of synthetic biology. Washington, DC: United States National Academies the Organisation for Economic Co-operation and Development, and the Royal Society.

Parekh, S.R. (2004) *The GMO Handbook: Genetically Modified Animals, Microbes, and Plants in Biotechnology*. Totowa, NJ: Humana Press.

Pleiss, J. (2006) The promise of synthetic biology. *Applied Microbiology and Biotechnology*, 73(4), 735–739.

van der Ploeg, R.R., Bohm, W. and Kirkham, M.B. (1999) On the origin of the theory of mineral nutrition of plants and the Law of the Minimum. *Soil Science Society of America Journal*, 63(5), 1055–1062.

Qin, S., Müller, W.E.G. and Cooper, E.L. (2011) Marine biotechnology. Evidence based complementary and alternative medicine. *eCAM*, September, 2.

Ramakrishnan, V. (2002) Ribosome structure and the mechanism of translation. *Cell*, 108(4), 557–572.

Rasmussen, S. et al. (2003) Bridging nonliving and living matter. *Artificial Life*, 9(3), 269–316.

Rawls, R.L. (2000) 'Synthetic biology' makes its debut. *Chemical and Engineering News*, 78(17), 49–53.

Ro, D.K. et al. (2006) Production of the antimalarial drug precursor artemisinic acid in engineered yeast. *Nature*, 440(7086), 940–943.

Roberts, S.M., Turner, N.J., Willetts, A.J. and Turner, M.K. (1995) *Introduction to Biocatalysis using Enzymes and Microorganisms*. Cambridge: Cambridge University Press.

Rothwell, R. and Robertson, A.B. (1973) The role of communications in technological innovation. *Research Policy*, 2(3), 204–225.

Sapsed, J. and Salter, A. (2004) Postcards from the edge: local communities, global programs and boundary objects. *Organization Studies*, 25(9), 1515–1534.

Sauro, H.M. (2008) Modularity defined. *Molecular Systems Biology*, 4(166), 2.

Scheper, T.H. and Reardon, K.F. (2008) Sensors in biotechnology. In Göpel, W., Hesse, J. and Zemel, J.N. (eds), *Sensors Set: A Comprehensive Survey*. Weinheim: Wiley-VCH Verlag, pp. 1023–1046.

Schmidt, M. (2010) Xenobiology: a new form of life as the ultimate biosafety tool. *BioEssays*, 32(4), 322–331.

Schwille, P. (2011) Bottom-up synthetic biology: engineering in a tinkerer's world. *Science*, 333(6047), 1252–1254.

Serrano, L. (2007) Synthetic biology: promises and challenges. *Molecular Systems Biology*, 3, 5.

Shiva, V. (1991) The Green Revolution in the Punjab. *The Ecologist*, 21(2), 57–60.

Shortland, M. and Gregory, J. (1991) *Communicating Science: A Handbook*. New York: Longman.
Sims III, R.J., Mandal, S.S. and Reinberg, D. (2004) Recent highlights of RNA-polymerase-II-mediated transcription. *Current Opinion in Cell Biology*, 16(3), 263–271.
Smith, V.H., Tilman, G.D. and Nekola, J.C. (1999) Eutrophication: impacts of excess nutrient inputs on freshwater, marine, and terrestrial ecosystems. *Environmental Pollution*, 100(1–3), 179–196.
Springham, D.G., Moses, V. and Cape, R.E. (1999) *Biotechnology: The Science and the Business*. Amsterdam: Harwood Academic Publishers.
Sprinzak, D. and Elowitz, M.B. (2005) Reconstruction of genetic circuits. *Nature*, 438(7067), 443–448.
Star, S.L. (2010) This is not a boundary object: reflections on the origin of a concept. *Science, Technology and Human Values*, 35(5), 601–617.
Star, S.L. and Griesemer, J. (1989) Institutional ecology, 'translations', and boundary objects: amateurs and professionals on Berkeley's museum of vertebrate zoology. *Social Studies of Science* 19(3), 387–420.
Szostak, J.W., Bartel, D.P. and Luisi, P.L. (2001) Synthesizing life. *Nature*, 409(6818), 387–390.
Szybalski, W. (1974) *In vivo* and *in vitro* initiation of transcription. Panel discussion, in Kohn, A. and Shatkay, A. (eds), *Control of Gene Expression*. New York: Plenum Press, pp. 404–405, 411–412, 415, 417.
Tassey, G. (2000) Standardization in technology-based markets. *Research Policy*, 29(4–5), 587–602.
Thieman, W.J. and Palladino, M.A. (2003) *Introduction to Biotechnology*. San Francisco, CA: Benjamin Cummings (Pearson).
Tornier, G. (1901) Ueberzahlige Bildungen und die Bedeutung der Pathologie fur die Biontechnik (mit Demonstrationen). *Verhandlungen des V. Internationalen Zoologen-Congressen*, Berlin.
Umenhoffer, K. et al. (2010) Reduced evolvability of *Escherichia coli* MDS42, an IS-less cellular chassis for molecular and synthetic biology applications. *Microbial Cell Factories*, 9(38), 12.
US Congress, Office of Technology Assessment (1988) New developments in biotechnology: US investment in biotechnology. Special Report, OTA-BA-360. Washington, DC: US Government Printing Office.
Utterback, J.M. (1971) The process of technological innovation within the firm. *Academy of Management Journal*, 14(1), 75–88.
Van Zuylen, J. (1981) The microscopes of Antoni van Leeuwenhoek. *Journal of Microscopy*, 121(3), 309–328.
Vancompernolle, K. and Ball, P. (2005) Synthetic biology applying engineering to biology. Area, Report of a NEST High-Level Expert Group.
Veening, J.W., Smits, W.K. and Kuipers, O.P. (2008) Bistability, epigenetics, and bet-hedging in bacteria. *Annual Review of Microbiology*, 62, 193–210.
Vinuselvi, P., Park, S., Kim, M., Park, J.M., Kim, T. and Lee, S.K. (2011) Microfluidic technologies for synthetic biology. *International Journal of Molecular Sciences*, 12(6), 3576–3593.
Wegberg, M. (2004) Standardization process of systems technologies: creating a balance between competition and cooperation. *Technology Analysis and Strategic Management*, 16(4), 457–478.

Weiss, R. and Panke, S. (2009) Synthetic biology-paths to moving forward. *Current Opinion in Biotechnology*, 20(4), 447–448.
Wilford, J.N. (1992) Jar in Iranian ruins betrays beer drinkers of 3500 BC. *New York Times*.
Wrotnowska, D. (1981) Pasteur. First investigations on fermentation (1855–1857). *Clio medica*, 15(3–4), 191–199.
Yakura, E.K. (2002) Charting time: timelines as temporal boundary objects. *Academy of Management Journal*, 45(5), 956–970.
Yeager, J.F. and Munson, S.C. (1945) Physiological evidence of a site of action of DDT in an insect. *Science*, 102(2647), 305–307.
Yi, D. (2008) Cancer, viruses, and mass migration: Paul Berg's venture into eukaryotic biology and the advent of recombinant DNA research and technology, 1967–1980. *Journal of the History of Biology*, 41(4), 589–636.
Zheng, Y.T. and Sriram, G. (2010) Mathematical modeling: bridging the gap between concept and realization in synthetic biology. *Journal of Biomedicine and Biotechnology*, 16pp., http://www.hindawi.com/journals/bmri/2010/541609/.

11. Methodological challenges of capturing innovation in developing countries

Marija Drenkovska

INTRODUCTION

Innovation tends to be considered as a major driver of both economic growth and competitiveness of companies and industries. However, measurement and analysis of innovative activities and their impacts at micro, meso and macro levels have often been burdened with conceptual and applicative difficulties (Aralica et al., 2008).

Another major challenge in studying the relation between growth and innovation (apart from the methodological one), especially in developing countries, is the collection and the quality of data. In obtaining data for empirical works, studies usually either exploit official statistical sources (Van Ark et al., 2009; Fukao et al., 2007; Majcen et al., 2011) or obtain them through surveys (Miyagawa et al., 2010; Prašnikar, 2010). Both approaches have their own advantages and disadvantages. The main benefits of the use of official data are in the number of data points and easy (or easier) cross-section and time series analyses. Such studies describe the phenomenon by capturing the expenditure on research and development (R&D), employment in R&D, number of patents of the company, and so on. Nevertheless, there is an acute problem regarding official data in developing countries, where the data on intangibles are often either completely lacking, or only partial datasets exist, often of poor quality and/or unsuitable in terms of variables. In these cases the survey methodology, if properly designed, could be the only way to conduct serious research. Standardized (internationally comparable) questionnaires have become the most commonly exploited source of data in the recent literature on innovation. And while innovation surveys have substantially improved the existing knowledge on innovation, they still face some challenges in their diffusion in developing countries (e.g. Mairesse and Mohnen, 2010; OECD, 2005; Mytelka et al., 2004). This is due to different reasons that range from collection and reliability issues of the data to selecting the most appropriate approach and measurements of the types of innovation and R&D in these countries.

My interest in developing countries relates to the proposition that a different type of R&D is needed in technology followers (Forbes and Wield, 2000). Namely, existing data show that the size and characteristics of R&D and innovation are slightly different in technology followers in comparison to leaders. Given these observations, as well as the different 'quality of exposure' that technological follower companies face, together with a research team at the Faculty of Economics, University of Ljubljana I have designed a questionnaire which is largely based on the Community Innovation Survey (CIS) questionnaire and attempts to comprehensively capture the innovation and the R&D activity of the developing countries of the Western Balkans.

This chapter is structured as follows. It first offers a short discussion on the existing methodological and data challenges pertaining to innovation studies. Then I briefly present and discuss the key adaptation in the questionnaire, and conclude with an agenda for future work and implications for innovation policy makers in developing countries.

METHODOLOGICAL AND DATA CHALLENGES

There is unanimous consensus among developmental economists regarding the relationship between technological change and development. The agreement among the different approaches that have investigated this relationship is that the origin of the development process is rooted in the growth of productivity, and not in a simple factor accumulation process (Bogliacino et al., 2009).

Many studies in the conceptualization of technology, however, have used either R&D expenditures or patents as indicators. These are two common traditional sources of data used in empirical works. The data on patents for different countries has been available for much longer, coinciding with the development of intellectual property rights and the institution of national patent offices in the nineteenth century. There are however certain concerns regarding this kind of data. The main issue is that patents do not cover all innovations within a company; that is, they only cover innovations that are relatively new, deemed worthy to be patented by the company applicant, and they yet may never be marketed. Additionally, the characteristics of the patent systems vary among countries, which makes this kind of data inadequate for comparative studies. In the end, the patent data may be non-existent in the first place for some emerging economies (since they do not have national patent offices, as found in Uvalic, 2006).

Data on R&D expenditures, on the other hand, have been collected on a regular basis at an international level since the 1950s (Mairesse and Mohnen, 2010). And while a number of empirical studies have succeeded in establishing a positive relationship between growth of productivity and growth of R&D investment (Goto and Suzuki, 1989, for supplying industries; Wakelin, 1998, for a sample of UK firms), a large number of recent studies warn that the indicators based on R&D bear several limitations as a measure of technological change. First, they underestimate technological activities in manufacturing and service industries where much of the technical change takes place around design and manufacturing, which is not captured by the concept of R&D. Second, small and medium-sized firms often do not possess a separate business unit devoted to R&D. Using R&D expenditure or R&D personnel as a measure will therefore underestimate their innovation activities. And finally, as an input measure, R&D expenditures are only loosely connected with the output of technological change (Aralica et al., 2008).

Apart from the methodological debate, another persisting issue in studying innovation and technological change in developing countries remains the lack of data. In many cases the R&D data are not collected in the first place, and in others only partial datasets exist, often of poor quality and/or unsuitable in terms of variables.

In light of the aforementioned measurement and data challenges that characterize the two sources of data, a third source that has become especially popular in modern literature is the innovation survey. Innovation surveys have substantially improved the existing knowledge on innovation and have provided more precise and comprehensive measures of both innovative inputs and outputs. They have enabled investigations of phenomena that were previously impossible to study and have allowed previously unsubstantiated ideas to be confirmed (Guellec and Pattinson, 2002). Europe has pioneered the surveys on innovation in firms, developing a common methodology, described in the *Oslo Manual* (OECD/Eurostat, 2005). This practice has progressively extended to other countries and represents the current international standard for the methodology of innovation surveys (Bogliacino et al., 2010). Although there have been attempts to include the non-OECD countries in the surveying process, the analysis done on the state of innovation in these countries does not go beyond a superficial examination of innovation indicators and trend analyses (e.g., see European Commission, 2011). The findings of such analyses often indicate inadequate innovation performance of the countries as compared to the European Union (EU) average; however so far they have not been supported by comprehensive firm-level innovation surveys.

A key reason for this state of affairs can be found in issues pertaining to the obtained survey outputs. As many studies have already brought to our attention (e.g., Mairesse and Mohnen, 2010; OECD/Eurostat, 2005; Mytelka et al., 2004), the survey questionnaires, given their attempt to be simplistic, standardized and comparative, often risk only partially capturing the phenomenon, especially in technology-follower countries, and often neglect the relevant aspects of innovation that might be important for (just) developing countries. Another challenge to be kept in mind when preparing a survey, that experience in studying intangibles in developing countries also confirms, is the wording and definition of concepts, which need to be carefully formulated to match the knowledge and experience of the respondent. Thus, in order to address the reliability issue, the language, the structure, the scope and so on of the questionnaires may need to be adapted before launching a survey in a developing country. One of the ways in which I hope to achieve this in the questionnaire is by systematic description of the concepts through statements that leave little room for misunderstanding. This, and other key aspects of my approach, I discuss in the following section.

Proposed Methodology

The proposed questionnaire is largely based on the CIS (Community Innovation Survey) questionnaire and follows the recommendations of the *Oslo Manual*. It has been adjusted and carefully constructed in a way that aspires to address the above-mentioned issues of non-comparable, non-reliable and often non-existing data (e.g., see Edvardsson et al., 2000).

In continuation I briefly present the questionnaire and discuss a few important modifications (introduction of new questions and putting more weight on others) that address innovation aspects that concern developing countries more specifically. The approach I propose allows me to analyse the differences between better- and less-well-performing firms in relation to their capabilities (and subsequently their competences), and to better understand the importance of the relative proximity of the markets where companies operate, the utilization of the potential sources of knowledge, as well as the company's strategic orientation towards R&D. These, I believe, are crucial elements for understanding the innovation gap between technological leaders and technological followers.

KEY ASPECTS OF THE PROPOSED QUESTIONNAIRE

The questionnaire is characterized by ten main sections: (1) target markets and characteristics of the industry (q1[1]); (2) types of new products (q2 to q6, q16); (3) types of new processes (q7 to q8); (4) technology development as a source of innovative products (q9); (5) innovation linkages and sources (q10 to q12); (6) organization of R&D activity and R&D expenditure (q13 to q15); (7) examination of competences (q17 to q19); (8) examination of capabilities (q20 to q22); (9) hard data (q23); and (10) external financial aid for innovation (q24).

In this chapter I discuss three main aspects that the questionnaire emphasizes. These are: the importance of competencies (and capabilities) in a technological follower; the different 'quality of exposure' (exposure to markets with different level of development); and the specificity of both external and internal (own R&D) knowledge sources.

Competencies

Competencies of a company refer to the ability to utilize resources that spread across multiple functions, products and markets in a sustainable and synchronized manner (Rajkovič and Prašnikar, 2009). Capabilities are repeatable patterns of actions in the use of assets to create, produce and/or offer products to a market (Grant, 1991). Only those key capabilities that are relatively unique and common to various business functions, products and business units are likely to form competencies of a company (Sanchez and Heene, 2004).

Different authors examine the importance and measurement of different types of competencies. According to some (Belderbos et al., 2004; Danneels, 2002; Griffin and Hauser, 1992), in order to innovate successfully, companies have to possess key competencies relating to technology and customers. Technological competencies refer to a firm's ability to generate, as well as assimilate, transform and exploit, the acquired knowledge (Zahra and George, 2002). Marketing competencies (sometimes referred to as customer competencies) stem from the marketing concept that puts customer needs first. Activities that refer to marketing competencies include, among others, obtaining real-time information about changes in customer needs; communicating with customers about their potential and current demands; acquiring real-time information regarding competitors' evolution of strength and weaknesses; building and enhancing marketing channels; establishing and managing long-term relations with suppliers; and creating and managing long-term customer relationships.

Another network of competencies whose importance is emphasized by Rajkovic and Prašnikar (2009) are the complementary competencies. Although some authors define them as an interaction between technological and marketing capabilities and competencies (Song et al., 2005), various studies have, however, identified them as an independent group and have warned that a failure to value them properly can lead to a deficient identification of key capabilities. There is a substantial amount of literature arguing that a combination of technological and marketing capabilities and competencies can create a competitive advantage (Chang, 1996; Dutta et al., 1999; Song et al., 2005). The role of complementary competencies, according to Wang et al. (2004), is to: (1) integrate different technological specialties; (2) combine different functional specialties; (3) exploit synergies across business units; (4) combine in-house resources with external capabilities required; and (5) integrate the dynamic competence-building process for superior performance. I argue that these three types of capabilities and competences (technological, marketing and complementary) are the foundation for managing innovative activities in technological follower company and that all other capabilities are derived from them. They play a central role in the innovation value chain in developing countries.

An example of operationalization of competencies is given in Table 11.1. The question set in the table measures respondents' perception of their competencies in relation to their main competitors' competencies. The set measures them on a five-point Likert scale ranging from 'considerably worse than the main competitors' to 'considerably better than the main competitors'. These question sets in the questionnaire familiarize the respondent with the concepts of the technological, marketing and complementary competencies. Only then is the respondent asked to rate

Table 11.1 Example of a set of questions on competencies

Technological competences	1	2	3	4	5
Research and development in the firm is advanced					
Number of available technological capabilities inside the firm or through strategic partnership is quite large					
We are good at predicting technological trends					
Research and development in the firm is advanced					

Note: Respondents are asked to evaluate the performance of the company compared to their main competitors in particular aspects on a scale from 1 to 5 (1 = considerably worse than the main competitors; 5 = considerably better than the main competitors)

Source: Intangibles questionnaire for Slovenia, Prašnikar et al. (2012b).

Table 11.2 Example of a set of questions on capabilities

Technological capabilities	YES	NO
Our technological capabilities exceed those of average companies in the industry		
Technological capabilities have been developed to such an extent that we can claim to be more technologically competent than our competitors		
Dynamically, outdated technological capabilities are being continuously replaced by new		

Note: An example of a yes/no cascading set explained below.

Source: Intangibles questionnaire for Slovenia, Prašnikar et al. (2012b).

the company's capabilities in respect to the competitive environment. The capabilities (technological, marketing and complementary), as opposed to the competencies, are measured in sets of cascading questions (which I will discuss in greater detail later on), as shown in Table 11.2.

Quality of Exposure

The open innovation paradigm recommends that the company, instead of 'doing it all itself', it should be able to utilize the knowledge coming from its environment in its innovation process. Part of this knowledge will reach the firm from external sources (Cassiman and Veugelers, 2002) and part of it is pre-existing within the company. A successful firm possesses the abilities to identify, transform, disseminate and utilize this knowledge in the creation of new products and processes. The utilization of these sources is especially important for a company that is still distant from the technological frontier, but aspires to build strong foundations for long-term competitiveness on the market. Coming from the idea that different 'quality of exposure' has a different impact on the company's general behaviour and development of capabilities (also capabilities to innovate), two questions that are grounded in the CIS questionnaire are expanded to capture the different exposures. More precisely, the question on the geographical market orientation and the question regarding innovation partners have been expanded to include the markets of the Western Balkans (as a control for exposure to less developed markets).

Internal R&D

The discussion on sources of knowledge leads us to the question on the internal R&D of a technological follower. In their argument about different types of R&D and innovation activities in technological followers, as opposed to those in technological leaders, Forbes and Wield (2000) discuss the specificity of the learning mechanisms of the R&D department of follower companies. These departments can become the locations for organized learning, the problem-solvers of last resort in production, the in-house knowledge stores and gatekeepers, and the focus for independent design and product development capacity. They have the task to continuously monitor the company's environment and quickly identify and exploit external knowledge (see e.g. Fritsch and Lukas, 2001; Katz et al., 1995). To examine the role of the R&D department in my questionnaire I use the carefully constructed yes/no set of questions that ask whether the R&D department, if existing, systematically supports problem solving on the shop floor; whether it builds the absorption capacity of the company (and has the role of its gatekeeper[2]), stores the information and spreads it through the company; or whether it has developed to a level of an agent of change that sets guidelines for further technological development.

The R&D infrastructure of a company is key to acquiring and developing the knowledge and competences needed to approach the technology frontier. In the questionnaire I incorporate two question sets, regarding both innovation of products and innovation of processes, that capture the degree to which a company relies on its internal sources of knowledge (own research) and how much it is involved in innovative cooperation with other entities (that is, whether the novelty relies mostly on imitation, independent development or innovation cooperation). From a strategic perspective, I feel it is also important to examine the company's attitude towards R&D: does the company believe that R&D is an important source of competitive advantage or is R&D perceived just as an unnecessary cost? The latter offers an insight into the innovativeness logic in developing countries, where often companies are rather 'pushed' to innovate in order to be able to face the competition on their main markets. In many cases in these countries, the organization of R&D departments seems less straightforward, depending mainly on the size of the company, the target markets and the country's level of development (for organizational design of R&D, see Ambec and Poitevin, 2001).

These are only the central adaptations introduced in the questionnaire that I have selected for discussion (for a more detailed presentation, see Prašnikar et al., 2012b). In constructing the questionnaire I followed the existing recommendations in the literature concerning designing a survey for developing countries, and relied heavily on our own experience.

EXPECTED CONTRIBUTIONS

In this section I examine the contributions that the questionnaire can offer with the usage of the simple and straightforward yes/no questions, the cascading approach and the perception questions. Regarding the structure and type of the questions used in my questionnaire, a key place is held by the so-called cascading sets of questions that address the core aspects of the innovation process, from innovation inputs (R&D expenditures, innovation culture in the company, degree to which the company relies on own R&D, knowledge base of the company) through capabilities (technological, marketing and complementary) to innovation outcome (number of new products or services, company performance on the basis of newly introduced products) and, finally, revenues (from new products). The cascading approach is a novelty in the innovation survey methodology, introduced by Miyagawa et al. (2010). A typical cascading question set is composed of three sub-questions (Table 11.2), all three being of the type, 'answer with yes or no'. The yes/no approach mainly attempts to simplify the answering process, which also contributes to greater trust in the data obtained due to two major reasons. First, the statements are simple, clear and straightforward, short and easy to understand. Thus, the noise stemming from the respondent's lack of understanding or lack of interest (lack of time) is minimized. One alternative to this approach is the simple 'choose your characteristics from a list' type of questions, which offer a vast choice of answers. These questions are exposed to the 'choose-fast risk' that can impede the reliability of the answer. The other rationale for using the yes/no questions lies in the construction of the cascading sets. Namely, each consecutive question from the cascading set represents a dimension that is superior to the one in the previous sub-question. The respondent's choice is thus limited, as the margin between two sub-questions is big enough so that no doubt can arise about which 'category' the company fits into. This approach can contribute to greater reliability of the survey data.

Using the cascading technique, one can identify whether a certain problem or aspect of development is present in the company and at what development level. This is possible since the cascade set of questions allows us to construct a variable on a scale of 1–4, with 1 being the lowest development level and 4 the highest. The transformation to a 'factor' is already initiated by assigning a value 1 for the answer 'no' in the first sub-question, and a value 2 for the 'yes' answer. If the answer in the next sub-question is 'no', then the value remains the same, for example it has a value of 2 if the first sub-question is answered 'yes'. If the answer is 'yes' also in the second sub-question, the value increases to 3. Similarly if the answer is 'yes' also to the third sub-question, the final value of the variable for a certain question, composed of three sub-questions, has an end value of 4. As an end result we obtain a numerical variable which can be used in further empirical research. This kind of measure opens the possibility of different empirical analyses compared to the use of simple dummy variables.

The other type of questions used in the questionnaire is the so-called perceptional Likert scale type of questions. Their importance is threefold. First, they enable the analysis of the relationship between the actions taken by companies for innovation and their market performance (competitiveness). Second, this approach allows me to develop a more comprehensive proxy of the company's capabilities and competencies that are of key importance for the development of its innovation ability. Given that specific perceptional questions complement some of the cascading sets, they allow for testing of the consistency of the provided answers by measuring the concepts differently. And last but not least, they aspire to circumvent the subjectivity and pretension of the respondents.

Another attempt to circumvent subjectivity of responses[3] is the deliberate order the questions follow in the questionnaire. For example, the perception questions follow the questions on the number and level of originality of newly introduced products and processes.

IMPLICATIONS AND FUTURE WORK

Once the output of the survey is obtained, the first step will be to examine the link between the environment, corporate performance and innovation; a link that is of key importance for developing countries. Subsequently, the innovation activity will be examined in relationship to other elements of intangible capital (human resources, organizational and relational capital), as defined by Prašnikar (2010). That will allow me to shift the focus into the firm, and by following the literature from Porter (1985) to Pisano and Shih (2009) to examine whether firm cohesion, focus on strategy and internal cooperation contribute to investments in human capital, firm dynamic abilities (capabilities and competences) and, finally, innovation. Another direction in which the study on innovation and the intangible capital will be extended is by including other countries from the region (predominately the emerging markets in the Western Balkans). The questionnaire(s) will be adapted to the development and structural characteristics of these countries, as was done also for Slovenia. For now, the research has been extended to the Republic of Srpska (Bosnia and Herzegovina) and Albania.

The developing countries in the majority of cases (Palley, 2011) benefit from export-led growth. With access to global information, global ideas, new knowledge, better processes, and so on, the learning opportunity is immense (the learning-by-exporting hypothesis). However, some of these countries are still more 'inward'-oriented (see Prašnikar and Knežević Cvelbar, 2012; Prašnikar et al., 2012a). In this case the innovation processes are enabled chiefly through developed complementary capabilities and competences of the firms. They facilitate the learning process and hence the progress from a pure imitative strategy to the domestic generation of technological knowledge. Strong technological capabilities mean adequate systematic interactions necessary to generate, exchange and recombine knowledge. Those integrated with developed marketing capabilities, which work towards reducing the risks associated with the market introduction of innovations, become antecedents of innovative performance for an inwardly oriented technological follower.

However, given that the developing countries (at least those of the Western Balkans) will eventually follow the Slovenian path of economic development, the power of learning by exporting should be acknowledged, not just in companies but also at the policy level. I expect that the findings from my study will be able to offer a broader understanding of the types and aspect of innovativeness in developing countries and a better comprehension of the influence of the global markets on companies' and countries' ability to innovate.

ACKNOWLEDGEMENTS

The work presented in this chapter is part of a broader study in which a number of academic workers participated. Drenkovska's work has been supported by the ManETEI: Management of Emerging Technologies for Economic Impact project, grant agreement number PITN-GA-2009-238382, funded by the Seventh Framework Programme, theme Marie-Curie Actions, identifier FP7-PEOPLE-2008-ITN.

NOTES

1. The designations q1 to q24 represent sets of questions on the same topic. The number of questions within a set varies, being at most 11, apart from the cascading sets where the

number of questions is always three. The sets are discussed in greater detail in Prašnikar et al. (2012b).
2. Gatekeepers can translate technical knowledge into terms meaningful to managers, and their presence in the organization has been found to correlate positively with the performance of development projects (Katz et al., 1995).
3. Prudent managers are more likely to maintain higher levels of perception about the degree of competition and to undertake innovation efforts against potential competitors.

REFERENCES

Ambec, S. and Poitevin, M. (2001). Organizational design of R&D activities. Centre for Studies in Economics and Finance, Working Paper No. 60.

Aralica, Z., Domagoj, R. and Dubravko, R. (2008). Innovation propensity in Croatian enterprises: results of the community innovation survey. *South East European Journal of Economics and Business*, 3(1), 77–88.

Belderbos, R., Carree, M. and Lokshin, B. (2004). Cooperative R&D and firm performance. *Research Policy*, 33(10), 1477–1492.

Bogliacino, F., Perani, G., Pianta, M. and Supino, S. (2009). Innovation in developing countries: the evidence from innovation surveys. Paper prepared for the FIRB conference Research and Entrepreneurship in the Knowledge-based Economy, Universita L. Bocconi, Milano.

Bogliacino, F., Perani, G., Pianta, M. and Supino, S. (2010). Innovation and development: the evidence from innovation surveys. BETA, Document de Travail no 2010 – 13, Université de Strasbourg.

Cassiman, B. and Veugelers, R. (2002). R&D cooperation and spillovers: some empirical evidence from Belgium. *American Economic Review*, 92, 1169–1184.

Chang, T. (1996). Cultivating global experience curve advantage on technology and marketing capabilities. *International Marketing Review*, 13, 22–42.

Danneels, E. (2002). The dynamics of product innovation and firm competences. *Strategic Management Journal*, 23, 1095–1121.

Dutta, S., Narasimhan, O. and Rajiv, S. (1999). 'Success in high-technology markets: Is marketing capability critical?' *Marketing Science*, 18, 547–568.

Edvardsson, B., Gustavsson, A., Johnson, M.D. and Sandén, B. (2000). *New Service Development and Innovation in the New Economy*. Lund: Studentlitteratur.

European Commission (2011). Innovation Union Scoreboard 2011: Research and Innovation Union Scoreboard. http://ec.europa.eu/enterprise/policies/innovation/files/ius-2011_en.pdf.

Forbes, N. and Wield, D. (2000). Managing R&D in technology-followers. *Research Policy*, 29, 1095–1109.

Fritsch, M. and Lukas, R. (2001). Who cooperates on R&D? *Research Policy*, 30, 297–312.

Fukao, K., Hamagata, S., Miyagawa, T. and Tonogi, K. (2007). Intangible investment in Japan: measurement and contribution to economic growth. Rieti Discussion Paper, 07-E-034.

Goto, A. and Suzuki, K. (1989). R&D capital, rate of return on R&D investment and spillover of R&D in Japanese manufacturing industries. *Review of Economic and Statistics*, November, 555–564.

Grant, R.M. (1991). The resource-based theory of competitive advantage: implications for strategy formulation. *California Management Review*, 33, 114–135.

Griffin, A. and Hauser, J.R. (1992). Patterns of communication among marketing, engineering, and manufacturing – a comparison between two new product teams. *Management Science*, 38(3), 360–373.

Guellec, D. and Pattinson, B. (2002). Innovation surveys: lessons from OECD countries' experience. *Science Technology Industry Review*, 27, 77–102.

Katz, R., Tushman, M. and Allen, T.J. (1995). The influence of supervisory promotion and network location on subordinate careers in dual ladder RD&E setting. *Management Science*, 41(5), 848–863.

Mairesse, J. and Mohnen, P. (2010). Using innovations surveys for econometric analysis. NBER Working Paper No. w15857.

Majcen, B., Verbic, M., Ivanova, O. and Cok, M. (2011). R&D and economic growth in Slovenia: a dynamic general equilibrium approach with endogenous growth. *Panoeconomicus, Savez ekonomista Vojvodine, Novi Sad, Serbia*, 58(1), 67–89.

Miyagawa, T., Lee, K., Kabe, S., Lee, J., Kim, H., Kim, Y. and Edamura, K. (2010). Management practices and firm performance in Japanese and Korean firms – an empirical study using interview surveys. RIETI Discussion Paper, 10-E-013.

Mytelka, L., Goedhuys, M., Arundel, A. and Gachino, G. (2004). Designing a policy-relevant innovation survey for NEPAD, a study prepared for the Development Bank of Southern Africa Limited. United Nations University for New Technologies.

OECD/Eurostat (2005). *Oslo Manual – Guidelines for Collecting and Interpreting Innovation Data*, Paris: OECD.

Palley, T. (2011). The rise and fall of export-led growth. Levy Economic Institute, Working Paper No. 675.

Pisano, G. and Shih, W. (2009). Restoring American competitiveness, *Harvard Business Review*, 87(7–8), 114–125.

Porter, M.E. (1985). *Competitive Advantage: Creating and Sustaining Superior Performance*. New York: Free Press.

Prašnikar, J. (ed.) (2010). The role of intangible assets to exit the crisis. Ljubljana: Časnik Finance.

Prašnikar, J. and Knežević Cvelbar, L. (eds) (2012). Intangible assets as a potential for growth in Republic of Srpska. Ljubljana: Ekonomska fakulteta.

Prašnikar, J., Memaj, F. and Redek, T. (eds) (2012a). Albania: the role of intangibles for the future growth. Ljubljana: Ekonomska fakulteta.

Prašnikar, J., Redek, T. and Drenkovska, M. (2012b). Methodological challenges of measuring intangible capital in developing countries: the case of innovation activity. Ljubljana: Ekonomska fakulteta.

Rajkovič, T. and Prašnikar, J. (2009). Technological, marketing, and complementary competencies driving innovative performance of Slovenian manufacturing firms. *Organizacija*, 42: 77–86.

Sanchez, R. and Heene, A. (2004). *The New Strategic Management: Organization, Competition and Competence*. New York: Wiley.

Song, M., Droge, C., Hanvanich, S. and Calantone, R. (2005). Marketing and technology resource complementarity: an analysis of their interaction effect in two environmental contexts. *Strategic Management Journal*, 26, 259–276.

Uvalic, M. (2006). National systems of research and development in the Western Balkan countries. See-ERA.NET-Consortium, February.

Van Ark, B., Hao, J.X., Corrado, C. and Hulten, C. (2009). Measuring intangible capital and its contribution to economic growth in Europe. *EIB Papers*, 14(1), 62–93.
Wakelin, K. (1998). Innovation and export behaviour at the firm level. *Research Policy*, 26, 829–841.
Wang, Y., Lo, H.P. and Yang, Y. (2004). The constituents of core competencies and firm performance: evidence from high-technology firms in China. *Journal of Engineering and Technology Management*, 21, 249–280.
Zahra, S.A. and George, G. (2002). Absorptive capacity: a review, reconceptualization, and extension. *Academy of Management Review*, 27(2), 185–203.

12. Renewable energy policies for eco-innovation

Renata Dombrovski

INTRODUCTION

This chapter summarizes recent approaches in evaluating the policy instruments used to promote eco-innovations. The value added of the chapter is, firstly, in revealing the possible research avenues for scientists who are conducting theoretical or empirical (econometric or case study) analyses. Secondly, and differently from other reviews dealing with similar issues, this chapter links all these different research approaches with the impact of the main financial and fiscal instruments on both technological innovations and their diffusion. Although, so far, these instruments have usually been examined separately, or comparisons made including two or three different ones, the comprehensive overview of their impacts definitely provides a broader picture that could support policy-making decisions in the field of renewable energy.

It is well known that the pressure on global energy resources is high. Therefore, current climate policy should adopt financial and fiscal support measures that will limit harmful emissions and promote clean energy. Emerging from the identification of this issue, the main purpose of this chapter is to critically examine the recent advances in researchers' efforts to identify the effectiveness and efficiency of these support measures in fostering the process of technological change. The term 'effective' denotes the extent to which a policy instrument achieves policy objectives (that is, achieves an increase in the number of sustainable renewable energy (SRE) innovations or/and renewable energy capacity installations). The term 'efficient' refers to the ability of a policy instrument to reduce energy generation costs (EC, 2008). A general finding is that SRE support instruments promote new technologies (OECD, 2010). However, different instruments do not equally promote hydro, geothermal, solar, wind and biomass technologies (Johnstone et al., 2010; Falconett and Nagasaka, 2010; Dong, 2012; EEA, 2011). Therefore, the chapter focuses on a detailed overview of theoretical and empirical studies that address these issues, determining gaps in the literature and revealing possible further research directions. It has been found that researchers mostly examine the effects of one particular policy instrument on a technological change, or compare the impacts of two or three instruments (Morris and Read, 2001; Gagelmann and Frondel, 2005; Oikonomou et al., 2008; Butler and Neuhoff, 2008; Peretto, 2009; Commins et al., 2009; Coria, 2009; Markandya et al., 2009; Falconett and Nagasaka, 2010; Rogge and Hoffmann, 2010; OECD, 2010; Noailly, 2012; Bodas et al., 2012; Dong, 2012; Kažukauskas and Jaraite, 2011; Antoci et al., 2012). While trying to determine why one instrument is better than another, many papers do not focus exclusively on one particular type of renewable technology (Dinica, 2006; Bodas et al., 2012; Marques and Fuinhas, 2011). On the other hand, if a distinction between SRE technologies is made, the research focus is mostly on wind or solar technologies solely (Dong, 2012; Groba et al., 2011). Moreover, the majority of studies focused on various support instruments do not cover all 27 European Union (EU) countries. The lack of empirical analysis, especially on the micro level, in addressing this emergent topic is also identified. Further research should provide a new methodological framework in order to include all main support instruments and their interaction terms while identifying their impact on eco-innovations. Technological change has positive impacts on energy efficiency, competitiveness and economic growth. Most countries have similar environmental and economic goals, namely, greenhouse gas (GHG) emission reductions, energy savings, money savings and economic growth. However, their choice of financial and fiscal instruments varies, depending mainly on the national priorities. Investments in renewable energy sources (solar, wind, geothermal, hydro, biomass and biofuel) significantly foster the process of reaching the environmental and economic targets. Unfortunately, the share of renewables in total electricity net generation is still low, only 21.7 per cent according to the data for 2010 (EIA, 2012). Moreover, electricity generated from SRE sources is more expensive than conventionally generated electricity. That indicates the need for future rigorous empirical analysis that would identify the most effective and efficient SRE technology-specific instruments, necessary to support the policy decision-making process.

The chapter is structured as follows. The second section below documents the main steps towards the development of these technologies referring to related legislation. The third section discusses the link between financial and fiscal instruments and different stages and types of technology. The fourth section addresses the key characteristics and effectiveness and efficiency of main policy instruments with respect to promotion of the technology changes. The fifth section identifies further research opportunities and policy implications. The final section concludes.

MAIN STEPS TOWARD THE PROMOTION OF THE TECHNOLOGY CHANGES

The climate change debate focused on renewable energy has attracted attention for decades and it certainly continues to grow. The process towards clean technologies development and diffusion started with the Vienna Convention for the Protection of the Ozone Layer from 1985 (UNEP, 2001) and the United Nations Framework Convention on Climate Change (UNFCCC, 1992), that recognized, among other important issues, that all countries and especially developing ones need to achieve sustainable social and economic development. To fulfil these requirements countries have to implement new technologies which will achieve greater energy efficiency and reduce GHG emissions. The next important step was the Montreal Protocol on Substances that Deplete the Ozone Layer from 1987 (UNEP, 2000). A few years later, by signing the Kyoto Protocol (UNFCCC, 1998) – adopted in 1997 and entered into force in 2005 – the contracting countries agreed to reduce their collective GHG emissions by at least 5 per cent from the 1990 level in the commitment period from 2008 to 2012. In order to help countries to meet their targets by reducing GHG emissions in a cost-effective way, the Kyoto Protocol introduced three market-based mechanisms: emissions trading, the Clean Development Mechanism (CDM) and Joint Implementation (JI). The 2007 United Nations Climate Change Conference held in Bali (UNFCCC, 2007), the 2009 United Nations Climate Change Conference held in Copenhagen, Denmark (UNFCCC, 2009) and the 2010 United Nations Climate Change Conference which took place in Cancun, Mexico (UNFCCC, 2010) delivered important decisions toward a cleaner environment future. All these conferences contributed significantly to fighting climate change, but some EU countries still need to revise their SRE policy in order to foster the SRE technological change process. In March 2011 the European Commission (EC) adopted a 'Roadmap for Moving to a Competitive Low Carbon Economy in 2050' with the goal of keeping the climate change below 2 degrees Celsius. Another accepted target is to obtain between 80 per cent and 95 per cent GHG emissions reductions by 2050 from a 1990 baseline. With the aim of reaching this target, a European Commission report (EC, 2011) points out that more focus is needed on energy efficiency policies. Innovative solutions are required in order to meet the EU's low carbon targets by 2020. The EU is currently on the way to accomplishing two of three proposed objectives, namely GHG emissions reductions of 20 per cent and an increase in the share of renewable energy by the same percentage by 2020. The third EU target, achieving a 20 per cent increase in energy efficiency by 2020, will not be accomplished unless additional fiscal policy solutions are found. According to the European Commission, increased energy efficiency is the most cost-effective and the fastest way to reduce GHG emissions. Therefore, examining and implementing the most effective and efficient policy instruments has become one of the greatest challenges for researchers and policy makers worldwide. The spread of eco-policies over the years indicates the emerging need to face the environmental and economic challenges (Busch and Jörgens, 2005).

Before addressing the barriers to and opportunities for finding an effective and efficient financial and fiscal policy framework, it is necessary to understand the relationship between policy instruments and technological change. Therefore, the next section will contribute to a deeper understanding of these instruments with respect to different stages and types of SRE technologies.

THE LINK BETWEEN RENEWABLE ENERGY POLICIES AND TECHNOLOGICAL CHANGE

The research findings reveal that it is very important for countries to develop and implement new environmentally friendly technologies and use cleaner energy in terms of reducing greenhouse gas emissions (Gerlagh, 2008; Zhai et al., 2008; Peretto, 2009; Nixon et al., 2010). Therefore, a large number of financial and fiscal instruments have been introduced to promote eco-innovation. According to the EEA (2011), the term 'eco-innovation' includes all three stages of the technological change process: invention, innovation and diffusion. The distinction between these three stages of technological change economic theory is credited to Schumpeter (1942). Following his classification, invention encompasses the development of new products or processes, innovation relates to their commercialization, and diffusion implies the spread of new products and processes across the market. However, researchers mostly make a distinction only between innovation and the implementation of emerging technologies. At this point, the link between technological innovation and diffusion becomes clear. Innovations are a precondition for implementation activities and a part of the climate change fight. Optimal fiscal solutions are needed in both cases. The majority of studies have confirmed that successful renewable energy policies foster eco-innovations. However, the level of success of policy instruments varies across the different types of SRE technologies (Markandya et al., 2009; Johnstone et al., 2010; Falconett and Nagasaka, 2010). In addition, various instruments are more or less effective and efficient for different technological development stages (EEA, 2011). The main problem identified is that the literature does not provide unique conclusions regarding the impact of policy-specific instruments on technological innovation and diffusion. The emerging debate on the role of energy taxes in the technological change process gives mixed results. While a group of studies (OECD, 2010; Næss-Schmidt et al., 2010; Johnstone et al., 2010) confirm that energy taxation drives technological innovation, Noailly (2012) does not find a significant impact of energy taxes (captured by energy prices) on innovations. In the case of technological diffusion, researchers (Coria, 2009; Liu and Espínola-Arredondo, 2013) determine scenarios under which environmental taxes promote diffusion processes. Research by Johnstone et al. (2010) on fiscal incentives shows for example that targeted tax credits do not have an impact on innovation in renewable technologies. Later, focusing on research and development (R&D) tax credits, Czarnitzki et al. (2011) identify their positive effect on the innovation output. Regarding the EU Emissions Trading System (ETS), Anderson et al. (2011) show that emissions trading encourages moderate technological change. Later, Calel and Dechezlepretre (2012) do not find that the EU ETS has impacted upon the direction of technological change at all. Focusing on innovations, limited innovation effects of the pilot phase of the EU ETS are acknowledged in the research by Gagelmann and Frondel (2005). But not in line with these results, Rogge and Hoffmann (2010) argue that the EU ETS influenced the sectoral innovation system of power generation technologies in Germany. When examining the impact of Kyoto instruments on technological diffusion, Bodas et al. (2012) emphasize that these instruments are not encouraging SRE technologies in the BRICS countries (Brazil, Russia, India, China and South Africa). On the other hand, Antoci et al. (2012) identify how the EU ETS can promote the diffusion of new clean technologies. Focusing on comparisons between feed-in tariffs (FITs) and renewable portfolio standards (RPSs), Johnstone et al. (2010) emphasize that FITs better promote innovations in solar technologies, while obligations and tradable certificates more effectively induce innovations in wind technologies. A different scenario is documented in the case of technological diffusion, where FITs increase wind capacity more than an RPS (Dong, 2012).

It should be noticed that one group of studies more explicitly analyses the effectiveness and/or efficiency of policy instruments in promoting technological innovation, while other researchers link the instruments with technological diffusion. Studies are elaborated in more depth below. Further research efforts are needed in order to clarify those mixed results.

The development of emerging clean energy technologies demands SRE policy systems at the country level (Dinica, 2006; Lipp, 2007; Siriwardena et al., 2007; Strand, 2007; Hart, 2008; Vollebergh, 2008; Delucchi and Jacobson, 2011). According to the EC (2007), the EU has a commitment to reduce GHG emissions to a level that would limit the global temperature increase to 2 degrees Celsius compared to pre-industrial levels. However, as is emphasized in the EC (2007) report, with the energy and transport policies that are currently in force, European carbon dioxide (CO_2) emissions would increase by around 5 per cent by 2030 and global emissions would rise by 55 per cent. That further implies that the present system of renewable energy policies within the EU is not sustainable. The Ministry of Economic Affairs, Agriculture and Innovation (2011) pointed out that achieving a more sustainable energy system is a huge challenge, but also an opportunity for energy policy to reduce the dependence on fossil fuels and to make a step towards a low-carbon economy target by 2050.

Overall, renewable energy policies differ in their degree of success. If effective and efficient, they are a driving force in promoting renewable technologies and in reaching environmental and economic goals. Therefore, there is a current need to examine the impact of energy taxes and other financial instruments on the innovation and diffusion of clean technologies.

SUMMARY OF KEY CHARACTERISTICS, EFFECTIVENESS AND EFFICIENCY OF RENEWABLE ENERGY POLICIES

The literature review identified two relevant groups of papers: theoretical analyses of the impact of renewable energy polices on eco-innovation, and empirical studies of the relationship. This section first presents key characteristics of policy instruments that support innovation and implementation of SRE technologies and then it focuses on the stage of technological development process to which they relate, on the methodology used, on the main findings and on the further research implications to foster the technological change. The most relevant, recent studies are examined in more detail (see Appendix Table 12A.1). The selected studies consist of the most comprehensive econometric analyses of panel data that address the impacts of policy instruments on innovation and implementation of the renewable energy technologies, and reveal future research directions.

This analysis addresses the following key market based instruments: environmental taxes, fiscal incentives, emissions trading, the Clean Development Mechanism, Joint Implementation, feed-in tariffs, renewable energy portfolio standards, government grants and carbon credits. In the process of examining the link between policy instruments and eco-innovation, policies can be classified as those that regulate either the price of renewable electricity or the quantity produced (Weitzman, 1974).

Price-based policy instruments, such as fiscal incentives (R&D tax incentives and tax credits), and quantity-based Kyoto instruments can directly support investments in SRE technologies. Another group of price-based policy instruments such as fixed or premium feed-in tariffs, and quantity-based policy instruments such as renewable portfolio standards, can directly support generation from these technologies. Apart from policy instruments that directly promote SRE technologies there are other instruments, such as environmental taxes, that may have an indirect impact on technological change (Haas et al., 2004; Held et al., 2006). However, all these energy policy instruments could impact upon all stages of the technological change process (detailed innovation and diffusion analyses are provided below). Along with those market-based instruments, command-and-control approaches, such as emissions standards, process or equipment specifications, address the technological change.

Renewable energy policies should be carefully designed to foster the innovation process and the diffusion of new technologies. Supported by a number of existing findings, taxation of non-renewable technologies is a good starting point.

Environmental Taxes and Fiscal Incentives

The economic rationale behind environmental taxes is that these taxes raise the price of environmentally non-friendly activities while providing incentives to mitigate the effects of the pollution. More precisely, the studies confirm that environmental taxes and investment subsidies increase the innovation and diffusion of energy technologies (for a detailed review see EEA, 2011). However, their effect varies across sectors, technologies and types of innovation. The studies also reveal that the impact of these taxes cannot be examined without taking into consideration different design characteristics of country-specific sustainable renewable energy policies. These studies are later examined in more detail.

Four types of environmental taxes are energy, transport, pollution and resource taxes. Environmental taxes should be examined and implemented carefully. If they are too rigid in a certain country, they can become counterproductive, causing the 'carbon leakage' effect (Babiker, 2005; Barker et al., 2007; Næss-Schmidt et al., 2010). This means that firms can change location to countries with lower taxes and use non-renewable technologies instead of making the decision to innovate or implement clean alternatives. Barker et al. (2007) examine potential carbon leakage from six EU countries that implemented environmental tax reform (ETR) unilaterally in the period from 1995 to 2005. Their study contributes to the literature in a way that it measures potential leakage from historical actions and uses a Cambridge Econometrics energy–environment–economy model (E3ME) econometric dynamic model of Europe. Their results show that carbon leakage is small or negative due to technological spillover effects.

The literature has dealt with the analysis of different environmental taxes and their impacts on the technological change. The first group of studies (Næss-Schmidt et al., 2010; EEA, 2011; Noailly, 2012) assess the impact of these taxes on innovation, while the second group (Commins et al., 2009; Coria, 2009) assess their impact on diffusion of SRE technologies.

Næss-Schmidt et al. (2010) focus on the role of taxes in the innovation of energy technologies. They conducted econometric analysis using the European patent data from 1978–2007. Their study covers 33 countries, 13 energy products and seven technologies. They used quasi-static panel regression with fixed effects at the country level and dynamic panel regression due to robustness concerns, to find the link between energy taxes, prices and patents. The Copenhagen Economics Renewables Innovation Model was applied. Their main findings indicate that energy taxation is an efficient driver of innovation. R&D policies that support long-term innovation need to be a supplement to energy taxation.

One of the major findings of a recent OECD study of taxation, innovation and environment (OECD, 2010) is that environmental taxes can stimulate innovation. The EEA (2011) report examines different options under which the revenues collected from carbon taxes can be redirected back to the economy. The results based on macroeconometric models show that using some of the revenues to support renewable energy would have positive impacts on the economy. However, further research is needed in considering country-specific conditions that could not be fully examined by modelling framework presented in the study. In addition, CDM and JI measures were not taken into account. Therefore, the reports conclude by noting that the discussion on market-based instruments should be intensified. In addition, the EU needs a mix of policies to reach GHG emissions targets and to maximize prosperity.

Noailly (2012) investigates the impacts of regulatory energy standards in buildings codes, energy taxes captured by energy prices, and specific governmental energy R&D expenditures on technological innovation intending to improve energy efficiency in buildings. The econometric analysis of patent data estimates the impact of different policy instruments on technological innovation for seven EU countries from 1989 to 2004. A conditional fixed effect Poisson model with robust standard errors clustered at the country level is used. The main results reveal that a strengthening of 10 per cent of the minimum insulation standards for walls would increase the likelihood to file additional patents by about 3 per cent. The study finds that energy prices have no significant impact on patenting activities, while the effect of governmental energy R&D support is small but positive. The author also highlights future research possibilities. Firstly, further work should use patent data at the firm level to study how policies can influence firm behaviour. Secondly, attributes such as the stability or flexibility of the measures might be relevant. Thirdly, it should be empirically examined how innovations and patents effectively contribute to the reduction of GHG emissions. Finally, the author suggests that how different policy instruments contribute to higher energy efficiency through the diffusion of technologies should also be examined.

Commins et al. (2009) examine the impact of energy taxes and the EU ETS on the employment, investment behaviour and productivity of the EU firms in different sectors, using firm-level panel data for the period 1996 to 2007. The main results of their study indicate that energy taxes increased total factor productivity and returns to capital, but decreased employment. The effects on investments are mixed. The results of their analysis also reveal large variations between sectors, not just in the size of the estimated effects but also in their signs. Additionally, findings show a negative effect of the EU ETS on productivity and profits, while the effect on labour and investment is insignificant. However, the dataset used covered only an experimental phase of the EU ETS. As indicated in the study, further research should reinvestigate that issue, using data from the second phase of the EU ETS.

Coria (2009) develops the adoption model to examine how the choice between taxes and permits impacts upon the clean technology adoption patterns, when firms engage in imperfect competition in the output market. It is assumed that the regulator commits to the *ex post* optimal amount of emissions before firms start with the technology adoption process. The author uses a dynamic approach to analyse the impact of the policy instrument choice over the periods prior to technological diffusion. The results show that when the output demand is more elastic, auctioned permits dominate over emissions taxes in accelerating the technological diffusion. Further analyses on technological diffusion should consider different settings as well.

On the other hand, fiscal incentives (R&D tax incentives, tax credits, tax exemptions or reductions) are also used to promote technological innovations and the diffusion of renewable technologies. In some countries they are the main support instrument, while others most often apply them in combination with other renewable energy instruments. Fiscal incentives could be sufficient to stimulate SRE technologies in countries with higher energy tax rates; otherwise, a mix of instruments is needed (European

Commission, 2008). The first group of studies (Johnstone et al., 2010; Czarnitzki et al., 2011) link fiscal incentives with the technological innovations, while the second group (Markandya et al., 2009; Cansino et al., 2010) examine their impact on the diffusion of SRE technologies.

Johnstone et al. (2010) find, among others, that investment incentives have a positive effect on all renewable energy technologies except wind. Targeted tax credits and voluntary programmes do not have an impact on technological innovations (detailed analysis of their research is provided in the subsection below, 'Feed-in tariff and renewable energy portfolio standards'). On the other hand, Czarnitzki et al. (2011) find that R&D tax credits increase the innovation output of Canadian manufacturing firms. The authors conduct their analysis employing cross-sectional data, based on the Canadian 1999 Survey of Innovation, and use a non-parametric matching approach. For future research, the use of panel data, and objectively measured instead of self-reported performance indicators, are suggested.

Markandya et al. (2009) developed a simple economic engineering type model of consumers' behaviour to analyse the tax incentives aimed at promoting the production and consumption of energy-efficient appliances in different European countries. From the cases considered, they found that tax credits on boilers are a cost-effective option in Denmark and Italy. Subsidies on integrated compact fluorescent light bulbs (CFLi bulbs) are cost-effective in France and Poland in terms of euros per tonne of CO_2 abated. In addition, their results inform that the subsidies are in most cases less cost-effective than energy taxes. The authors emphasize that further research should take distributional factors into account because of the concern that an increase in energy prices might hurt the very poor. Secondly, the proposed model does not fully allow spillover effects of the incentives. Finally, a dynamic modelling approach would be a valuable extension to the model presented.

Cansino et al. (2010) present an overview of the main tax incentives aiming to promote green electricity within the European Union countries. They find that 16 EU countries use tax incentives together with other instruments, mainly with quota obligations and price regulation. However, the problem lies in the fact that not all available technologies are promoted. The authors additionally identify that an exemption on the payments of excise duties for electricity when it is generated from renewable energy sources is most widely used. The EU goal is to provide a renewable energy policy framework. This goal is hard to achieve because of the great diversity of tax incentives used to promote green electricity. Despite this problem, authors argue that different case studies can be used to inform the EU policy development.

Emissions Trading, Clean Development Mechanism and Joint Implementation

The Kyoto Protocol introduced three market-based instruments – emissions trading, the Clean Development Mechanism (CDM) and Joint Implementation (JI) – to help countries to achieve their emission requirements under the Protocol. Emissions trading, as determined in Article 17 of the Kyoto Protocol, allows countries that exceed their emissions reduction commitments under the Protocol to buy emissions units from other countries that are under their Kyoto targets. The Clean Development Mechanism – as set out in Article 12 of the Protocol – allows a country with an emission reduction or limitation targets under the Kyoto to implement an emissions reduction project in developing countries. Such projects can earn Certified Emission Reduction (CER) credits, each equivalent to 1 tonne of CO_2. CERs can be traded and sold and can give industrialized countries some flexibility in how they meet their Kyoto targets. Joint Implementation – defined in Article 6 of the Kyoto Protocol – allows a country with an emissions reduction or limitation commitment under the Kyoto Protocol to earn Emission Reduction Units (ERUs) from an emissions reduction or removal project in any other country with targets under the Protocol. An ERU is equivalent to 1 tonne of CO_2.

In general, the underpinning economic rationale of the Kyoto instruments is that the emissions reductions and low-carbon technology installations should be achieved in a cost-effective way. However, it was argued (Gagelmann and Frondel, 2005; EC, 2013) that during the first EU ETS trading period (2005–2007) too many emissions allowances were issued and distributed free of charge. Another reason for the limited impact of the EU ETS was the lenient cap on emissions and unlimited usage of CDM and JI credits. Consequently, the price of first-period allowances decreased to zero in 2007 and firms did not have enough incentives to reduce emissions and develop SRE technologies. According to the European Commission (EC, 2013), based on 'learning by doing', the number of allowances was reduced by 6.5 per cent in the second trading period (2008–2012). Major reform, including the introduction of an EU-wide cap on emissions, reduced by 1.74 per cent each year, and auctioning of allowances, will affect the third trading period (2013–2020). Despite these improvements, researchers emphasize that further research on preventing the carbon leakage caused by Kyoto should be intensified (Antimiani et al., 2013).

Recent studies have begun to examine the impact of Kyoto instruments on technological innovation (Gagelmann and Frondel, 2005; Rogge and Hoffmann, 2010) and diffusion processes (Schneider et al., 2008; Bodas et al., 2012; Antoci et al., 2012). Gagelmann and Frondel (2005) emphasize that the literature does not provide a unique answer to the question of whether the ETS causes more innovation than other policy instruments, such as command-and-control. Based on the studies reviewed, the authors conclude that the innovation effects of the EU ETS in its pilot phase from 2005 to 2007 were limited, especially due to lenient emissions targets in these first years. Later, Rogge and Hoffmann (2010) empirically investigate the actual innovation impacts of the EU ETS in the initial years after its implementation. The authors conduct analyses for Germany, based on 42 interviews with experts in the field of the EU ETS, the power sector and technological innovation. Despite low expectations of the innovation impacts of the EU ETS, triggered by Gagelmann and Frondel's (2005) findings, the authors argue that the EU ETS has influenced the sectoral innovation system of power generation technologies in Germany in several areas. However, based on their research results, it is not expected that the revised EU ETS will be sufficient to decarbonize the power sector. Future research areas are envisaged as well. Firstly, the findings for Germany should be compared with results from other countries. Secondly, since the amounts of emission allowances issued under the EU ETS, free allocation rules and innovation patterns vary across sectors, the future analysis should include other sectors covered by the EU ETS as well. Thirdly, in-depth company case studies focused on technological and firm-level differences might be a valuable contribution to the topic investigated. Finally, a survey of firms in several countries would allow identification of the extent of innovation effects triggered by the ETS.

By assessing empirical studies and conducting expert interviews, Schneider et al. (2008) find that the CDM contributes to low-carbon technology transfer. However, its performance is affected by geography, technology and project size. The impact of firm size or other firm characteristics still needs to be analysed.

Bodas et al. (2012) conduct an empirical analysis using aggregated national data to examine whether the Kyoto incentives have stimulated the diffusion of renewable energy technologies in the BRICS countries (Brazil, Russia, India, China and South Africa). The results suggest that the Kyoto mechanisms are not creating incentives for the use of sustainable variants of renewable energy technologies in the BRICS countries. Instead, these instruments are encouraging installed technological capacity. The authors also address some limitations of their study that are mostly related to its reliance on macro-aggregated and secondary sources of data which do not allow them to investigate differences on the levels of the sustainability of energy-producing technologies based on renewable sources. In order to further investigate the impact of Kyoto instruments on technological diffusion, project-level data should be collected to identify, among others, the critical elements that attract buyers to invest in CDM and JI projects. In addition, research should distinguish whether projects are private buyer-driven or government-driven.

Antoci et al. (2012) examine the impact of the EU ETS on the diffusion of new clean technologies, taking into account firms' strategic behaviours and penalties to firms that go above their emissions allowance limits. The authors analyse the evolutionary game model with pairwise random matching. They assumed that each firm has to decide whether it will adopt new clean technology, or meet the emission targets using the old technology and pollution permits. The results show that properly defined penalties to non-compliant firms and an increase in permits trade can promote non-polluting scenarios and the diffusion of new clean technologies. In order to strengthen the realism of the model, the evolutionary game should be extended to the case of *n* firm possible matching.

Further research on the impact of Kyoto instruments on technological diffusion should be intensified in order to clarify the dilemma of whether these instruments actually stimulate the diffusion of sustainable renewable energy technologies.

Feed-In Tariff and Renewable Energy Portfolio Standards

Feed-in tariff (FIT) refers to a policy instrument that reduces the risk of a long-term investment in renewable technologies. Under a FIT, renewable electricity producers are paid a fixed rate or premium for each kilowatt-hour generated and prices are guaranteed for a certain period of time. Quota obligations, also referred to as renewable portfolio standards (RPSs), oblige suppliers to produce a certain percentage of their electricity from renewable energy sources. Certified renewable energy generators get certificates for units of electricity they produce (renewable energy certificates – RECs). Renewable energy producers can sell RECs which prove the renewable source of the electricity.

The basic economic rationale behind feed-in tariffs and renewable portfolio standards is to reduce the costs of renewable electricity production. Pros and cons of these instruments related to technology change are identified in the research paper by Johnstone et al. (2010). The authors conduct panel analysis using patent data for 25 countries over the period 1978–2003 to examine the effect of environmental policies on technological innovation in the case of renewable energy. The contribution of this paper to the recent literature is twofold: (1) the cross-country focus allows them to examine the effects of feed-in tariffs, production quotas and investment subsidies; and (2) they study patent activity with respect to different sources of renewable energy. Their results indicate that different support instruments are effective for different SRE sources. More precisely, feed-in tariffs are effective in promoting innovation in more costly energy technologies, such as solar power. On the other hand, obligations and tradable certificates are effective in inducing additional innovations in technologies that are close to being competitive with fossil fuels, such as wind power. The authors also draw attention to further research possibilities, emphasizing that accounting for variation in natural conditions as determinants of patenting in SRE technologies is needed.

Dong (2012) examines the effectiveness of feed-in tariffs and renewable portfolio standards in promoting wind capacity development. The author designs four different modelling scenarios and conducts a regression analysis using a panel dataset for 53 countries for the period 2005–2009. The main findings indicate that FITs increase total wind capacity more than RPSs across countries. He does not find a significant difference in annual wind capacity between FITs and RPSs. The analysis also reveals that wind energy development acts in response to high electricity demand and oil reliance. Further research should, among other issues, investigate how wind energy development impacts upon other renewable energy industries.

Falconett and Nagasaka (2010) develop a probabilistic model to calculate the effects of governmental grants, feed-in tariffs and renewable energy certificates on the financial returns of small-scale hydroelectric, wind energy and solar photovoltaic (PV) systems. The authors also examine the impact of carbon credits on the net present value of renewable projects, and compare it with other instruments. Carbon credit is a permit that gives the right to the holder to emit 1 tonne of carbon dioxide. If countries or groups reduce GHG emissions below their emission allowances, they get the credit. Others that pollute above their emission targets have to buy it. Carbon credit is a tradable financial instrument the purpose of which is to reduce harmful emissions. The results of the Falconett and Nagasaka (2010) study indicate that the FIT is the best mechanism to increase the profitability of solar PV systems and wind energy projects. The green certificate mechanism is the most suitable one to increase the profitability of hydroelectric projects. They also find that governmental grants and carbon credits are secondary support instruments in comparison with FITs and RECs. Finally, they conclude that the efficiency of the support mechanisms varies, depending on the stage of development of the renewable technologies that are being implemented.

One of the few micro empirical analyses is the research paper by Kažukauskas and Jaraite (2011) who investigate how Tradable Green Certificates (TGCs) and FITs affect power-generating firms, using cross-country micro-level data for 27 EU countries for the period 2002–2007.

The main finding indicates that power-generating firms that implemented TGCs were more profitable than firms that implemented FITs. However, the exact sources of the higher profitability are not identified and further research is required on that aspect.

IMPORTANCE OF FURTHER RESEARCH FOR THE RENEWABLE ENERGY POLICY DECISION-MAKING PROCESS

Researchers might utilize insights from this review in order to contribute to the process of technology changes. Details of gaps in the theoretical, econometric and case study literature are provided, organized according to the policy instruments used to support eco-innovations. However, major possible research opportunities could be summarized as follows.

In a frequently cited research paper, Jaffe et al. (2002) emphasize that empirical evidence generally supports theoretical findings that market-based instruments have a greater significant positive impact on eco-innovation than command-and-control approaches. However, the authors also identify empirical studies that are not consistent with the theory. Therefore, further work is needed in order to resolve the obvious dilemma.

In general, further research should include all main financial and fiscal policy support instruments into analysis to empirically identify the most effective and efficient ones for fostering technological innovation and diffusion. In addition, other elements of country-specific renewable energy policy such as political, economic, social, environmental, technical should be evaluated. All these renewable policy elements should be examined in more depth through different case studies, and then used to support comprehensive and more general econometric findings. Despite various approaches to this emerging topic, there is a lack of evidence that focuses on SRE policy design.

Research to date reveals that further efforts are needed in the examination of the stability and flexibility of country- and technology-specific instruments. Moreover, the impact of support instruments on technology-specific innovation activities and consequent reductions of GHG emissions due to firms' patenting activities should be further examined using historical panel datasets. Another interesting task would be to test how policy instruments increase energy efficiency through technological diffusion. Theoretical models should consider different settings (see e.g. Liu and Espinola-Arredondo, 2013 for more detailed recommendations) when examining the impacts of policy instruments on technological change. In addition, they should fully account for spillover impacts of the incentives.

Focusing on Kyoto Protocol instruments, the EU ETS is the right tool for GHG emissions reductions as this is the primary goal of the emissions trading. The theoretical and empirical (although mainly case study) findings confirm that the Emissions Trading System can trigger innovation as well. The strength of its impact on technological change should be further examined, incorporating the changes in the EU ETS design that are or will be included in the second (2008–2012), third (2013–2020) and fourth (2021–2028) trading periods (EC, 2013). The studies so far have focused mainly on its pilot phase (2005–2007). Due to the fact that the amounts of emission allowances issued under the EU ETS, free allocations and innovation patterns differ across sectors, further analyses should address different sectors covered by the scheme. In addition, further research should combine historical patent, environmental R&D, EU ETS and firm financial data in order to conduct econometric analyses and provide more comprehensive results that cannot be obtained using case studies alone. However, in-depth case studies at the firm level focused on technological and firm-level differences in the EU countries could significantly contribute to conclusions on the effects of the EU ETS. Comprehensive analyses of the impact of Kyoto instruments on technological diffusion should use project-level data to clarify the unsolved dilemma: do these instruments really promote the diffusion of clean technologies? The policy decision-making process needs all these findings in order to secure sustainable renewable energy policies within countries.

CONCLUDING REMARKS

This chapter provides an up-to-date critical literature review analysis that considers the overall promotion of technology changes and determines further research suggestions. It may be summarized that policy support instruments have a positive influence on technological innovation and implementation in the area of renewable energy. However, technology-specific analyses emphasize the importance of instrument choice, pointing to differences in the effectiveness and efficiency of instrument types when promoting different renewables. Moreover, when examining their impact on eco-innovation, researchers predominantly compare two policy support instruments, for example FIT versus RPS or energy taxes versus incentives. If referring to a specific technology, especially when using the number of capacity installations as a dependent variable, they most often focus solely on wind or solar SRE sources. A summary of the existing findings and identification of gaps in the literature are certainly valuable steps for further research contributions to this field.

Focusing on the research gaps identified, our future work aims to support the debate on the effectiveness of financial and fiscal instruments in supporting SRE technologies, providing empirical results. The value added of that econometric study is in its comprehensiveness due to the examination of (among others) FIT (premium and fixed for geothermal, wind, solar and biomass technologies), RPS, tender, tax incentives and investment grants and their impact on technology-specific SRE generation and additional SRE production, using a panel dataset for 27 EU countries from 1990 to 2010. Additional SRE production is defined as the ratio between technology-specific renewable generation and the differences between total net electricity generation and the net electricity imports. The main results that emerge from the analysis using the pooled OLS approach, with country and time fixed effects included, show that in the case of SRE generation, fixed FIT, RPS and tender have a positive and significant impact only on wind electricity generation. In the case of additional SRE production, fixed and premium FIT and RPS positively and significantly impact upon additional geothermal production, premium FIT has this impact on additional wind electricity production, while tax incentives and investments grants have a positive significant impact on additional solar production. Further analysis would identify the countries that might be driving the results.

ACKNOWLEDGEMENTS

The author gratefully acknowledges the support of the Marie Curie FP7 programme: Management of Emerging Technologies for Economic Impact, grant no. PITN–GA–2009–238382. I am grateful to two anonymous reviewers for their very constructive comments and suggestions, which enabled substantial improvement of the chapter. The chapter has also benefited from discussion with participants of the EURAM 2012 conference (Rotterdam School of Management, Erasmus University).

REFERENCES

Anderson, B., Convery, F. and Di Maria, C. 2011. Technological change and the EU ETS: the case of Ireland. IEFE Working Paper No. 43.

Antimiani, A., Costantini, V., Martini, C., Salvatici, L. and Tommasino, M.C. 2013. Assessing alternative solutions to carbon leakage. *Energy Economics* 36: 299–311.

Antoci, A., Borghesi, S. and Sodini, M. 2012. ETS and technological innovation: a random matching model. FEEM Working Paper No. 79.

Babiker, M.H. 2005. Climate change policy, market structure, and carbon leakage. *Journal of International Economics* 65(2): 421–445.

Barker, T., Junankar, S., Pollitt, H. and Summerton, P. 2007. Carbon leakage from unilateral environmental tax reforms in Europe, 1995–2005. *Energy Policy* 35(12): 6281–6292.

Bodas, F.I.M., Dantas, E. and Iizuka, M. 2012. The Kyoto mechanisms and the diffusion of renewable energy technologies in the BRICS. *Energy Economics* 42(3): 118–128.

Busch, P.O. and Jörgens, H. 2005. The international spread of environmental policy innovation. FFU-Report No. 02-2005, Berlin.

Butler, L. and Neuhoff, K. 2008. Comparison of feed-in tariff, quota and auction mechanisms to support wind power development. *Energy Policy* 33(8): 1854–1867.

Calel, R. and Dechezlepretre, A. 2012. Environmental policy and directed technological change: evidence from the European carbon market. FEEM Working Paper No. 22, Milan.

Cansino, J.M., Pablo-Romero, M., Román, R. and Yñiguez, R. 2010. Tax incentives to promote green electricity: an overview of EU-27 countries. *Energy Policy* 38(10): 6000–6008.

Commins, N., Lyons, S., Schiffbauer, M. and Tol, R.S.J. 2009. Climate policy and corporate behaviour. ESRI Working Paper 329, Economic and Social Research Institute, Dublin. http://www.researchgate.net/publication/46460461_Climate_Policy_and_Corporate_Behaviour.

Coria, J. 2009. Taxes, permits, and the diffusion of a new technology. *Resource and Energy Economics* 31(4): 249–271.

Czarnitzki, D., Hanel, P. and Rosa, J.M. 2011. Evaluating the impact of R&D tax credits on innovation: a microeconometric study on Canadian firms. *Research Policy* 40(2): 217–229.

Delucchi, M.A. and Jacobson, M.Z. 2011. Providing all global energy with wind, water and solar power, Part II: Reliability, system and transmission costs and policies. *Energy Policy* 39(3): 1170–1190.

Dinica, V. 2006. Support systems for the diffusion of renewable energy technologies – an investor perspective. *Energy Policy* 34(4): 461–480.

Dong, C.G. 2012. Feed in tariff vs. Renewable portfolio standard: an empirical test of their relative effectiveness in promoting wind capacity development. *Energy Policy* 42(3): 476–485.

EIA. 2012. International Energy Statistics. http://www.eia.gov/cfapps/ipdbproject/IEDIndex3.cfm (accessed 14 September 2012).

European Commission (EC). 2007. An energy policy for Europe. COM 1 Final, Brussels, Belgium.

European Commission (EC). 2008. The support of electricity from renewable energy sources. Commission Staff Working Document SEC 57, Brussels.

European Commission (EC). 2011. A roadmap for moving to a competitive low carbon economy in 2050. COM 112 final, Brussels.

European Commission (EC). 2013. The EU Emission Trading System (EU ETS). EC, Brussels.

European Environment Agency (EEA). 2011. Environmental tax reform in Europe: opportunities for eco-innovation. EEA Technical Report 17, Copenhagen.

Falconett, I. and Nagasaka, K. 2010. Comparative analysis of support mechanisms

for renewable energy technologies using probability distributions. *Renewable Energy* 35(6): 1135–1144.
Gagelmann, F. and Frondel, M. 2005. The impact of emission trading on innovation – science fiction or reality? *Environmental Policy and Governance* 15(4): 203–211.
Gerlagh, R. 2008. A climate-change policy induced shift from innovations in carbon energy-production to carbon-energy savings. *Energy Economics* 30(2): 425–448.
Groba, F., Indvik, J. and Jenner, S. 2011. Assessing the strength and effectiveness of renewable electricity feed-in tariffs in European Union countries. DIW Discussion Paper 1176, Deutsches Institut für Wirtschaftsforschung, Berlin.
Haas, R., Eichhammer, W., Huber, C., Langniss, O., Lorenzoni, A., Madlener, R., Menanteau, P., Morthorst, P.E., Martins, A., Oniszk, A., Schleich, J., Smith, A., Vass, Z. and Verbruggen, A. 2004. How to promote renewable energy systems successfully and effectively. *Energy Policy* 32(6): 833–839.
Hart, R. 2008. The timing of taxes on CO_2 emissions when technological change is endogenous. *Journal of Environmental Economics and Management* 55(2): 194–212.
Held, A., Haas, R. and Ragwitz, M. 2006. On the success of policy strategies for the promotion of electricity from renewable energy sources in the EU. *Energy and Environment* 17(6): 849–868.
Jaffe, A.B., Newell, R.G. and Stavins, R.N. 2002. Environmental policy and technological change. *Environmental and Resource Economics* 22(1–2): 41–69.
Johnstone, N., Haščič, I. and Popp, D. 2010. Renewable energy policies and technological innovation: evidence based on patent counts. *Environmental and Resource Economics* 45(1): 133–155.
Kažukauskas, A. and Jaraite, J. (2011). The profitability of power generating firms and policies promoting renewable energy. CERE Working Paper 14, Centre for Environmental and Resource Economics, Umea, Sweden. http://papers.ssrn.com/sol3/papers.cfm?abstract_id=1964224.
Lipp, J. 2007. Lessons for effective renewable electricity policy from Denmark, Germany and the United Kingdom. *Energy Policy* 35(11): 5481–5495.
Liu, B. and Espínola-Arredondo, A. 2013. The impact of environmental taxes on firms' technology and entry decisions. Working Paper No. 2, Washington State University.
Markandya, A., Ortiz, R.A., Mudgal, S. and Tinetti, B. 2009. Analysis of tax incentives for energy-efficient durables in the EU. *Energy Policy* 37(12): 5662–5674.
Marques, A.C. and Fuinhas, J.A. 2011. Do energy efficiency measures promote the use of renewable sources? *Environmental Science and Policy* 14(4): 471–481.
Ministry of Economic Affairs, Agriculture and Innovation. 2011. Energy Report 2011. http://www.government.nl/documents-and-publications/reports/2011/11/01/energy-report-2011.html (accessed 5 October 2012).
Morris, J.R. and Read, A.D. 2001. The UK landfill tax and the landfill tax credit scheme: operational weaknesses. *Resources, Conservation and Recycling* 32(3–4): 375–387.
Næss-Schmidt, S., Hansen, E.S., Tops, J., Jensen, H.N. and Jespersen, S.T. 2010. Innovation of energy technologies: the role of taxes. Copenhagen Economics – Final Report, Copenhagen.
Nixon, J.D., Dey, P.K. and Davies, P.A. 2010. Which is the best solar thermal collection technology for electricity generation in north-west India? Evaluation of options using the analytical hierarchy process. *Energy* 35(12): 5230–5240.
Noailly, J. 2012. Improving the energy efficiency of buildings: the impact of

environmental policy on technological innovation. *Energy Economics* 34(3): 795–806.

OECD. 2010. Taxation, innovation and the environment. Organisation for Economic Co-operation and Development.

Oikonomou, V., Jepma, C., Becchis, F. and Russolillo, D. 2008. White Certificates for energy efficiency improvement with energy taxes: a theoretical economic model. *Energy Economics* 30(6): 30443–30462.

Peretto, P.F. 2009. Energy taxes and endogenous technological change. *Journal of Environmental Economics and Management* 57(3): 269–283.

Rogge, K.S. and Hoffmann, V.H. 2010. The impact of the EU ETS on the sectoral innovation system for power generation technologies – findings for Germany. *Energy Policy* 38(12): 7639–7652.

Schneider, M., Holzer, A. and Hoffmann, V.H. 2008. Understanding the CDM's contribution to technology transfer. *Energy Policy* 36(8): 2930–2938.

Schumpeter, J.A. 1942. *Capitalism, Socialism and Democracy*. New York: Harper & Brothers.

Siriwardena, K., Wijayatunga, P.D.C., Fernando, W.J.L.S., Shrestha, R.M. and Attalage, R.A. 2007. Economy wide emission impacts of carbon and energy tax in electricity supply industry: a case study on Sri Lanka. *Energy Conversion and Management* 48(7): 1975–1982.

Strand, J. 2007. Energy efficiency and renewable energy supply for the G-7 countries, with Emphasis on Germany. IMF Working Paper 299, International Monetary Fund.

UNEP. 2000. The Montreal Protocol on Substances that Deplete the Ozone Layer. http://www.unep.org/ozone/pdf/Montreal-Protocol2000.pdf (accessed 24 August 2010).

UNEP. 2001. Vienna Convention for the Protection of the Ozone Layer. http://ozone.unep.org/pdfs/viennaconvention2002.pdf (accessed 24 August 2010).

UNFCCC. 1992. United Nations Framework Convention on Climate Change. http://unfccc.int/resource/docs/convkp/conveng.pdf (accessed 24 August 2010).

UNFCCC. 1998. Kyoto Protocol to the United Nations Framework Convention on Climate Change. http://unfccc.int/resource/docs/convkp/kpeng.pdf (accessed 24 August 2010).

UNFCCC. 2007. The United Nations Climate Change Conference in Bali. http://unfccc.int/meetings/cop_13/items/4049.php (accessed 5 January 2010).

UNFCCC. 2009. The United Nations Climate Change Conference in Copenhagen. http://unfccc.int/meetings/cop_15/items/5257.php (accessed 5 January 2010).

UNFCCC. 2010. The United Nations Climate Change Conference in Cancun. http://unfccc.int/meetings/cancun_nov_2010/meeting/6266.php (accessed 5 January 2011).

Vollebergh, H.R.J. 2008. Lessons from the polder: energy tax design in The Netherlands from a climate change perspective. *Ecological Economics* 64(3): 660–672.

Weitzman, M.L. 1974. Prices vs. quantities. *Review of Economic Studies* 41(4): 477–491.

Zhai, X.Q., Wang, R.Z., Dai, Y.J., Wu, J.Y. and Ma, Q. 2008. Experience on integration of solar thermal technologies with green buildings. *Renewable Energy* 33(8): 1904–1910.

APPENDIX

Table 12A.1 Analyses of the most relevant econometric eco-innovation studies

Study	Data sources	Variables	Methodology	Key findings	Gaps in the literature and further research possibilities
Noailly (2012)	MURE database (Mesures d'Utilisation Rationnelle de l'Energie), a European project collecting information on measures for the rational use of energy and for renewables in Europe – data on the stringency of the national building codes; energy prices and taxes database from the International Energy Agency (IEA) – data on energy prices;	DEPENDENT VARIABLE: patent counts for 8 specific technologies related to energy efficiency in buildings INDEPENDENT VARIABLES: regulatory energy standards in building codes, energy taxes (captured by energy prices) and specific governmental energy R&D expenditures	The econometric analysis estimates the impact of different policy instruments on technological innovation for 7 EU countries over the 1989–2004 period; a conditional fixed effect Poisson model with robust standard errors clustered at the country level is used	– For the specific case of the building sector, strengthening regulatory standards would have a greater impact on innovation than energy prices or R&D support – Energy prices have no significant effect on patenting activities, while the effect of governmental energy R&D support is small, but positive and significant	– Future work should (using patent data at the firm level) study how policies can influence firm behaviour – Beyond the types of policy measures, other attributes such as stability or flexibility of the measures may be particularly relevant – More work is needed to measure how innovations and patents effectively contribute to reducing environmental impacts – A very interesting

	patents data were extracted from EPODOC, an internal database from the European Patent Office (EPO)				and worthwhile issue to consider is how various policy measures contribute to higher energy efficiency through the diffusion of technologies
Commins et al. (2009)	AMADEUS database, OECD, Eurostat and the IEA	DEPENDENT VARIABLES: total factor productivity (TFP), number of employees, investment and return on capital employed INDEPENDENT VARIABLES: tax rate, lagged tax rate, import intensity, education, output gap, electricity price, labour cost, ETS	This paper examines the impact of energy taxes and the EU ETS on a large number of firms in Europe between 1996 and 2007 using: – Olley–Pakes method – OLS panel regression in first differences	– Energy taxes increased total factor productivity (TFP) and returns to capital but decreased employment, with a mixed effect on investment, results vary between sectors in the size of the estimated effects and also in their signs	Future work should reinvestigate this topic, using data from the second phase of the ETS
Næss-Schmidt, et al. (2010)	IEA Data Services, EPO database	DEPENDENT VARIABLE: number of patents within a certain technology class	Multivariate econometric analysis using the: – quasi-static panel regressions with	– Energy taxation is a very effective driver of innovation – R&D policy supporting long-	– This study focuses on energy taxes and R&D policies solely – The impact of other support instruments

Table 12A.1 (continued)

Study	Data sources	Variables	Methodology	Key findings	Gaps in the literature and further research possibilities
Næss-Schmidt, et al. (2010)		(a measure of innovation) INDEPENDENT VARIABLES: total number of EPO patents (this variable serves to capture common trends in patenting which are unrelated to patenting incentives for environmentally relevant technologies), government R&D in the technology of interest, price of a relevant energy input, possible policy interventions, energy taxes	fixed effects at the country level – dynamic panel regression too (due to robustness concerns) to find the link between energy prices, taxes and patents – CERIM (Copenhagen Economics Renewables Innovation Model); the model can show the necessary carbon taxes to achieve a certain emission target – European patent data (30 years: 1978–2007, 33 countries, 13 energy products, 7 technologies)	term innovation needs to be a supplement to energy taxation	on innovation should also be empirically tested

Marques and Fuinhas (2011)	OECD Factbook 2009, MURE (Mesures d' Utilisation Rationnelle de l'Energie) II Database, Coordinated by the Institute of Studies for the Integration of Systems and the Fraunhofer Institute for Systems and Innovation Research ISI), Eurostat, World Bank World Development Indicators, International Financial Statistics of the IMF, BP Statistical Review of World Energy 2009	DEPENDENT VARIABLE: contribution of renewable sources to total energy supply EXPLANATORY VARIABLES: *policy* (number of energy efficiency policies and measures), *environment* (carbon dioxide emissions per capita), *socio-economic characteristic* (energy consumption per capita, income/GDP, energy security, prices of oil, natural gas and coal), *electricity generation* (contributions of coal, oil, natural gas and nuclear to electricity generation)	OLS and the quantile regression technique applied to a set of 21 EU countries in 2 time spans: 1990–1998 and 1999–2006	– For the period 1999–2006, the variable *efficiency energy policies and measures* was highly statistically significant in explaining the use of renewable sources – This paper also shows that the lobbying effect of traditional energy industries depends on the period under analysis and on the type of traditional energy source	– It might be helpful to identify and categorize countries with similar characteristics of renewable energy use by using cluster analysis

Table 12A.1 (continued)

Study	Data sources	Variables	Methodology	Key findings	Gaps in the literature and further research possibilities
Dong (2012)	World Wind Energy Report 2008 and 2009, Renewables 2010 Global Status Report, World Development Indicator Database, International Energy Statistics, World Bank	DEPENDENT VARIABLE: total/ cumulative wind capacity installed at the end of each year, annual wind capacity INDEPENDENT VARIABLES: FIT and RPS, GDP per capita, electricity net, consumption, net oil imports, wind resources, CO2 intensity, information about other promotion policies that each country has DUMMY VARIABLES: FIT and RPS – a country is coded as 1 if adopts the policy, an interaction term of these two policies is also created	Regression analysis (OLS, fixed effects model) using panel data for 53 countries for 5 years from 2005 to 2009, four different scenarios of modelling are designed	FIT is more effective than RPS in promoting wind capacity development across countries, since 2005 there is no significant difference in annual wind capacity between these two policies, developing wind power responds to high electricity demand and high oil dependence	Further studies should examine how wind energy development impacts other renewable energy industries, and what are the structural reasons for a country to adopt one of these two policies (FIT or RPS)

13. Managing structural ambiguity in collaborative R&D policy making

Miriam Wolf and Terence A. Wilkins

INTRODUCTION

Fostering innovation and supporting management of emerging technologies is not only a challenge on the company level but becomes even more complex when it comes to policy making. Throughout recent decades, states have been stepping back from their traditional role as a supplier of policies, and actors from outside the traditional policy-making system are increasingly engaging in the policy-making process (Hajer and Wagenaar, 2003). With this, companies, universities and civil society organizations are becoming political actors, directly involving in the policy elaboration process (Fischer, 2003; Rhodes 1996). Especially complex and technical policy areas such as research and development (R&D) policy are no longer subject to top-down decision making but have become processes of negotiation and consultation among a variety of actors (Kuhlmann, 2001; Jørgensen et al., 2006) who draw on different and sometimes competing frames of meaning, diverse interests and backgrounds (Hoffmann and Ventresca, 1999; Hajer and Wagenaar, 2003).

The European Union (EU) aims to foster European scientific and technological development and integration through its Framework Programmes for Research and Technological Development (FP), incentivizing cross-border partnerships, creating European 'togetherness' in science and technology, and interweaving and coordinating the efforts of actors and nation states (Caloghirou et al., 2004). To meet the needs of the heterogeneous community of actors, the European Commission (EC) depends on input from companies, universities, research institutes and other actors on developments and priorities in R&D throughout European sectors and nations (Jørgensen et al., 2006). Therefore, in the development of the Framework Programmes, the European Commission draws heavily on interaction with stakeholders from a variety of backgrounds. These interactions are often organized within relatively open and diverse policy networks (Rhodes, 1996; Jørgensen et al., 2006). Due to this, political activities of firms within Europe have changed considerably (Scherer and Palazzo, 2007). Stepping back from traditional, unilateral lobbying activities (Coen, 2007) they provide sources of information and expertise to policy makers and are involved in collaborative processes to elaborate upon policy priorities together with a broad range of other stakeholders (Boswell, 2008). Dealing with this heterogeneous environment poses new challenges for managers. They leave the comfort zone of organizations where the rules of the game and the context for action are relatively clear, and engage in polycentric and multilateral environments where multiple interests and backgrounds come together.

THEORETICAL BACKGROUND

Managers, their work, actions and decision making, have long constituted a pillar for understanding the behaviour of organizations (Simon, 1945). Managers lead by creating strategies and reshaping organizational structures (Chandler, 1962). They exercise their entrepreneurial competency by recombining resources that enable the growth of organizations (Penrose, 1959). Their actions are embedded in the wider patterns of organizational learning (Minzberg, 1979). Inspirational leaders set the strategic direction for companies (Burgelman, 2002) and middle managers exercise their autonomy to develop new organizational capabilities (Burgelman, 1994; Floyd and Lane, 2000). This literature and the consecutive scholarly work explicitly or implicitly assumes that managers operate within well-defined boundaries of organizations and that the consequences of their actions impact upon competitive dynamics in a landscape constituted of buyers, suppliers and rival firms.

Nevertheless, a growing stream of literature asserts that managers have to engage with a wider set of stakeholders that goes well beyond the competitive market environment (Donaldson and Preston, 1995). This scholarly work emphasizes the importance of a deeper engagement with non-market stakeholders and explores how firms engage with governments in order to influence policy development and regulatory frameworks (Hillman et al., 2004). Recently, the notion of a sharp division between firms and the non-market environment in general, and governments in particular, has been challenged by Scherer and Palazzo (2008, 2011). They argue that we need a fresh view of the role of firms in society and describe this phenomenon as 'politicization of the corporation' (Scherer and Palazzo, 2007). Relations between firms and the non-market environment have become polycentric and multilateral, and hierarchical relations are often replaced by network-based processes of coordination (Newman, 2005). Although this literature suggests that firms are becoming political agents, it remains mostly silent on the almost inevitable consequence that not only organizations as collective entities, but also their managers, take a role of political actors. Managers engaging in a political context see themselves confronted by an ambiguous structural environment where 'there are no clear rules and norms according to which politics is to be conducted and policy measures are to be agreed upon. To be more precise, there are no generally accepted rules and norms according to which policy making and politics is to be conducted' (Hajer, 2003: 175). Action takes place next to or across established social orders, challenging rules, norms and the logic of appropriateness (March, 1995) of the actors involved. There are no predefined rules determining who is responsible, who has authority over whom, what sort of accountability is to be expected, and thus, not only the policy content itself, but also the rules of the game themselves are called into question (Fischer, 2003; Hajer and Wagenaar, 2003).

Prevalent theories of management hardly provide a framework to understand managerial action in ambiguous structural environments. Focusing on the individual organization there is little space to consider collaboration, multi-party and multi-signification aspects (Lawrence et al., 1999), institutional structures other than homogeneous, stable and coercive (DiMaggio and Powell, 1983; Meyer and Rowan, 1977), and mechanisms of coordination other than hierarchy or autonomy (Lawrence et al., 1999; Siggelkow and Rivkin, 2005). Authority structures are assumed to be defined, interests to be congruent within the organization and thus, action is conceptualized as guided by a rather stable structural environment. The manager in this context acts as a system controller, who plans, organizes and leads within an organization (Klijn, 1996).

In this chapter, we explore management under structural ambiguity and address the question of how we may understand managerial agency differently in this context. Under structural ambiguity, guidance for action is low, increasing the likelihood of creative and strategic agency (Hajer, 2009; Kostova et al., 2008). As actors are exposed to a multiple, fluid and nascent structural environment, they develop, reshape and manipulate shared rules of the game in ongoing processes of social construction (Berger and Luckmann, 1966). Within such an environment, it may be more beneficial to conceptualize managers as political actors, who deliberately try to influence, direct and shape interactions and thus the construction of the social reality. Table 13.1 shows the frictions between prevalent theories of management, presupposing stable structural environments and management under structural ambiguity. While organizational and management theories in general and neo-institutional theory in particular

Table 13.1 Frictions between management under stable institutional environments and management under structural ambiguity

	Management under stable structural conditions	Management under structural ambiguity
Structural rules	Homogeneous, stable, coercive	Multiple, fluid, nascent
Agency	Agent-centric	Distributed
Process	Punctuated equilibrium	Ongoing construction
Coordination	Hierarchy, autonomy	Political negotiation
Interest	Congruent	Pluralistic
Construction of meaning	Homogenization	Bricolage

provide a very sound basis to understand column one, they struggle to conceptualize the same categories in the context of structural ambiguity.

Building on neo-institutional theory and its foundations in structuration theory (Giddens, 1984) and social constructionism (Berger and Luckmann, 1966), we explore the relationship between structure and agency as a fluid, ongoing process linked through a level of interaction (Blumer, 1966). Based on this, we develop a process model exploring managerial agency under heterogeneous and ambiguous conditions.

METHODOLOGY

Research Setting

In 2004, the European Commission emphasized the need to define R&D priorities by coordinating across European, national and regional levels and getting together a wide range of public and private stakeholders to work on 'a number of strategically important issues with high societal relevance [for] achieving Europe's future growth, competitiveness and sustainable development objectives' (European Commission, 2004). This, they pointed out, should be the task of so-called European Technology Platforms (ETPs) which would unite stakeholders around a common vision to define a strategic research agenda and mobilize a critical mass of research and innovation effort for particular industries. The key elements of European Technology Platforms were addressing challenging issues where Europe's future growth, competitiveness and sustainable development depend on major research and technological advances and producing a shared vision of all stakeholders (industry, public authorities, research community, regulators, civil society, operators, users and consumers) (European Commission, 2004).

Although Technology Platforms were initiated by the EC, they were independent entities and neither owned nor managed by the European Commission. Some of them evolved as loose networks gathering in annual meetings, while others have established legal structures with membership fees (CORDIS, 2012). The elaboration of priorities and action plans to address them was expected to provide a major input for the development of the Commission's proposals for FP7 and the Commission was 'convinced of the great potential benefits from the activities of the technology platforms and, in particular, of the appropriateness of defining industrial research needs in this consensual and bottom-up manner' (European Commission, 2005). Thus, ETPs were initiated to represent entire sectors and their stakeholders, rather than individual actor groups or specific policy issues within the sector. This complexity makes ETPs particularly interesting and raises questions about how they developed and how they managed to work together across differences in interests and backgrounds.

Research Design and Data Analysis

In order to understand managerial agency under structural ambiguity we adopt a qualitative case study approach about a particular Technology Platform, which we shall call WeETP. We gathered data from multiple sources: interviews, archival documents and observations, triangulating them within and across levels of agency, interaction and structural rules. We conducted 40 semi-structured interviews with actors involved in the Technology Platform. The interviewees were initially sampled to represent the breadth of actors involved (for example, academia, industry, policy makers). At a later stage, we focused on the actors appearing as key actors and coordinators within the platform. We reviewed approximately 2000 pages of documents (for example, commission documents, platform documents, social media postings, PowerPoint presentations). Additionally, Mirian Wolf shadowed the platform in action for three months from March to May 2012 (Czarniawska, 2007) and attended stakeholder events over a time period of two years.

Our analysis was conducted by travelling back and forth between data, literature and emerging concepts (Locke, 2001). The raw data were categorized, linked to emerging themes and aggregated into a theoretical framework in an iterative manner.

FINDINGS

The Development of WeETP

As one of the first European Technology Platforms, WeETP was launched in 2004 as a joint initiative between two major European industry associations, and four national science and industry associations. As representative bodies for industry and research the associations had been approached by the Commission and received two-year funding for setting up a representative network to elaborate priorities and action plans.

The European Industry Association took a lead in setting up the platform, building on an existing R&D group within the association where a number of industry representatives were meeting regularly to discuss common concerns in R&D. While the industry association had previously been strongly focused on lobbying activities, taking a clear industry position and a rather reactive mode of action, the commitment to WeETP demanded change towards broader representation, collaboration and consensus building. A manager from one of the leading companies of the sector was seconded into the industry association as a platform secretary. He and the R&D group within the association formed a core group that started to push the development of the platform by building up a broader network and propelling the elaboration of the policy documents.

As the representation of a broad and diverse number of actors was a basic prerequisite to gain political legitimacy with the Commission, a variety of stakeholders were targeted by this group and encouraged to join the platform, emphasizing the opportunities to influence policy, acquire funding or build up networks. Responsibilities were divided according to the existing industry structures and focus areas of the members of the R&D group. Each of them engaged a 'Sherpa', responsible for the elaboration of the respective topic areas. From this, a working group structure and an editorial team emerged. Each of the working groups engaged in building up a network which operated quite autonomously, although linked through close cooperation between the editorial team who agreed on common approaches to organize the input. Based on this working structure, the key documents were elaborated: the Vision Document, the Strategic Research Agenda and the Implementation Action Plan.

While the Strategic Research Agenda and the Implementation Action Plan were very technically oriented documents, the Vision Document aimed to define the direction of the sector until 2025. The elaboration of the vision was complicated through conflicts of interest and opinions between the core group members, which led to a final document which was very broadly written to accommodate multiple interpretations. The other two documents were more technically specialized and, therefore, responsibilities could be more easily split between the working groups. Through this, conflict was avoided as the topic areas could exist alongside each other. The working groups were rather self-sufficient, each setting up their own network, organizing meetings for brainstorming, collecting ideas and finally setting priorities for policy. With this, the members of the core team had a certain control over the development of the priorities, while at the same time the collaborative work provided them with a stock of knowledge about the developments and interests of a broad range of other actors in the field. Thus, the elaboration of the documents was a constant process of expansion and contraction between consulting broader groups and decision making within a smaller group. The documents were presented to the broader community at stakeholder events, and by 2007 all three documents had been officially published. Shortly after, FP7 started and the funding from the Commission to organize the Platform ran out.

The core group decided to keep the platform up and running with their own funds, as the project calls from the Commission were now being published and the stakeholders involved aimed at applying for the project calls that were related to the proposals made in the Implementation Action Plan. Three major visionary projects with a strong core group involvement, and a large number of other projects, emerged from the proposal and between 2007 and 2009 research projects funded to the tune of over €900 million were 'inspired by WeETP thinking'. With this, WeETP had accomplished a considerable impact on the funding priorities within FP7. Throughout the years of its operation, WeETP had acquired a quite recognized position as a responsible and representative partner for the Commission and a source of information about the development, interests and priorities of the sector. Interaction between the Commissioners and the core group actors had become quite frequent.

Building on these successes, the core group aimed to push the platform further towards the development of the next Framework Programme, Horizon 2020. At that time, some considerable changes occurred. As the funding had run out the role the platforms would play in the future was quite undefined. Many of the people who had played a major role in the first phase of the platform changed jobs and also the team within the association went through changes in staff, although the organizational composition stayed considerably stable. Most of the companies which were engaged in the core group had seconded new staff into the association to continue WeETPs work, although under uncertainty as to how its future would look. With this, a significant change in the targeted stakeholders as well as in the strategy occurred. In the first phase of WeETP emphasis was set on research, and academic and research actors were playing a quite significant role. The new phase was dominated by industry actors emphasizing innovation and market potential rather than the role of knowledge and research. While the earlier WeETP had been 'technology driven, concentrating mainly on knowledge creation, and measuring success from the number of grants received or how much funding support has taken place', the intention was now 'to emphasise much more strongly innovative outputs, such as demonstrator projects that highlight novel options and technologies' (Interview, Manager WeETP). With this, the creation of economic value was put at the top of the agenda and research in this context was a means to corresponding ends.

While this change in emphasis was taking place, it also became clear that the working structure WeETP had developed may not be operable in the context of the new Framework Programme. Therefore, the time between 2009 and 2011 was characterized by strong uncertainty. The core group was closely observing the Commission's work towards the new Framework Programme. As the orientation of the new FP was beginning to crystallize, it became clear that the new funding period would be oriented much more along what have been called societal challenges and multidisciplinary as well as cross-sectoral collaboration. With this, the sector-driven approach the platform had developed was incompatible with the new developments. The core group engaged in analysing the new policy directions, and building working groups around initiatives emerging with the new policy directions. Those were now rather issue-driven than sector-driven and required to build up a new network that would be able to integrate with the new developments. Similarly to when the platform was first set up, the core group strategically partnered with stakeholders having links with the particular issues emerging, elaborating on topics and priorities and finally elaborating proposals in small teams, closely linked to the core group.

While the new Framework Programme was still in preparation at time of writing (March 2013), it is yet uncertain how strongly WeETP will impact upon the new Framework Programme. Certainly, the platform has produced significant policy impacts over the last eight years of its existence (2004–2012), and was strongly driven by a core group coordinating and representing a broad range of actors.

Actor Groups

As illustrated above, two major actor groups were part of the platform. On the one hand, the core group was made up of actors who coordinated stakeholders as well as the development of the specific policy issues in a rather permanent manner. Members of this group were in close contact with each other and belonged to the most influential organizations in the field.

The broader stakeholder groups were involved to varying degrees. Some of them were operating as permanent members of working groups, while others were only sporadically attending stakeholder meetings. They were less familiar with each other and with WeETP as a whole, and joined and left the platform activities based on their own, often specific or technical interests. The broader stakeholder group was made up of industry actors (often smaller, less influential companies), academics and research organizations and other actors with an interest in specific topics addressed by WeETP.

While the core group actors were very well informed about the composition of the field and where expertise and interests are located, the members of the broader stakeholder group were principally specialized experts. This differentiation in knowledge served as a basis for exchange within the platform. While the core group aimed to gain political legitimacy by representing the field, members of the broader group often asserted that they were less interested in influencing policy but rather more in networking, knowing who does what, understanding the funding opportunities or involving in the development of particular technologies. Thus, the core group was trading field-level expertise to the broader stakeholder groups as well as the policy makers. This relationship is shown in Figure 13.1.

We also found that the division into two actor groups was a basis for

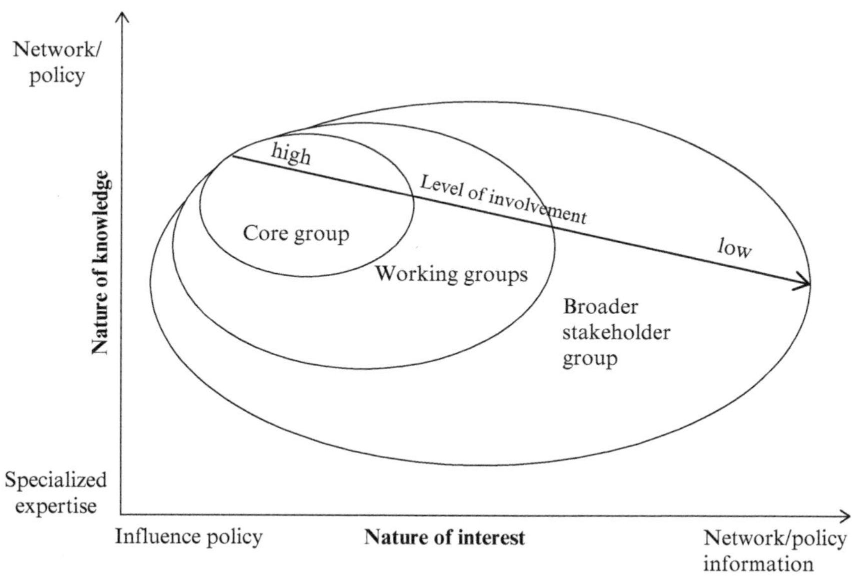

Figure 13.1 Stakeholder groups, knowledge and interest

the emergence of organizational structures as well as issue interpretations within the platform. While the core group was actively involved in carving out organizing structures, the broader stakeholders mirrored them by playing according to the 'rules of the game' set by the core group. In the following, we particularly focus on theorizing upon how the core group coordinates this complex and heterogeneous environment.

TOWARDS UNDERSTANDING MANAGERIAL AGENCY UNDER STRUCTURAL AMBIGUITY

Orchestrating Backgrounds and Interests

As the previous account has shown, the industry association took the role of an institutional entrepreneur, as an actor having an 'interest in particular institutional arrangements and who leverage resources to create new institutions or to transform existing ones' (Garud et al., 2007; Battilana et al., 2009). The association engaged as a creative agent whose social position and skills enabled it to identify opportunities in the field and to mobilize resources to effect the desired change (Fligstein, 1997). The association initiated a fundamental change in how the sector was dealing with policy making. This transformation was initiated by the targeted involvement of actor groups which had previously not, or only marginally, been working with the association. As those actors came from a variety of backgrounds, they had different 'socially constructed, historical patterns of material practices, assumptions, values, beliefs, and rules by which [they] produce and reproduce their material subsistence, organize time and space, and provide meaning to their social reality' (Thornton and Ocasio, 2008: 101) – different institutional logics. In order to recruit the potential allies, institutional entrepreneurs must be able to theorize the project (Figure 13.2, 1a) in such a way that it matches the values, interests and problems of actors from different institutional backgrounds. They must be able to manage meaning and influence 'how others understand, frame and make sense of established or new structures, practises and behaviours' (Zilber, 2007: 1036). For this, they need considerable knowledge about the backgrounds of these actors and how involvement would resonate with their goals and values.

At the same time, the entrepreneur is involved in a process of political selection (Figure 13.2, 1b) based on knowledge about the interests of the different actors. As political legitimacy is gained through the representation of a broad spectrum of interests (Jørgensen et al., 2006), institutional entrepreneurs in a political environment may pick and mobilize actors

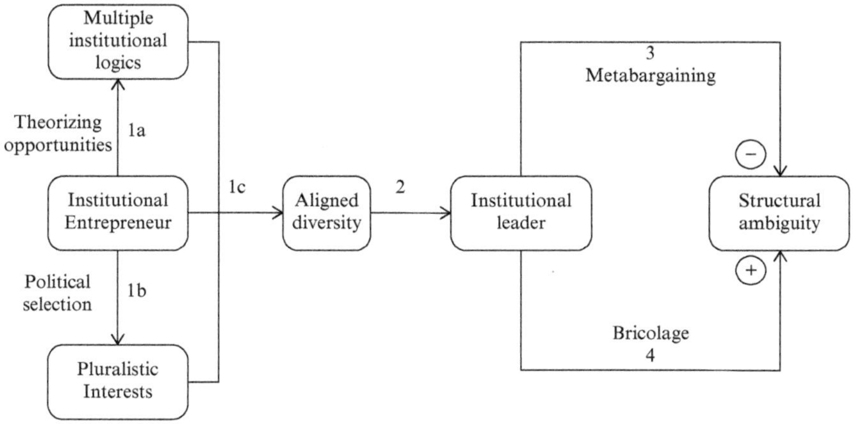

Figure 13.2 Process model for understanding managerial agency under structural ambiguity

representing a variety of different (or even conflicting) interests. With interests being pluralistic the actors may have an adversarial relationship to one another (Ansell and Gash, 2007). Awareness about contradictions is crucial for the entrepreneur to appraise which actors may, despite conflicting interests, be able to engage in collaboration. Policy-making collaborations are consensus-oriented and thus the likelihood of reaching some degree of consensus between the different actors is fundamental for the collaboration to emerge (Ansell and Gash, 2007; Innes and Booher, 2007; Rhodes, 1996).

Through theorization of opportunities and political selection the institutional entrepreneur engages in the orchestration of what we call an 'aligned diversity'. The latter maximizes the representation of interests and backgrounds but, at the same time, maximizes the likelihood of the corresponding actors to become aligned or strive to discover areas of agreement and interdependency (Ansell and Gash, 2007; Rhodes, 1996). In contrast to neo-institutional social movement accounts, where a certain homogeneity of interests within a collective and a mobilization of bias against another group is the basis of collective action (Hensmans, 2003; Levy and Scully, 2007), in political collaboration the plurality of interests is indispensable and thus, assimilation of interests is not necessarily desirable as it affects representativeness and thus political legitimacy. The collaborative, political environment presupposes diversity and, as shown above, diversity is a precondition for its existence. Thus, the institutional entrepreneur engages in the orchestration (Figure 13.2, 1c) of an aligned diversity which initiates interaction between varieties of actors.

From an Aligned Diversity to Institutional Leaders

Once the aligned diversity starts interacting, new practices emerge which the institutional entrepreneur can influence only to a limited degree. Agency at this stage becomes more distributed (Garud and Karnoe, 2003) and the engagement of the institutional entrepreneur becomes one of multiple strands of action within the group (Schütz, 1970). As we cannot presuppose an institutional vacuum before the orchestration, the aligned diversity reproduces pre-existing patterns of power, influence and interaction. Nevertheless, due to the reassembling of backgrounds and interests, the collaborative setting starts to produce new interaction patterns and, thus, a process of social structuration (Berger and Luckmann, 1966; Giddens, 1984). The inherent structural ambiguity forces the actors to imagine alternative possibilities and practices (Emirbayer and Mische, 1998). Each of the actors aims to satisfy their background and interests while aligning with the other actors to produce a common output. Thus, they engage in a conscious creation process (Barley and Tolbert, 1997) where they deliberately negotiate the rules of the game through playful interaction: trying out what works and what does not work in this social context. With this, they build up a shared reality (Weber and Glynn, 2006) of how interaction within the platform can be organized. The common ground for interaction emerging within the group gives them the capacity to socialize further actors into the emerging system (Berger and Luckmann, 1966). Those are orchestrated in a similar political process, as described above. Nevertheless, the actors now taken in by the institutional leaders are not confronted with the same degree of structural ambiguity. They are socialized into a system of rules that is passed on, and thus those actors tend to engage in a 'taken-for-granted' mode of interaction. They work within the given structures and, with this, mirror the organizational structures, roles and rules established by the group of institutional leaders. This leads to a 'thickening' of the system (Berger and Luckmann, 1966). Thus, the group of institutional leaders initiates a feedback loop engaging in structuring social rules, while a broader group of actors engages in mirroring them.

Managing Ambiguity: Metabargaining and Bricolage

Drawing on a conception of conscious and intentional agency (Blumer, 1966; Emirbayer and Mische 1998), we assume that the broader group of actors engaging in mirroring behaviour will only take this role as long as they see their interests satisfied. Therefore, knowledge about interests and backgrounds again becomes crucial for the group of institutional leaders in order to maintain control over the rule-setting within the group.

Klijn (1996) and Quack (2007) propose to differentiate actions in multi-actor settings in strategic and more incidental actions in relation to the rules of the game. In this context, meta-bargaining is defined as deliberate strategizing to influence the multiplex institutional environment through practices aiming to change interaction rules and organizational arrangements while also influencing goals and perceptions. On the other hand, bargaining refers to 'everyday muddling through' (Quack, 2007) in a complex environment. Actors engaging in bargaining take the rules of the game, such as decision making, general goals and organizational structure as given, and strategize within this given set of rules. Thus, while the group of institutional leaders will engage predominantly in creative metabargaining practices, another, larger group of actors within the collaboration will engage predominantly in more iterative and practical evaluative bargaining (Emirbayer and Mische, 1998) and, with this, mirror the social structuration processes driven by the metabargaining leaders. The interlocking of metabargaining and bargaining actions leads to the emergence of shared rules of the game and thus to a consolidation of structural rules (Figure 13.2, 3). As the basic rules of interaction become more and more taken for granted, roles become detached from the individuals who established and negotiated them and can thus be taken over by new representatives (Berger and Luckmann, 1966). Structural ambiguity is decreased through this process.

Nevertheless, management under structural ambiguity implies managing divergent and even rival backgrounds and interests (Hajer, 2003) as each party within the political collaboration maintains its distinctive identity. The distinctiveness is important not only for political legitimacy as such but also for the institutional leaders in securing their position. They build up a toolkit (Swidler, 1986) of knowledge about the specific interests and backgrounds that the actors bring to the table. This toolkit is a principal source of power in the process. In this context, complexity, diversity and contradictions provide the institutional leaders with material to manage the ambiguous environment. Drawing on their knowledge of the context, they continuously rearrange and recombine the different structural backgrounds in order to meet the needs of day-to-day interactions within the collaboration. In our model, we refer to this as ongoing bricolage (Figure 13.2, 4) (Levi-Strauss, 1966; Innes and Booher, 2007). Bricoleurs play with heterogeneous concepts, strategies and actions brought forward by different actors and rearrange them in order to fit a particular context at hand. Although the outcome of this may be qualitatively different from the material used for bricolage, the latter is not manipulated as such, and still recognizable within the new arrangement. Thus, it can be disassembled and rearranged and actors within the collaboration will be able to recognize their 'part' in the arrangement. While the Levi-Straussian bricoleur is a quite solitary figure who collects his material without a specific purpose, our view is more strategic and interactive (Garud and Karnøe, 2003). Bricolage is hard to foresee or plan, and thus, if practised on an ongoing basis, it sustains structural ambiguity. It is, nevertheless, highly dependent on the involvement of a diversity of actors who provide local knowledge (Garud and Karnøe, 2003). Rather than a clear end, the bricoleur has only a rough idea what the final outcome would be as the latter depends on the different inputs and how they can be arranged. The more diverse the material, the more resources the bricoleur has to draw on (Innes and Booher, 2007).

Thus, the group of institutional leaders acts strategically by accumulating knowledge about the interests and backgrounds of the other actors involved. Rather than as an obstacle, the diversity of interests and backgrounds serves as a resource for them and the constant reassembling of this accumulated material sustains structural ambiguity, and thus space for agency. This allows adjusting to changes and safeguarding the powerful role of the group of institutional leaders as managers of ambiguity.

Implications

As our model suggests, engaging in collaborative R&D policy making confronts managers with an ambiguous environment and the challenge to collaborate across different backgrounds and interests. This has consequences for theory as well as for practice.

On the one hand, the 'politicization of the corporation' leads to a need to theorize differently about the environment managers are operating in, from a focus on stability and homogeneity towards a more dynamic, distributed and pluralistic environment where agency enfolds in distributed action (Garud and Karnøe, 2005), constantly assimilating and at the same time sustaining ambiguities. On the other hand, it brings new practical challenges for managers. In order to gain influence in political processes, they have to engage in new practices and build up knowledge about other actors' interests and backgrounds, about where they overlap, where they contradict and how they can be coordinated. The ability of managers to engage in bricolage and metabargaining is highly likely to determine their success in influencing the political process. Therefore, managers engaging in policy initiatives should, while providing specialized expertise, raise their awareness of backgrounds and interests in their broader environment and emphasize overlaps and compatibilities in interests rather than conflicts and incompatibilities. Last but not least, engagement of private actors in policy making also brings new challenges for policy makers in how to monitor and understand how power and influence unfold in such bottom-up initiatives. As other authors have noted, accountability for decision making within policy networks is dispersed and thus hard to trace (Rhodes, 1996), and actors who are skilled in metabargaining and bricolage may have considerable advantages in the process. However, these advantages are blurred through the collaborative setting and therefore hard to trace back from final documents. Thus, representativeness of policy networks should be measured not only in terms of the diversity of actors but also in terms of their potential ability to influence. This capacity has a social dimension determined by social position and skills for metabargaining and bricolage, as well as a material dimension determined by the access to financial means to guarantee ongoing engagement in policy initiatives. As our study has shown, industry actors may be more likely to maintain involvement through their own funding, while academic actors may depend on public support. Last but not least, the set-up of networks by private actors tends to reproduce existing patterns of power and influence. Therefore, it is important to monitor the process of stakeholder involvement and to make sure that stakeholders from potentially weaker positions are assisted not only financially but also socially, by means of building up skills for metabargaining and bricolage. Through this, less powerful actors may gain considerable influence in the core group and thus the setting of the rules of the game of the collaboration.

CONCLUSION

While this chapter draws on a single case study and is thus limited in terms of generalizability, our aim was to direct attention towards theoretical and practical challenges and questions emerging from collaborative R&D policy making, which is gaining major importance on the European (Hajer and Wagenaar, 2003) but also the international level (Hardy and Maguire, 2010; Hoffman and Ventresca, 2002). As our empirical and theoretical findings indicate, the changing engagement of companies with their non-market environment, and in particular the increasing involvement of managers in collaborative processes to elaborate R&D policy priorities with a wider set of stakeholders, constitutes a variety of new challenges. Further research in this area, we believe, will contribute to a better understanding of the practical, normative and theoretical implications of this phenomenon.

ACKNOWLEDGEMENT

This chapter benefits from the authors' involvement in the Marie Curie Initial Training Network funded by the European Commission through the 7th Framework Programme (ManETEI, GA PITN-GA-2009-238382).

REFERENCES

Ansell, C. and Gash, A. 2007. Collaborative governance in theory and practice. *Journal of Public Administration Research and Theory* 18: 543–571.

Barley, S.R. and Tolbert, P.S. 1997. Institutionalization and Structuration: Studying the Links between Action and Institution. *Organisation Studies* 18(1): 93–117.

Battilana, J., Leca, B. and Boxenbaum, E. 2009. How actors change institutions: towards a theory of institutional entrepreneurship. *Academy of Management Annals* 3(1): 65–107.

Berger, P. and Luckmann, T. 1966. *The Social Construction of Reality: A Treatise in the Sociology of Knowledge*. New York: Doubleday Anchor.

Blumer, H. 1966. Sociological implications of the thought of George Herbert Mead. *American Journal of Sociology* 71(5): 535–544.

Blumer, H. 1996. *Symbolic Interactionism: Perspective and Method*. Englewood Cliffs, NJ: Prentice-Hall.

Boswell, C. 2008. The political functions of expert knowledge: knowledge and legitimation in European Union immigration policy. *Journal of European Public Policy* 15(4): 471–488.

Burgelman, R.A. 1994. Fading memories: a process theory of strategic business exit in dynamic environments. *Administrative Science Quarterly* 39(1): 24–56.

Burgelman, R.A. 2002. Strategy as a vector and the inertia of coevolutionary lock-in. *Administrative Science Quarterly* 47: 325–357.

Caloghirou, Y., Vonortas, N.S. and Ioannides, S. (eds). 2004. *European Collaboration in Research and Development: Business Strategies and Public Policy*. Cheltenham, UK and Northampton, MA, USA: Edward Elgar.

Chandler, A.D. 1962. *Strategy and Structure: Chapters in the History of the American Industrial Enterprise*, Cambridge, MA: MIT Press.

Coen, D. 2007. Empirical and theoretical studies in EU lobbying. *Journal of European Public Policy* 14(3): 333–345.

CORDIS. 2012. European Technology Platforms. Available at: http://cordis.europa.eu/technology-platforms/home_en.html (accessed 20 December 2012).

Czarniawska, B. 2007. *Shadowing and Other Techniques for Doing Fieldwork in Modern Societies*. Malmö: Liber.

DiMaggio, P.J. and Powell, W.W. 1983. The iron cage revisited: institutional isomorphism and collective rationality in organizational fields. *American Sociological Review* 48(2): 147–160.

Donaldson, T. and Preston, L.E. 1995. The stakeholder theory of the corporation: concepts, evidence, and implications. *Academy of Management Review* 20(1): 65–91.

Emirbayer, M. and Mische, A. 1998. What is agency? *American Journal of Sociology* 103(4): 962–1023.
European Commission. 2004. European Technology Platforms. Knowledge for growth. Available at ftp://ftp.cordis.europa.eu/pub/technology-platforms/docs/tp_leaflet_en.pdf (accessed 20 December 2012).
European Commission. 2005. Status Report. Development of Technology Platforms. Available at ftp://ftp.cordis.europa.eu/pub/etp/docs/etp1streport_en.pdf (accessed 20 December 2012).
Fischer, F. 2003. *Reframing Public Policy: Discursive Politics and Deliberative Practices*. New York: Oxford University Press.
Fligstein, N. 1997. Social skill and institutional theory. *American Behavioral Scientist* 40(4): 397–405.
Floyd, S.W. and Lane, P.J. 2000. Strategizing throughout the organization: managing role conflict in strategic renewal. *Academy of Management Review* 25(1): 154–177.
Garud, R., Hardy, C. and Maguire, S. 2007. Institutional entrepreneurship as embedded agency: an introduction to the Special Issue. *Organization Studies* 28(7): 957–969.
Garud, R. and Karnøe, P. 2003. Bricolage versus breakthrough: distributed and embedded agency in technology entrepreneurship. *Research Policy* 32: 277–300.
Garud, R. and Karnøe, P. 2005. Distributed agency and interactive emergence. In Floyd, S.W., Roos, J., Jacobs, C.D. and Kellermanns, F.W. (eds), *Innovating Strategy Processes*. Oxford: Blackwell Publishing.
Giddens, A. 1984. *The Constitution of Society*. Oxford: Polity Press.
Hajer, M.A. 2003. Policy without polity? Policy analysis and the institutional void. *Policy Sciences* 36: 175–195.
Hajer, M.A. 2009. *Authoritative Governance: Policy Making in the Age of Mediatization*. Oxford: Oxford University Press.
Hajer, M.A. and Wagenaar, H. 2003. *Deliberative Policy Analysis: Understanding Governance in the Network Society*. Theories of Institutional Design. Cambridge: Cambridge University Press.
Hardy, C. and Maguire, S. 2010. Discourse, field-configuring events, and change in organizations and institutional fields: narratives of DDT and the Stockholm Convention. *Academy of Management Journal* 53(6): 1365–1392.
Hensmans, M. 2003. Social movement organizations: a metaphor for strategic actors in institutional fields. *Organization Studies* 24: 355–381
Hillman, A.J., Keim, G.D. and Schuler, D. 2004. Corporate political activity: a review and research agenda. *Journal of Management* 30(6): 837–857.
Hoffmann, A.J. and Ventresca, M.J. 1999. The institutional framing of policy debates: economics versus the environment. *American Behavioural Scientist* 42(8): 1369–1391.
Hoffman, A.J. and Ventresca, M. 2002. *Organizations, Policy and the Natural Environment. Institutional and Strategic Perspectives*. Stanford, CA: Stanford University Press.
Innes, J.E. and Booher, D.E. 2007. Consensus building as role playing and bricolage. *Journal of the American Planning Association* 65(1): 9–26.
Jørgensen, K.E., Pollack, M. and Rosamond, B. (eds). 2006. *The Sage Handbook of European Union Politics*. Thousand Oaks, CA: Sage.
Klijn, E.H. 1996. Analyzing and managing policy processes in complex networks:

a theoretical examination of the concept policy network and its problems. *Administration and Society* 28(1): 90–119.
Kostova, T., Roth, K. and Dacin, M.K. 2008. Institutional theory in the study of multinational corporations: a critique and new directions. *Academy of Management Review* 33(4): 994–1006.
Kuhlmann, S. 2001. Future governance of innovation policy in Europe – three scenarios. *Research Policy* 30: 953–976.
Lawrence, T.B., Phillips, N. and Hardy, C. 1999. Watching whale watching: exploring the discursive foundations of collaborative relationships. *Journal of Applied Behavioral Science* 35(4): 479–502.
Levi-Strauss, C. 1966. *The Savage Mind*. New York: Oxford University Press.
Levy, D.L. and Scully, M. 2007. The institutional entrepreneur as modern prince: the strategic face of power in contested fields. *Organization Studies* 28(7): 971–991.
Locke, K. 2001. *Grounded Theory in Management Research*. Thousand Oaks, CA: Sage.
March, J.G. 1995. The future, disposable organizations and the rigidities of imagination. *Organization* 2(3): 427–440.
Meyer, J.W. and Rowan, B. 1977. Institutionalized organizations: formal structure as myth and ceremony. *American Journal of Sociology* 83(2): 340–363.
Minzberg, H. 1979. An emergent strategy of direct research. *Administrative Science Quarterly* 24: 582–589.
Newman, J. 2005. *Remaking Governance: Peoples, Politics and the Public Sphere*. Bristol: Policy Press.
Penrose, E. 1959. *The Theory of the Growth of the Firm*. Oxford: Oxford University Press.
Quack, S. 2007. Legal professionals and transnational law-making: a case of distributed agency. *Organization* 14(5): 643–666.
Rhodes, R.A.W. 1996. The new governance: governing without government. *Political Studies* 44(4): 652–667.
Scherer, A.G. and Palazzo, G. 2007. Towards a political conception of corporate responsibility: business and society seen from a Habermasian perspective. *Academy of Management Review* 32: 1096–1120.
Scherer, A.G. and Palazzo, G. 2008. Corporate social responsibility, democracy, and the politicization of the corporation. *Academy of Management Review* 33: 773–775.
Scherer, A.G. and Palazzo, G. 2011. The new political role of business in a globalized world: a review of a new perspective on CSR and its implications for the firm, governance, and democracy. *Journal of Management Studies* 48(4): 899–931.
Schütz, A. 1970. *On Phenomenology and Social Relations*. Chicago, IL: University of Chicago Press.
Siggelkow, N. and Rivkin, J.W. 2005. Speed and search: designing organizations for turbulence and complexity. *Organization Science* 16(2): 101–122.
Simon, H. 1945. *Administrative Behavior*. New York: Free Press.
Swidler, A. 1986. Culture in action: symbols and strategies. *American Sociological Review* 51(2): 273–286.
Thornton, P.H. and Ocasio, W. 2008. Institutional logics. In Greenwood, R., Oliver, C., Sahlin-Andersen, K. and Suddaby, R. (eds), *Organisational Institutionalism*. London: Sage.

Weber, K. and Glynn, M.A. 2006. Making sense with institutions: context, thought and action in Karl Weicks theory. *Organization Studies* 27(11): 1639–1660.
Zilber, T.B. 2007. Stories and the discursive dynamics of institutional entrepreneurship: the case of Israeli high-tech after the bubble. *Organization Studies* 28(7): 1035–1054.
Zilber, T.B. 2011. Institutional multiplicity in practice: a tale of two high-tech conferences in Israel. *Organization Science* 22(6): 1539–1559.

PART IV

Utilizing new technologies for innovation and learning

INTRODUCTION

Recent years have witnessed a growing interest in management through advanced communication technologies. In this part we have chapters that report how utilization of the new technologies can provide new capabilities that support management in both creating innovation and managing the dissemination of knowledge in organizations and innovation through these technologies. A growing stream of studies takes interest in how new concepts and technologies affect work within organizations and their performance. Among the key concepts in the academic literature are open innovation (Chesbrough, 2003) and co-creation (Vargo et al., 2008). Indeed, many organizations are experimenting with open innovation; however, the vast majority of them are large entities, often with a global presence, such as multinational consulting companies or national agencies.

In this regard Madiedo in Chapter 14 challenges our understanding of open innovation by examining its relevance for family-owned firms. In this chapter, the author explores the effect of two important factors, namely ownership and control, on the propensity of the family business to pursue open innovation initiatives. Indeed, open innovation is a risky proposition as it brings with it high coordination costs as well as the need to restructure the innovation process within the firm to accommodate idea initiation from external parties. In a family business where resources are scarce and often governed by family members rather than recruited managers, pursuing an open innovation strategy may increase the risk involved for both the management of the innovation process as well as appropriating the benefits from such engagement. It is suggested in this chapter that a unique characteristic of a family-controlled firm is to pursue decisions based on a socio-emotional element that incorporates the value the owner endows to their controlling position. The chapter offers a set of propositions that advance our understanding of the factors affecting the use and application of open innovation in family business.

Another key concept the academic literature has been interested in recently is the idea of co-creation of value (Füller, 2010). In Chapter 15, Pacauskas goes against the common conceptualization of co-located co-creation to examine the concept in virtual environments. In particular, the chapter takes interest in understanding how various collaborative and co-creation platforms help to bring in innovation from a wide audience of end-users (Peffers et al., 2003), and the factors affecting their performance. Indeed, while co-creation with co-located contributors has been studied rather extensively, research on virtual contributor communities is still limited. The chapter reviews the literature on the acceptance of online tools for co-creation, the motivating factors that incentivize individuals to contribute, and the effect of the firm on the contributor's involvement. The chapter concludes by drawing on the individual's intention versus willingness to contribute, subject to the hedonic nature of the technology. Furthermore, the chapter suggests that a scheme of incentives can in fact change the individual's tendency to contribute to an online co-creation platform. Last but not least, the firm's brand name has a major effect on the individual's tendency to contribute.

New technologies have also played a key role in examining how organizations cope with technological change. While numerous studies highlight the challenges organizations face by introducing new technological platforms to their employees, Alexiou in Chapter 16 explores the potential positive effects of changes. The chapter seeks to advance our understanding of how adversarial changes in the environment or organization can be made into opportunities, and how organizations can be made resilient to changes. The chapter develops the concept of resilience to change as a meta-capability through a three-phase model of change.

Recent developments in Web tools and social media platforms have affected the nature and intensity of interactions between organizations and their end-users and suppliers. Many firms have incorporated Facebook and LinkedIn into their Customer Support function in order to seek to benefit from ongoing interactions with their stakeholders. The effect of such transformations in the technologies supporting business functions on the firm's value chain and its response to market demand is as yet unclear. In the last chapter in this part, Chapter 17, Durgam assesses such transformation by exploring the emerging concept of social customer relationship management (SCRM) (Baird and Parasnis, 2011). The author builds on the recent transformation of Web tools including the introduction of Web 2.0 to argue that recent developments in Web tools are in fact creating a 'Social CRM' platform. The chapter examines the elementary conditions possessed by organizations required to benefit from a Social CRM platform. The chapter reviews the emergence of Web 2.0 as a social platform to accommodate far more complex interactions. The chapter discusses the process, people and technology aspects involved in shaping the involvement in the CRM strategy and goes on to offer a framework that accommodates Social CRM as an emerging platform.

In total the chapters offer an interesting view into the changes to be brought in order to speed up the dissemination of innovation and crowdsourcing innovation through new digital technologies. Many of the technologies are still in early phases of utilization, but offer interesting new possibilities for the learning and innovative organizations.

Ilan Oshri and Matti Rossi

REFERENCES

Baird, C.H. and Parasnis, G. (2011). From social media to social customer relationship management. *Strategy and Leadership*, 39(5), 30–37.

Chesbrough, H.W. (2003). *Open Innovation: The New Imperative for Creating and Profiting from Technology*. Cambridge, MA: Harvard Business Press.

Füller, J. (2010). Refining virtual co-creation from a consumer perspective. *California Management Review*, 52(52), 2.

Peffers, K., Gengler, C.E. and Tuunanen, T. (2003). Extending critical success factors methodology to facilitate broadly participative information systems planning. *Journal of Management Information Systems*, 20(1), 51–85.

Vargo, S.L., Maglio, P.P. and Akaka, M.A. (2008). On value and value co-creation: a service systems and service logic perspective. *European Management Journal*, 26(3), 145–152.

14. Open innovation in family firms: a matter of socio-emotional wealth

Juan Pablo Madiedo M.

INTRODUCTION

Competitive capabilities and resources shape the firm's market position and competitiveness (Barney, 1991; Teece et al., 1997; Krajewski et al., 2010). Practitioners and academics have recognized cost, quality, speed and flexibility as key elements for success in dynamic market situations (Petersen et al., 2003; Aberdeen Group, 2004). Still, because of specialization and constrained resource endowments, for developing and delivering a new successful product or service offer, firms cannot work by themselves and need to rely on business partners (Benton and Maloni, 2005; Koufteros et al., 2007; Mitra and Singal, 2008).

The literature has stated that working within an open innovation framework with suppliers and customers has a positive effect on the company's performance (Chesbrough, 2003, 2006a; Enkel et al., 2009). High levels of openness and the corresponding integration with business partners are usually related to greater responsiveness and resilience due to the increase of information visibility and of operational knowledge, and lower costs of doing business because of the stronger links that allow firms to target their customer requirements effectively, while leveraging their resources for reducing the count of non-conformities and their magnitude (Dyer, 1996; Grant, 1996; Kim, 2009). These benefits, however, do not come easily. Open innovation requires information sharing and coordination between the different companies' processes (Chesbrough, 2003, 2004, 2006a; Dahlander and Gann, 2010; Huizingh, 2011). And when the right relationship management mechanisms are absent and the business models of the partners are unaligned, the chances soar for free-riding, for hold-ups and for leakages (Cassia et al., 2012; Chesbrough, 2006a; Chesbrough and Schwartz, 2007; De Massis et al., 2012; McCarter and Northcraft 2007).

Regardless of the position of the company in the value chain (or network), as a buyer or as a supplier, most companies have to deal not only with the substantial potential benefits of open innovation but also with its potential downside. Benefit assessment and control mechanisms should be developed and set in place (Chesbrough, 2003, 2006b; Dahlander and Gann, 2010; Petersen et al., 2005).

In this analysis I build on the idea that although beneficial, the risky nature of open innovation may inhibit certain types of companies from working closely with other firms for developing such kind of innovation arrangements. The focus of my argument is the family firm (De Massis et al., 2012).

A family firm as defined by Chua et al. (1999) is a 'business governed or managed by members of the same family or a small number of families in a manner that is potentially sustainable across generations of the family or families'. Acknowledging that 'family firms are not simply a unique phenomenological setting but are significantly different from non-family firms' (Berrone et al., 2012: 258), I contend that the associated risk of sharing with, licensing or even selling to a business partner, relevant information about the market and the own needs, as well as that of setting up an organizational structure in order to support the business relationship, may have a different effect on the level of openness that characterizes the innovation process of family and non-family firms.

Management of family controlled firms is not necessarily driven exclusively by financial goals (Gómez-Mejía et al., 2007; Gómez-Mejía et al., 2010; Gómez-Mejía et al., 2012; Zellweger and Astrachan, 2008). The literature has provided evidence that family-controlled firms exhibit a preference to make decisions on the basis of the preservation of what several authors regard as socio-emotional endowment, meaning the preservation of 'the stock of affected-related value that a family derives from its controlling position in a particular firm' (Berrone et al., 2012: 259). In this sense, whether families care about control over the organization at the expense of business risk and better economic performance, looking forward to work in a close manner with other firms in the value chain (or network) and to jointly innovate for developing and delivering products to the customer may not necessarily be common practice for family firms.

While one might argue that success factors for open innovation as for several other managerial issues remain a subject of discovery, research and discussion, it is possible to consider that, in most cases, two major activities should take place when working within an open innovation framework. First, firms should accurately assess the business partner's technological capabilities and performance (including assets). And second, business partners should develop a cooperative relationship to assess and solve the information-sharing requirements of the identified business opportunities and business models (Chesbrough, 2003, 2006a; Chesbrough and

Schwartz, 2007; Gassmann et al., 2010; Petersen et al., 2005; Kim, 2009; Lee et al., 2010; Yeung et al., 2009; Jayaram and Tan, 2010).

Nevertheless, both activities create two particular situations that may hurt the family-controlled firms' socio-emotional endowment. First, families may lack the human capital to accurately evaluate the partners' potential and their subsequent performance. This would inevitably create the need to incorporate non-family managers into the firm's decision-making process. Delegating part of the authority and relying on the outsider's point of view and assessments diminishes the family's managerial control.

Second, strong collaboration between firms demands a fair amount of resources. Because of constrained endowments, companies may find their chances to successfully develop such kinds of relationships limited if they are to work with several business partners. Consequently, as non-family-controlled firms, family-controlled companies may find themselves dealing with fewer suppliers over time. If this is the case, firm dependency on these particular business relationships will grow stronger, and future endeavours will rely mostly on the supplier's or customer's agreement to pursue similar or supporting enterprises. Once the links are created and the operations are developed on the basis of the supplier–buyer association, family firms are likely to encounter 'exceptional challenges' (De Massis et al., forthcoming: 3) and the family's control over the company's fate would be significantly reduced.

On a different note, the impact of the loss of socio-emotional capital on the level of integration may be attenuated by the effect of trust and of the potential to exercise power. Increasing levels of trust and a substantial potential to exercise power can be associated with moderate to high levels of alignment between the business partners' interests and consequently of their business models. In any of these cases, strict control on the part of the family members is not necessary but only sufficient to involve other firms in the open innovation process.

Taking these ideas into consideration, and from the socio-emotional perspective, this analysis is set to explore how family-controlled firms face the challenge provided by such a new model of innovation and how they compare to their counterparts, the non-family-controlled companies. This research offers two key contributions to the literature on innovation management and family firms. First, it extends the previous innovation management literature related to open innovation, by including a behavioural factor related to family firms, the socio-emotional wealth effect. The analysis of this effect allows for developing collaboration relationships among business partners, following a different logic from that of the usual agreements that call for cost reduction on the basis of efficiency and task specialization. Second, it complements the literature on the management of family firms by including in the analysis of the relationship between business partners, the effect of perceived trust and power.

THEORETICAL DEVELOPMENT AND PROPOSITIONS

Both academics and practitioners agree that no matter what strategy firms pursue when trying to build a market position, resources and capabilities are fundamental for achieving sustainable competitive advantage (Barney, 1986, 1991; Jacobsen, 1988; Peteraf, 1993; Amit and Schoemaker, 1993; Teece et al., 1997). However, because of the increasing sophistication of customers and the growing demand for efficiency, companies have encountered difficulties in developing all the required activities that the final customer demands, in a profitable manner (Enkel et al., 2009; Koufteros et al., 2007; Lee et al., 2010). The pressure resulting from this situation has pushed companies towards specialization. Specialized companies are still required to deliver the product offer to their customers, and to do so they have to look for the necessary resources and capabilities outside the organizational boundaries (Chesbrough, 2003, 2006b; Chesbrough and Schwartz, 2007; Enkel et al., 2009; Das and Teng, 2000; Kang and Kang, 2009). In order to secure such resources in an economic, acceptable condition and to attain the competitive advantage that comes with their exploitation, suppliers and buyers need to open and to cooperate for creating new and radically improving all sorts of products and processes (Kopczak and Johnson, 2003; Yeung et al., 2009). By working together, companies expect to generate and appropriate higher value than those that decide not to pursue, or have no opportunities to engage in, such collaborative agreements (Dyer and Singh, 1998; Villena et al., 2009). Open innovation, then, defined by Chesbrough et al. (2006: 1) as 'the use of purposive inflows and outflows of knowledge to accelerate internal innovation, and to expand the markets for external use of innovation, respectively', is expected to have a positive effect on business performance.

Open Innovation and Family Firms

Because firms can only manage a constrained set of resources and engage in a limited number of business relations (Kouferos et al., 2007), collaborative innovation actions play a crucial role for obtaining high levels of technical and business performance (Chesbrough, 2003, 2006b; Chesbrough and Schwartz, 2007; Dahlander and Gann, 2010; Laursen and Salter, 2006; Lee et al., 2010; Song and Di Bendetto, 2008; Villena et al., 2009;

Flynn et al., 2010). The literature has established that collaborative innovation actions and integration with business partners results in faster product development processes (Dyer and Singh, 1998; Lee et al., 2010; Monckza et al., 1998; Petersen et al., 2003), higher quality levels (Henkel, 2006; Petersen et al., 2005; Tan, 2001), better relationship-building capabilities (Jayaram and Tan, 2010; Lee et al., 2010) and increased final customer satisfaction (Bleakley, 1995; De Jong and von Hippel, 2009).

While most firms would agree that the latter represent extraordinary benefits and strong mechanisms to deal with the challenges of a dynamic business environment in which innovation drives the corporate strategy (Aberdeen Group, 2004), the decision to develop open innovation activities and collaborate with other companies in the value chain (or network) may not be as straightforward for family-controlled firms.

According to previous research, family-controlled firms management tends to give priority to factors other than pure economic results (Berrone et al., 2010; Berrone et al., 2012; Cruz et al., 2010; Gómez-Mejía et al., 2007; Gómez Mejía et al., 2010; Gómez Mejía et al., 2012; Jones et al., 2008). In fact it is argued that family owners make decisions to protect what is referred to as their socio-emotional wealth (Gómez-Mejía et al., 2007; Gómez Mejía et al., 2010; Gómez Mejía et al., 2012; Jones et al., 2008; Zellweger and Astrachan, 2008). The socio-emotional wealth encompasses several factors, most of them strongly related to the 'family's affective needs' (Jones et al., 2008: 1009). As stated by Jones et al. (2008: 1009) in the literature review of their study, factors include:

> the unrestricted exercise of personal authority vested in family members (Schulze et al., 2003b); fulfillment of psychological needs for belonging, identification, and intimacy (Kepner, 1983; Westhead et al., 2001); the perpetuation of family emblems and values (Handler, 1990); maintaining the legacy of the founder and preservation of family dynasty (Casson, 1999); utility derived from placing trusted relatives in key positions; enhancing the family's social capital (Arregle et al., 2006); being part of a tight social group or clan; and the opportunity to be altruistic to family members (Schulze et al., 2003a)

Among those factors, and perhaps the most important of them, is the ability of family members to exercise control over the firm. Being able to do so allows family owners to use the firm as a mechanism to pursue their particular interests (Berrone et al., 2012; Gómez-Mejía et al., 2012; Zellweger et al., 2011).

As well as for non-family-controlled firms, open innovation may represent a large set of advantages for family-controlled firms. However, it also causes major threats to the family members' socio-emotional wealth because of potential losses of control over the firm's decision-making process. In most cases, the performance of the open innovation process depends on the right assessment of the business partner's capabilities and on the cooperative relation between the two firms to assess and solve the information-sharing requirements of the identified business opportunities (Chesbrough, 2003, 2006a; Chesbrough and Schwartz, 2007; Gassmann et al., 2010; Petersen et al., 2005; Koufteros et al., 2007; Jayaram and Tan, 2010). In that sense, the integration process menaces family control in two ways.

First, as a result of their own company's specialization, families may lack the human capital to accurately evaluate the potential benefits of adapting certain technologies and working more closely with certain business partners. Whether the family firm still decides to pursue the collaborative innovation process, it would be required to bring into the company non-family managers and to delegate to these managers part of, if not entirely, the decision-making responsibility regarding the relationship with the partner. Handing over authority to these family outsiders lessens the family's managerial control (Berrone et al., 2012; Gómez-Mejía et al., 2010). Second, cooperative efforts between the supplier and the buyer may require embedded ties to develop and support them. However, because of high demands and resource constraints, firms can develop only a few of these relationships (Laursen and Salter, 2006; Song and Di Bendetto, 2008). Family-controlled firms are no exception to this rule. Embedded ties are then related to business partners' base rationalization (Berrone et al., 2012; Koufteros et al., 2007). Dealing with fewer suppliers and customers reduces the company's negotiation power and increases company interdependencies. In that sense, the family firm would see its future endeavours subjected to those of a reduced number of its partners, and vice versa, diminishing the family's control over the company's fate.

Following these arguments on the threat that open innovation represents to the socio-emotional wealth of the family members that control the firm one could propose that:

Proposition 1: Innovation portfolios of family-controlled firms are characterized by lower levels of open innovation than those of non-family-controlled firms.

Effect of Moderators

Prior research has shown that trust and power play major roles in the development of business relationships. Both are thought of as affecting the perceived interest alignment of the exchange parties, and consequently as key factors for reducing transaction costs, inducing desirable behaviour and easing conflict resolution (Kanter, 1994; Doz, 1996; Leifer and Mills, 1996; Song and Di Bendetto, 2008).

On the one hand, and thinking of trust as 'the extent to which a firm believes that its exchange partner is honest and/or benevolent' (Geykens et al., 1998), one could argue that it 'would reduce the perception of risk associated with uncertainty and opportunistic behaviors' (Song and Di Bendetto, 2008). Following that rationale, when firms engage in business with a third party, the presence of trust eases the decision to share information for coordinating business processes and aligned business models, as expected when striving for open innovation. Trust levels will vary according to the outcome of the business transactions over time. One would expect that if trust levels increase, the buyer and the supplier will grow closer and the level of integration will be augmented. The opposite follows after the trust is somehow reduced or lost.

Given the interest in preserving their socio-emotional wealth, family members who control their own firms will tend to favour resource exchanges with the firms whose representatives they have developed trust in (Jones et al., 2008). In such a sense, one could propose that:

Proposition 2: The level of trust in a business partner positively moderates the relationship between family control and the level of open innovation that characterizes the business portfolio between both firms.

On the other hand, power, defined as the ability to influence a third party to modify its behaviour or decision-making processes in order to favour one's objectives (Benton and Maloni, 2005; Song and Di Bendetto, 2008; Yeung et al., 2009), can be employed to promote collaborative actions and to empower higher levels of performance (Benton and Maloni, 2005). Power obtained, whether from mediated sources (reward, coercive and legal legitimate) or from non-mediated sources (such as expert and referent-based, and traditional legitimate) (Maloni and Benton, 2000) may allow family firm owners to reshape the business exchange or the direction that the innovation process follows on a discretionary basis, and without the need to engage in complex negotiations with the supplier or the customer. If this is the case, the threats to their socio-emotional wealth are substantially reduced as the perception of loss of control over the firm's decisions and fate diminishes. One would expect, then, that as the family owners' perception of the potential to exercise power over the supplier–buyer relationship remains at considerably high levels, the level of collaboration between companies in an innovation business relationship will increase. Following that line of thought:

Proposition 3: The perceived power position over a business partner positively moderates the relationship between family control and the level of open innovation that characterizes the business portfolio between both firms.

Open Innovation and Performance

Although in several cases innovating in association with suppliers and customers has been proved to be beneficial, the features and the focus of the open innovation process that yield valuable results are contingent on the main goals that the company pursues (Chesbrough and Schwartz, 2007; Dahlander and Gann, 2010; Flynn et al., 2010; Laursen and Salter, 2006). In the end, however, and after taking into consideration factors such as the position of the company within the value chain (or network), the industry to which buyers and suppliers belong, the state of development of the product or the service matter of the business relationship (Christensen et al., 2005; Mitra and Singal, 2008), as well as the firm's capabilities and fallibilities (Chesbrough, 2003; Keupp and Gassmann, 2009), the concern remains about the actual effect of open innovation on the performance of the organization.

While, as described before, family firms' managers may be reluctant to set up open innovation initiatives in order to safeguard the socio-emotional endowment of the family group, their apprehension may well be justified, in addition, by the potential downside of such initiatives. Open innovation brings onto the firm an extra burden of complexity, and consequentially uncertainty, that makes it hard to predict its outcome. As described by Chesbrough (2003, 2006b) and several other authors, open innovation requires rethinking, and most times recomposing, if not the entire firm, then at least those parts of the business model that relate to the innovation process. However, often the firm's history and experience, as well as the limitations of those in charge to 'understand the totality of the task that the organization is performing' (Chesbrough, 2003: 69), lead the organization to maintain several and sometimes conflicting approaches to its business. Because of the existence of these multiple and divergent views, the firm might run into insufficient or poor allocation of managerial resources to its different innovation projects. If this is the case, and in connection with a constrained knowledge base, managers could find themselves overwhelmed by complexity and in a spiral of uninformed and poor decision making. This decision-making pattern can, swiftly and effortlessly, turn the benefits of open innovation into harmful losses.

Following this line of thought one would expect, for family firms and in line with the findings of Laursen and Salter (2006) in their study about UK manufacturing firms, that there is a curvilinear (inverted U-shape) relation between the level of open innovation in the firm's research portfolio and its innovation performance level. 'Family mental biases, maps and filters' (Berrone et al., 2012: 259), and a more risk-averse climate towards innovation compared to non-family-controlled firms (De Massis et al., forthcoming), would make of the latter a more stringent relationship in family-controlled firms. The degree of open innovation at which the relationship between open innovation and business performance achieves its maximum value is lower for family controlled firms than for non-family-controlled firms. In such a sense one could propose that:

Proposition 4: The level of open innovation in a firm's innovation portfolio is curvilinearly (taking an inverted U-shape) related to business performance.

Proposition 5: In family-controlled firms, compared to non-family-controlled organizations, the maximum of the curvilinear (inverted U-shape) function that links the level of open innovation in the firm's innovation portfolio and business performance is achieved at lower levels of open innovation.

Discussion

'Firms which do not cooperate and which do not exchange knowledge reduce their knowledge base on a long term basis and lose the ability to enter into exchange relations with other firms and organizations' (Koschatzky, 2001: 6, on Enkel et al., 2009: 311). Family firms do not escape such a rule.

As family firms try to deal with the fact that their resources and knowledge are limited, open innovation may prove to be of substantial importance, for it allows companies to leverage their knowledge and that of their business partners. The knowledge combination of internal sources with that of external contributors improves the firm's capability to respond to the challenges caused by evolution of the supply and demand markets, and most likely at a fraction of the cost compared to the firm working in isolation.

However, setting up the basis and developing the activities to achieve the expected performance and benefits of open innovation may not be as easy as it appears; in fact several factors should be taken into consideration while considering working with other companies on such initiatives. The ownership and control of firms may indeed represent a significant factor that either boosts or blocks the ideas pertaining openness and collaboration. Family-controlled firms, because of their different prioritization of goals, might even consider developing arm's-length relationships and closing themselves, rather than building up embedded ties that permit the exploitation of knowledge and expertise. Such behaviour tends to perpetuate the detrimental effects of the 'not invented here and the not sold here viruses' (Chesbrough, 2003) and affect the company's business model potential.

The purpose of this chapter was that of analysing how the idiosyncrasies of family firms, represented by the intention of managers to preserve the socio-emotional wealth of the family, might have an effect on the representativeness of open innovation in the innovation practices portfolio of the organization. In this analysis I also set out to explore how the influence of socio-emotional wealth could be moderated. The trust and power factors have been key to the extant literature in several other contexts; however, this is to my knowledge the first study that combines the topics of both open innovation and family firm management. Trust and power may well serve the purpose of modifying manager perceptions, as they are closely related to the emotional effect (Berdahl and Martorana, 2006; Slovic, 1999; Van Kleef et al., 2006), one of the main drivers of socio-emotional wealth.

While this chapter provides insightful information about the behaviour of family firms within their value chain (or network), and how perceptions of trust and power may affect the relation between business partners when it comes to open innovation, it also aims to provide some ideas about research design that may represent major opportunities for exploration in future academic endeavours. First, taking into consideration that strong business partnerships and consequently the open innovation and collaboration processes are developed over time, future studies may look at how the willingness of suppliers and customers to work on strengthening their relationship is affected by changes in the performance of the collaborative innovation. Second, future research may advance the analysis of the open innovation process by taking a look and collecting data from pairs or networks of companies that are able to assess and comment on their work with specific suppliers and specific customers; this approach would allow a more balanced view of the collaboration and the actual perception of trust and power that each of the companies hold, faced with each other. Finally, it would be interesting to study whether competition among suppliers and customers has a significant effect on the integration process; in the current economic environment, one may be able to find suppliers and customers competing in the same markets, directly or via alternative value network relationships; identifying networks of companies and how the competition among these to serve the final customer evolves over time, may provide additional information and better explain the behaviour of the business partners in certain stages of the open innovation process.

All in all, this study serves as a point of reference not only for the academic community, but also for managers who propose to take their companies along the open innovation path, and when major associations are to be developed with firms controlled by single family groups.

ACKNOWLEDGEMENT

This work has been developed with the support of the Marie Curie PF7 Programme within the framework of the Management of Emerging Technologies for Economic Impact (ManETEI) project, grant agreement number: 238382.

REFERENCES

Aberdeen Group (2004). 'Product development in consumer industries benchmark report'. Boston, MA.

Amit, R. and Schoemaker, P.J.H. (1993). 'Strategic assets and organizational rent'. *Strategic Management Journal*, 14 (1), 33–46.

Arregle, J-L., Herbert, L. and Beamish, P.W. (2006). 'Mode of international entry: the advantages of multi-level methods'. *Management International Review*, 46 (5), 597–618.

Barney, J. (1986). 'Strategic factor markets'. *Management Science*, 32 (10), 1231–1241.

Barney, J.B. (1991). 'Firm resources and sustained competitive advantage'. *Strategic Management Journal*, 17 (1), 99–120.

Benton, W.C. and Maloni, M. (2005). 'The influence of power driven buyer/seller relationships on supply chain satisfaction'. *Journal of Operations Management*, 23, 1–22.

Berdahl, J. and Martorana, P. (2006). 'Effects of power on emotion and expression during a controversial group discussion'. *European Journal of Social Psychology*, 36, 497–509.

Berrone, P., Cruz, C., Gomez-Mejia, L.R. and Larraza-Kintana, M. (2010). 'Socioemotional wealth and corporate responses to institutional pressures: do family controlled firms pollute less?' *Administrative Science Quarterly*, 55, 82–113.

Berrone, P., Cruz, C. and Gómez Mejía, L.R. (2012). 'Socioemotional wealth in family firms: theoretical dimensions, assessment approaches and agenda for future research'. *Family Business Review*, 25 (3), 258–279.

Bleakley, F.R. (1995). 'Some companies let suppliers work on site and even place orders', *Wall Street Journal*, 12 January, pp. A1, A6.

Cassia, L., De Massis, A. and Pizzurno, E. (2012). 'Strategic innovation and new product development in family firms: an empirically grounded theoretical framework'. *International Journal of Entrepreneurial Behavior and Research*, 18 (2), 198–232.

Casson, M. (1999). 'The economics of family firm'. *Scandinavian Economic History Review*, 47, 10–23.

Chesbrough, H. (2003). *Open Innovation: The New Imperative for Creating and Profiting from Technology*. Boston, MA: Harvard Business School Press.
Chesbrough, H. (2004). 'Managing open innovation'. *Research and Technology Management*, 47 (1), 23–26.
Chesbrough, H. (2006a) 'Open business models: how to thrive in the new innovation landscape'. Boston, MA: Harvard Business School Press.
Chesbrough, H. (2006b). 'Open innovation: a new paradigm for understanding industrial innovation'. In Chesbrough, H., Vanhaverbeke, W. and West, J. (eds), *Open Innovation: Researching a New Paradigm*. Oxford: Oxford University Press, pp. 1–12.
Chesbrough, H. and Schwartz, K. (2007). 'Innovation business models with co-development partnerships'. *Research Technology Management*, 50 (1), 55–59.
Chesbrough, H., Vanhaverbeke, W. and West, J. (eds) (2006). *Open Innovation: Researching a New Paradigm*. Oxford: Oxford University Press.
Christensen, J., Olesen, M. and Kjær, J. (2005). 'The industrial dynamics of open innovation – evidence from the transformation of consumer electronics'. *Research Policy*, 34, 1533–1549.
Chua, J.H., Chrissman, J.J. and Sharma, P. (1999). 'Defining the family business by behavior'. *Entrepreneurship Theory and Practice*, 23 (4), 19–32.
Cruz, C., Gomez-Mejia, L.R. and Becerra, M. (2010). 'Benevolence, family altruism, and chief executive – top management team agency relations in family owned firms'. *Academy of Management Journal*, 53 (1), 69–89.
Dahlander, L. and Gann, D.M. (2010). 'How open is innovation'. *Research Policy*, 39, 699–709.
Das, T.K. and Teng, B.S. (2000). 'A resource-based theory of strategic alliances'. *Journal of Management*, 26, 31–61.
De Jong, J.P.J. and von Hippel, E. (2009). 'Transfers of user process innovations to process equipment producers: a study of Dutch high-tech firms'. *Research Policy*, 38, 1181–1191.
De Massis, A., Frattini, F. and Lichtenthaler, U. (2012). 'Research on technological innovation in family firms: present debates and future directions'. *Family Business Review*, 20, 1–22.
De Massis, A., Frattini, F., Pizzurno, E. and Cassia, L. (forthcoming). 'Product innovation in family vs. non-family firms: an exploratory analysis'. *Journal of Small Business Management.*
Doz, Y.L. (1996). 'The evolution of cooperating in strategic alliances: initial conditions or learning process?' *Strategic Management Journal*, 17 (Summer Special Issue), 55–83.
Dyer, J.H. (1996). 'Specialized supplier networks as a competitive advantage: evidence from the auto industry'. *Strategic Management Journal*, 17, 271–291.
Dyer, J.H. and Singh, H. (1998). 'The relational view: cooperative strategy and sources of interorganizational competitive advantage'. *Academy of Management Review*, 23 (4), 345–367.
Enkel, E., Gassman, O. and Chesbrough, H. (2009). 'Open R&D and open innovation: exploring the phenomenon'. *R&D Management*, 39 (4), 311–316.
Flynn, B., Huo, B. and Xiande, Z. (2010). 'The impact of supply chain integration on performance: a contingency and configuration approach'. *Journal of Operations Management*, 28, 58–71.
Gassmann, O., Enkel, E. and Chesbrough, H. (2010). 'The future of open innovation'. *R&D Management*, 40 (3), 213–221.

Geykens, I., Steenkamp, J.E.M. and Kumar, N. (1998). 'Generalizations of trust in marketing channels relationship using meta-analysis'. *International Journal of Research in Marketing*, 15, 223–248.
Gómez-Mejía, L.R., Cruz, C., Berrone, P. and De Castro, J. (2012). 'The bind that ties: socioemotional wealth preservation in family firms'. *Academy of Management Annals*, 5 (1), 653–707.
Gomez-Mejia, L.R., Haynes, K., Nunez-Nickel, M., Jacobson, K. and Moyano, J. (2007). 'Socioemotional wealth and business risks in family-controlled firms'. *Administrative Science Quarterly*, 52 (1), 106–138.
Gomez-Mejia, L.R., Makri, M. and Larraza-Kintana, M. (2010). 'Diversification decisions in family controlled firms'. *Journal of Management Studies*, 47 (2), 223–252.
Grant, R.M. (1996). 'Prospering in dynamically competitive environments: organizational capability as knowledge integration'. *Organization Science*, 7 (4), 375–388.
Handler, W.C. (1990). 'Succession in family firms: a mutual role adjustment between family and entrepreneur and next-generation family members'. *Entrepreneurship Theory and Practice*, 15 (1), 37–51.
Henkel, J. (2006). 'Selective revealing in open innovation processes: the case of embedded Linux'. *Research Policy*, 35, 953–969.
Huizingh, E.K.R.E. (2011). 'Open innovation: state of the art and future perspectives'. *Technovation*, 31 (1), 2–9.
Jacobsen, R. (1988). 'The persistence of abnormal returns'. *Strategic Management Journal*, 9, 415–430.
Jayaram, J. and Tan, K.C. (2010). 'Supply chain integration with third party logistics providers'. *International Journal of Production Economics*, 125, 262–271.
Jones, C.D., Makri, M. and Gomez-Mejia, L.R. (2008). 'Affiliate directors and perceived risk bearing in publicly traded, family controlled firms: the case of diversification'. *Entrepreneurship Theory and Practice*, November, 1007–1026.
Kang, K.H. and Kang, J. (2009). 'How do firms source external knowledge for innovation? Analyzing effects of different knowledge sourcing methods'. *International Journal of Innovation Management*, 13 (1), 1–17.
Kanter, R.M. (1994). 'Collaborative advantage: the art of alliances'. *Harvard Business Review*, 72 (4), 96–108.
Kepner, E. (1983). 'The family and the firm: a coevolutionary perspective'. *Organizational Dynamics*, 12 (1), 57–70.
Keupp, M.M. and Gassmann, O. (2009). 'Determinants and archetype users of open innovation'. *R&D Management*, 39 (4), 331–341.
Kim, S.W. (2009). 'An investigation on the direct and indirect effect of supply chain integration on firm performance'. *International Journal of Production Economics*, 119 (2), 328–346.
Kopczak, L.R. and Johnson, M.E. (2003). 'The supply chain management effect'. *MIT Sloan Management Review*, 44 (3), 27–34.
Koschatzky, K. (2001). 'Networks in innovation research and innovation policy – an introduction'. In Koschatzky, K., Kuliche, M. and Zenker, A. (eds), *Innovation Networks: Concepts and Challenges in the European Perspective*. Heidelberg: Physica Verlag, pp. 3–23.
Koufteros, X., Cheng, T.C.E. and Lai, K-H. (2007). 'Black-box and gray-box supplier integration in product development: antecedents, consequences and the moderating role of firm size'. *Journal of Operations Management*, 25, 847–870.

Krajewski, L., Ritzman, L. and Malhotra, M.K. (2010). *Operations Management: Strategy and Analysis*, 9th edn. Upper Saddle River, NJ: Prentice Hall.
Laursen, K. and Salter, A. (2006). 'Open for innovation: the role of openness in explaining innovation performance among UK manufacturing firms'. *Strategic Management Journal*, 27 (2), 131–150.
Lee, S., Park, G., Yoon, B. and Park, J. (2010). 'Open innovation in SMES – an intermediated network model'. *Research Policy*, 39 (2), 290–300.
Leifer, R. and Mills, P.K. (1996). 'An information processing approach for deciding upon control strategies and reducing control loss in emerging organisations'. *Journal of Management*, 22, 113–137.
Maloni, M.J. and Benton, W.C. (2000). 'Power influences in the supply chain'. *Journal of Business Logistics*, 21(1), 42–73.
McCarter, M. and Northcraft, G. (2007). 'Happy together?: Insights and implications of viewing managed supply chains as a social dilemma'. *Journal of Operations Management*, 25, 498–511.
Mitra, S. and Singal, V. (2008). 'Supply chain integration and shareholder value: evidence from consortium based industry exchanges'. *Journal of Operations Management*, 26, 96–114.
Monckza, R., Petersen, K.J. and Ragatz, G.L. (1998). 'Success factors in strategic supplier alliances: the buying company perspective'. *Decision Sciences*, 29 (3), 553–577.
Peteraf, M.A. (1993). 'The cornerstones of competitive advantage: a resource-based view'. *Strategic Management Journal*, 14 (3), 179–191.
Petersen, K.J., Handfield, R.B. and Ragatz, G.L. (2003). 'A model of supplier integration into new product development.' *Journal of Product Innovation Management*, 20 (4), 284–299.
Petersen, K.J., Handfield, R.B. and Ragatz, G.L. (2005). 'Supplier integration into new product development: coordinating product, process and supply chain design'. *Journal of Operations Management*, 23, 371–388.
Schulze, W.S., Lubatkin, M.H. and Dino, R.N. (2003a). 'Exploring the agency consequences of ownership dispersion among the directors of private family firms'. *Academy of Management Journal*, 46 (2), 179–194.
Schulze, W.S., Lubatkin, M.H. and Dino, R.N. (2003b). 'Toward a theory of agency and altruism in family firms'. *Journal of Business Venturing*, 18 (4), 473–490.
Slovic, P. (1999). 'Trust, emotion, sex, politics, and science: surveying the risk-assessment battlefield'. *Risk Analysis*, 19 (4), 689–701.
Song, M. and Di Bendetto, C.A. (2008). 'Supplier's involvement and success of radical new product development in new ventures'. *Journal of Operations Management*, 26, 1–22.
Tan, K.C. (2001). 'A structural equation model of new product design and development'. *Decision Science Journal*, 32 (2), 195–226.
Teece, D.J., Pisano, G. and Shuen, A. (1997). 'Dynamic capabilities and strategy management'. *Strategic Management Journal*, 18 (7), 509–533.
Van Kleef, G., De Dreu, C.K.W., Pietroni, D. and Manstead, A.S.R. (2006). 'Power and emotion in negotiation: power moderates the interpersonal effects of anger and happiness on concession making'. *European Journal of Social Psychology*, 36, 557–581.
Villena, V., Gomez-Mejía, L.R. and Revilla, E. (2009). 'The decision of the supply chain executive to support or impede supply chain integration: a

multidisciplinary behavioral agency perspective'. *Decision Sciences*, 40 (4), 635–665.

Westhead, P., Cowling, M. and Howorth, C. (2001). 'The development of family companies: management and ownership imperatives'. *Family Business Review*, 14 (4), 369–401.

Yeung, J.H.Y., Selen, W., Zhang, M. and Huo, B. (2009). 'The effects of trust and coercive power on supplier integration'. *International Journal of Production Economics*, 120, 66–78.

Zellweger, T. and Astrachan, J. (2008). 'On thc emotional value of owning a firm'. *Family Business Review*, 21 (4), 347–363.

Zellweger, T., Kellermanns, F. and Chrisman, J. (2011). 'Family control and family firm valuation by family CEOs: the importance of intentions for transgenerational control'. Working Paper, University of Tennessee.

15. Engaging users in value co-creation activities

Darius Pacauskas

INTRODUCTION

Have you ever thought that customers will start creating products for themselves and will buy them afterwards? It sound unrealistic, but it is the truth. Do not get this wrong. Excited consumers are not breaking into factories, kicking out current workers and doing the manufacturing jobs themselves. Instead, they are giving a hard time to research and development departments, while identifying their needs, designing product prototypes, searching for bugs, providing technical support and advertising the product. This phenomenon, called 'value co-creation', entails companies approaching their clients, while clients provide a proper solution. For example, car enthusiasts and fans of Audi designed their own future infotainment systems (Füller et al., 2004).

However, sending letters, hanging posters or just screaming in the streets are not very effective ways when we want to approach as many people as possible. To solve this problem, digital technologies come to the rescue. With the help of online environments, companies can easily access current or future customers around the world. The easiest way to get a large number of customers is to address existing online communities (Nambisan, 2002).

If we look to the scientific literature for a definition of 'online community', we can find very different explanations. For example, Joseph Cothrel and Ruth Williams (1999) defined online community as 'a group of people who use computer networks as their primary mode of interaction'. Barry Wellman (1979) defined community 'as networks of interpersonal ties that provide sociability, support, information, a sense of belonging, and social identity', while Jake McKee, former Global Community Development Manager for the LEGO company, says, 'people often think that blogs, forums, wikis, and other tools are community. In actuality, those tools are just that – tools. They can help you to build community, but they aren't actually "community"' (McKee, 2005). We could find hundreds of different definitions, but most if not all of them can be classified under these requirements: shared interest, a platform and rules for communication (De Souza and Preece, 2004). To fulfil these criteria, we may gather a group of people in one virtual place, but we will not necessarily achieve the expected outcomes. Marketing guru Seth Godin in his book *Tribes* (2008) describes slightly different communities: groups of people who have a common sense of the future they are trying to build, connected with a leader and each other, and doing something meaningful. A value co-creation community is a real tribe, but to build it, we need to get users who are full of passion and unsatisfied needs. Finding such people is not as hard as it seems. Every person on Earth is dreaming about making something meaningful that they could be proud of, and most of them are not completing their daily duties. Tata Consultancy Services, an IT consulting company from India, launched an online forum where employees could post a problem and get help from other workers, for example to solve difficulties with the Java programming language. The most surprising thing was not that the community was thriving or that employees spent a lot of their time helping each other, but that the most contributing consultants in each category were people in an official position who had nothing to do with the problems solved (Kiron, 2012). For example, some of the people who were on the leader board in the Java category were not hired for their Java knowledge. It was not even something that Tata Consultancy Services knew they were good at. Digital technology makes knowledge transfer much easier nowadays.

Value co-creation activities can happen in different online environments, which are called virtual customer environments. It can be a simple bulletin board, an online forum, or a social networking tool such as Facebook, where users post their ideas and others comment on them. It can also be a more sophisticated platform, where companies provide a prepared tool with many functions for creating design or testing a platform with available product features, which users can then add to or take away from, and see how this might change the price of the product (Nambisan, 2002).

We can roughly describe the purpose of such platforms by looking at another concept that is interchangeably used with the value co-creation concept: crowdsourcing (Geiger et al., 2011). A crowdsourcing system is a crowd of humans to help solve a problem defined by the system owners (Doan et al., 2011). Even though crowdsourcing is more about taking advantage of cheaper labour with no interest in co-creators, the main components are roughly the same: there should be a problem defined by the company that is developing a product or a service; there should be a community that is dealing, or will be dealing, with a particular problem; and there should be the right approach to gather information from the user-innovator community. And whatever approach companies are using, users are investing their time, knowledge and efforts, making a contribution; and it is obvious that every company wants to see more contributions.

There are three main streams of scientific studies that relate to users' motivation for contributing in online communities. One stream studies how technology or information technology (IT)-related tools like online platforms affect users' willingness to contribute, and they hold that the technology they use can shape users' perception (Sawhney et al., 2005). A second stream divides users according to their roles, and holds that users behave in communities according to their own nature, and only some users can come up with creative solutions (Preece et al., 2004). The third stream studies brand communities, that is, companies and customer relationships in online communities, the outcome of shared activities, and ways to increase brand strength in their customers' eyes (Schau et al., 2009). This chapter will explore these topics, and look in more detail at these three main factors that affect users' willingness to co-create:

1. How users are accepting an online platform.
2. Motivation that drives users to contribute.
3. How a company's reputation can influence users.

In the next sections, I will look more deeply at each of these factors and illustrate them with examples from case studies. At the end of this chapter, I will generate propositions and suggest possibilities for future research studies.

ONLINE TOOL ACCEPTANCE

All the major scientific contributions in technology acceptance were made before the 1990s. Martin Fishbein and Icek Ajzen in 1975 developed the theory of reasoned action. On the basis of this theory, in 1985 Icek Ajzen developed the theory of planned behaviour. The third theory, the technology acceptance model, was developed in 1989 by Fred Davis, and is still unsurpassed in the information systems area. These three models are based on users' beliefs and are generally used to predict users' intention to perform a particular behaviour. These models can be decomposed into main components: ease of use, perceived usefulness, social influence, self-efficacy, attitude, and willingness. I discuss each of them below.

User-friendly systems have proven to be a leader in the IT field (Doan et al., 2011). Products from Apple Inc. do not attract attention just for looking fancy and having the best micro components; Apple devices are well known for their user-friendliness. Users are prepared to pay two or three times the price of other equipment just to have a stress-free environment. This is what ease of use predicts. If a user feels that a system will require less learning or effort to use, they will be more willing to adopt it.

The next component is perceived usefulness. The main question that could arise is: what advantages do users see in doing other people's work for free? A good example of this is open source software (OSS) communities. If you look at it peripherally, you can get the impression that there is no fun in programming software, but the reality is that OSS communities attract many skilful programmers. Lakhani and Wolf (2003) made a study of this phenomenon and found that hackers were participating in the OSS movement because they enjoyed it: their motivation was their feeling of creativity, their need for better software, intellectual stimulation and their desire to improve their programming skills. But co-creation should not be just something tangible that customers are doing, it can also be criticism. Customers in value co-creation environments like to complain, and most often they are using a harsher manner than they present in person. Consumers want to be heard, they want someone to listen to them, and it does not matter if a company changes the product according to their complaints. It is very important to reply to criticism and show that you care. McDonald's knows that. Seth Godin, in his book *Tribes* (2008), wrote about a woman whose child had not found a toy in their 'Happy Meal', and the woman immediately wrote a complaint to McDonald's. Managers reacted and sent her a toy. It appears that this lady was the leader of a local tribe, and she praised this act to her community. Now imagine what could have happened if the McDonald's managers had ignored her or responded negatively.

The main idea of these models is that users judge a system from their own perspective, and afterwards they form an attitude toward a particular system. How easy it is to use and how useful it is are tricky questions if you have just started using a system, and this is where social influence comes in. Users are searching for feedback from experienced users, and that information is influencing their attitude toward the system even before they use it (Pavlou, 2003).

The component that summarizes all the previous variables is attitude. After determining whether a particular system is user-friendly and useful, a user acquires a positive or a negative attitude towards the system. Attitude affects a user's willingness, that is, how enthusiastic a user is about using a particular system. This is a main predictor of a user's real behaviour. Even though a user thinks the system is useful and easy to use, there is one more factor that prevents forecasting their future behaviour with a high degree of certainty. That factor is self-efficacy. This describes how a user feels about their ability to arrive at a valuable solution. In other words, how confident they are and whether they have proper resources to accomplish a given task. Identifying the proper target group should solve this issue. If, for example, a company is targeting value co-creation activity to amateur and professional designers, users most probably will have their own tools and the company needs to concentrate on providing exciting tasks. But if the company intended to discover its customers' preferences around the world, customers should be equipped with simpler tools.

One good example of possible tools is the Swarovski case discussed in Füller et al.'s (2011) research article. Swarovski Enlightened™ invited designers and creative consumers from all over the world to engage in the 'jewellery design competition' community. To this end, a virtual social media platform was created, providing several tools. The community creation and the design competition were open not only to designers (for example, professional designers and design students) but also to people who were generally interested in jewellery, gemstones and related topics. Indeed 38.5 per cent of all participants were professional designers, 23.6 per cent were design students, and 23.6 per cent were hobby designers.

The competition was divided into two categories: With the 'jewellery configuration toolkit', participants were able to create their own items of jewellery, selecting from 24 components (for example, chain, pendant) in various colours and through the selection and free placement of 108 different Swarovski Enlightened™ gemstones. In the second category, freely created jewellery designs could be uploaded. More than 1700 participants joined the community to showcase their talent and submit their designs. In total, they created over 3000 pieces of jewellery: 980 freely designed items and more than 2000 using the provided toolkit. Designs were created in various styles, ranging from classic to sporty and from abstract to realistic. A lively community evolved, where participants uploaded their own avatars and pictures and gave feedback on both the configured designs and the freely created designs. The community platform allowed feedback either with short evaluations (1–5 points) or with 'vote and win' evaluations within different categories (from 'The design of jewellery looks good' to 'In particular I like the shape'). Comments on designs and suggestions for improvements were provided. Emotional but detailed voting took place. Community members made more than 23000 evaluations and contributed 2936 qualitative comments.

CHANGING THE ROLE OF USERS

Accepting a platform is only one factor that influences a user's contribution. Since a value co-creation platform is a virtual community used for product development, it has much in common with general online communities (Piller et al., 2005). The majority of online communities have one big problem: only a small number of users actively participate in the movement, while the others are free-riding (lurking), that is, collecting information but not communicating with other users by submitting information or contributing in any other way (Preece et al., 2004).

Despite the large numbers of registered users in online communities, participation drops to zero if we are referring to contribution. Butler (1999) found that 50 per cent of social, hobby and work mailing lists had no traffic over a period of four months. According to Adar and Huberman (2000), 10 per cent of the users provide 87 per cent of all the music in the popular peer-to-peer music-sharing service, Gnutella. In open source development communities, 4 per cent of the members account for 50 per cent of the answers on a user-to-user help site (Lakhani and Von Hippel, 2003), and 4 per cent of the developers contribute 88 per cent of the new code and 66 per cent of the code fixes (Mockus et al., 2002).

Lurking has traditionally been viewed as a negative social activity. Because lurkers dedicate a substantial amount of time to learning about the community, their knowledge about the group could be beneficial to the whole community. Passive users were also used as word-of-mouth marketers in open source software communities, convincing their friends to join the tribe or use the software (Lindman et al., 2010). Furthermore, companies are finding new ways to benefit from lurkers. One possibility is to use free-riders' activities in the forums to recognize common problems with their products. For example, if consumers detect errors in their devices, they probably go straight to the online community to search for solutions. Managers of communities can easily identify problems that need solving by viewing logs of the most visited topics and phrases used in search queries.

However, there are ways to change a user's behaviour from passive (lurking) to active (contributing). Preece et al. (2004) published a study in which they discovered the top reasons why users lurk:

1. Thought there was no obligation to post.
2. Needed time to find out about the group.
3. Thought they were helping by not posting.
4. Software problems.
5. Did not like the group.

Here are some quotes illustrating each reason: (1) 'Will start posting in the future if I feel I need to', 'I have gained the insight I needed from this community'; (2) 'If I am not completely confident about sharing my posts, then I dump it', 'I think that most of the members probably know more about the questions asked than I would'; (3) 'Never felt that I had enough to share or anything important enough to ask'; (4) 'Don't like the login requirements'; (5) 'This is not a very active community, and I belong to others where I am very well known now', 'Nobody replies!!!!', 'Many of the postings are either rude, offensive, off-topic or simply childish and ignorant', 'Fear of rejection, which has happened, or being insulted, which has also happened'.

From reading all these comments, it is not hard to realize that users are not motivated enough; they do not see benefits for active participation in the community (Preece et al., 2004). Online communities researchers came with the opinion that a key driver of knowledge contribution behaviour in an online community is the seeking of extrinsic and intrinsic incentives (Ma and Agarwal, 2007). Füller (2010) elaborated on extrinsic and intrinsic motives within co-creation platforms and found motivational factors for users' participation in value co-creation activities. Ten motives were identified: enjoyment (have fun), curiosity (try something new), altruism (help others), make friends (socialize), self-efficacy (a challenge to be mastered), information seeking, skill development (gain knowledge), recognition (satisfy ego), personal need (dissatisfaction with current products), and reward (money, prizes).

This does not mean that a value co-creation platform must provide all of these incentives. Managers should try different actions and search for a successful formula in their domain. One good example is a live experiment made by Farzan et al. (2010) on a large social network site deployed inside IBM (called 'Beehive'). During the experiment, Beehive had been live for over two years and had over 60000 registered users. Users came to the site to socialize and connect with their co-workers. Similarly to the majority of communities, 85 per cent of the content was created by 10 per cent of the users. Beehive had over 100000 photos, lists and events; and researchers every week selected a new group of 50 users to perform a small task – to promote one item from each category. Additionally, users were highlighted on the site as a way of thanking them for helping the community. As a result, 20 per cent of the free-riders who responded to the task during the next month actively participated in the community (previously 0 per cent), and 80 per cent actively browsed the content (previously 40 per cent).

If we look at another study made by Nambisan and Nambisan (2008), we can see that different types of value co-creation platforms provide different experiences for the user. According to the authors, platforms can be designed to support five customer roles in innovation and value co-creation: product conceptualizer, product designer, product tester, product support specialist and product marketer (Nambisan, 2002). Product conceptualizer: companies can encourage customers to interact among themselves to generate and advance product improvement and new product ideas. Product designer: customers can design their own versions of the 'ideal' product using virtual prototyping tools and design toolkits provided in the co-creation platforms. Product tester: users through virtual product technologies can be integrated in trying, searching for errors, or other ways of preparing the developed concept for customers. Product support specialist: this allows customers to leverage their product-related knowledge and expertise to extend support to peers. Product marketer: companies are able to pass along information about new products and shape peer perceptions through dialogue and discussions.

Depending on the customer innovation role, the nature of customer interactions and the technologies used in the value co-creation platforms will vary; and thus the nature of customers' experiences in the virtual environment will vary as well. For example, while Microsoft's Most Valuable Professional program concentrates primarily on product support activities, and Ducati's Tech Café focuses mainly on developing product concepts; users will have different experiences with each platform. Experience in value co-creation platforms is described by four components: pragmatic experience, sociability experience, usability experience and hedonic experience. The pragmatic component relates to customers' experiences in realizing product-related informational goals in the co-creation platforms (how useful the system is for their needs). The sociability component emphasizes the importance of community dialogue and the social policies (or rules of engagement) that frame such dialogue. The quality of the human–computer interaction defines the usability experience. Regardless of whether the technologies used are simple (for example, online discussion boards) or more complex (for example, 3-D product simulation tools), the ease with which customers can interact and perform tasks shapes their overall experience. Customers' interactions in the virtual environment can also be mentally stimulating or entertaining, a source of pleasure and enjoyment; this is the hedonic component. It can encompass the interaction with both other customers and with tools and technologies. Together, the four components provide a comprehensive view of the experience a customer participating in a co-creation activities can have.

However, based on the nature of the customer's interactions and value co-creation activities, some components are likely to be more important than others. In one of the studies, Nambisan and Nambisan

(2008) examined the virtual customer environments at four companies: a software company, a consumer electronics company, a high-technology company and a technology services company. The value co-creation platforms in the software company and the consumer electronics company encouraged customers to be product support specialists, while those in the high-technology company and the technology services company encouraged customers to conceptualize new products and services. Based on a survey of customers who visited these virtual environments, the authors evaluated the average customer experience profiles of the four companies and found some interesting results. In the first two companies, the customer experience profiles were skewed toward pragmatic and sociability experience components, while in the latter cases the emphasis was on pragmatic and hedonic components (Figure 15.1). These differences underline the different types of customer interactions.

To conclude, there are many reasons why people lurk in online discussion communities, and many lurkers are not selfish free-riders. There is much that we can do to make the community a more interesting, satisfying and comfortable environment for both lurkers and posters. Some solutions require improved software and better tools, but moderation and better interaction support are also critical to the overall success of online communities (Preece et al., 2004).

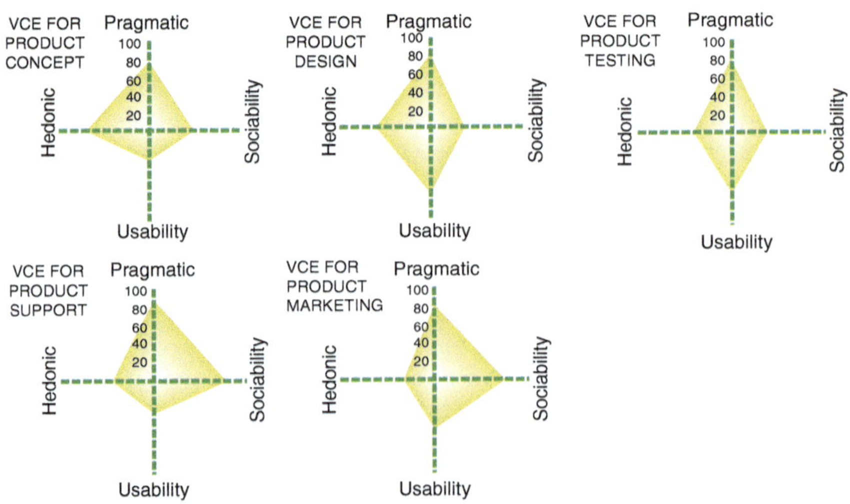

Figure 15.1 Users' experience in different virtual customers' environments (VCEs)

COMPANIES' INFLUENCE ON THE DECISIONS OF USERS

In value co-creation, the most desired outcome is widely adopted innovation (Sawhney et al., 2005). The previous innovation aim was new technology applications, while nowadays the focus has shifted to the delivery of meaningful value (Shneiderman, 2000). The innovation domain within an organization needs a common vision of what the organization is and how it is connected to its environment; in short, what is described by 'brand' (Abbing and Gessel, 2008). Brands are a product's logos, names, packaging design and other marketing mechanisms used to help connect consumers with the products sold by a company (Keller, 1993).

Customers like companies with innovative and meaningful strategies. When they are co-creating products for such organizations they have the feeling that they are changing the world. After all, customers will not just help to create a cheaper, faster or higher-quality solution, but will also develop psychological ownership and will be more willing to buy the product they were involved in developing it (Fuchs et al., 2010).

If we try to analyse the concept of 'brand', we will come to the conclusion that a brand is a collection of promises (Karjalainen and Snelders, 2010). 'Volvo' is promising a car for life, 'Audi' promises advanced technology, and 'BMW' promises a driving experience. Users create relationships toward a brand as they would toward another person, such as trust or attachment.

There are many variables that were widely used in researching brand impact, but Esch et al. (2006) tested the most important of them and distinguished three that describe users' future activities: (1) brand satisfaction; (2) brand trust; and (3) attachment to brand. Brand satisfaction is defined as a user's evaluation of whether or not the exchange relationship with the brand is rewarding. Satisfaction arises when a company at least fulfils promises for customers, or gives more value than expected. Even tiny details are important. Think about how consumers evaluate a product or service. For example, if it is an airline, consumers will not be able to tell whether the company does a good or a bad job maintaining the planes, but they can assess the food's tastiness, the crew's attitude and willingness to help, and the pleasantness of the environment inside the aeroplane.

There is one more important variable in the brand family: trust. Brand trust is the relationship between the confident expectations of the consumer and the resulting responsibility of the brand or retailer. Trust corresponds with brand satisfaction. If a brand meets the consumer's expectations, then along with a good reputation it will reinforce the consumer's trust in that brand, and vice versa. The same as with people: by doing the right things, you can earn a person's trust, but it is easy to lose that trust; and you can lose it quickly, practically overnight, regardless of your past reputation. As an example, let us consider 'Skoda'. It is a trusted brand name, but it was not always like this. Skodas in the 1990s can be illustrated with a joke: Why does a Skoda have a heated rear windscreen? Answer: To keep your hands warm when you push it. After Volkswagen's influence on the vehicles' manufacture, the most important issue was to convince consumers that Skoda is no longer a big load of problems, and this was achieved with the help of the humorous advertising message: 'The Fabia is a car so good that you won't believe it's a Skoda'.

To describe the overall consumer's feeling towards a brand, the term 'attachment' is used. Attachment is the longer-lasting, commitment-inducing bond between the brand and the consumer. People who are attached to a brand generally pay to wear the brand emblem or be associated with owning the brand. Starbucks is a good example. Consumers prefer to pay twice as much for a cup of coffee so that they can walk through the city holding a cup with the Starbucks logo. In other words, they feel that the brand describes their beliefs and is part of their lifestyle.

Let us take one more example of what can be achieved with proper brand–customer relationship management. This example is from a Bagozzi et al. (2012) research article. In their study, the authors explored virtual communities of Ducati motorcycle owners. Ties in a virtual brand community are indirect in the sense of being mediated online or through some other impersonal medium. Conversations are built around the topic of Ducati motorcycles. Users pose questions and give advice over the Internet to strangers to share knowledge about Ducati motorcycles and their ownership and use.

In online communities, 'Ducatisti' (a name used to characterize owners of Ducati bikes) discuss customization issues, engage in highly visible joint activities, and function as promoters of Ducati and serve as non-paid advocates, primarily through providing information and advice for beginners or long-time owners in personal ways, and with a sense of caring not possible via official corporate venues. The Ducati company gets a double win from this situation: a group of non-paid agents who are promoting Ducati bikes, and the trust of those who receive the messages, which is far from the trust in commercials that are spread through the company's official channels.

Consumers socially construct their experiences around the Ducati product and brand through membership in virtual brand communities. 'Ducatisti' expresses ownership of Ducati motorcycles in very personal terms as 'part of me'. Corporate Ducati can participate in such processes by purposefully managing its activities to foster the adoption and use of its products in what Scott and Lane (2000) termed 'organization identity construction'.

One way to strengthen the ties between brand and customer is by holding 'brandfests': 'corporate-sponsored events provided primarily for the benefit of current customers'. Each year, Ducati organized brandfest gatherings en masse, of upwards of 45 000 people in Italy. It was found that participation in such brand-intensive galas often leads consumers to extraordinary and memorable experiences with the brand. This experience changes pariticipants' attitudes toward the brand and strengthens intentions to purchase a particular product. Such brandfests also lead to greater trust in the corporation and brand. This is achieved as people experience a sense of kindness and caring from the organization and their agents, and learn about admirable attributes of an organization's products, practices, warranties, service, handling of complaints, and so forth. This develops confidence and positive expectations about future interactions with the organization concerning product quality, performance, and other aspects of an organization's offerings and relationships with consumers. Such activities have been suggested for moving away from 'employees' versus 'customers' and moving toward a more fine-grained approach that emphasizes the unique characteristics of individual–organizational relationships.

So what constitutes successful brand management? What needs to be concentrated on to ensure lasting success? For long-term brand success, relationships are the key. In branding terms, this means fostering brand trust, brand satisfaction and brand attachment among customers and target groups. Brand-building has never been about the short term, always about a longer view and about generating revenue over the long term. Brand managers need to think about the relationships they have with people. They are rarely easy, but they are often ultimately rewarding. People are people. They do not change when labelled 'consumer'. To be truly successful, marketers must move beyond the standard notions of the marketing mix and become obsessed with developing initiatives, both strategic and tactical, that ensure customer satisfaction with the brand, reinforce their trust in it, and foster a sense of meaningful emotional attachment. Customer relationship management has grown up and come of age (Esch et al., 2006).

DISCUSSION

After describing three main user contribution-related areas, I will now set out propositions that will be based on my analysis of each section and will suggest general implications that could lead the way for further studies.

In the second section, I discussed technology acceptance and its models. Unfortunately these models receive plenty of criticism (Bagozzi, 2007). Technology acceptance models were developed to understand users' intentions to use utilitarian (that is, work-related) systems. We can assume that users joining value co-creation platforms for hedonic reasons, or pleasure, are also users who do not intend to develop a product when they join communities; they are registering for more personal reasons. This behaviour can be explained by the term 'willingness' rather than 'intention' (Pomery et al., 2009). In this case, the proposition can be made that:

Models explaining users' willingness to join value co-creation activities or accept its tools should concentrate on the hedonic side of technology.

In the third section, it was seen that user motivation can vary, and it can be affected by proper acts within a community. Now the proposition can be derived that:

User roles are affected by the users' motives and incentives (which are provided by value co-creation platforms), and a user's role can be changed by providing sufficient incentives.

Further research could concentrate on exploring the nature and grounds of these phenomena, as well as empirically proving their existence, while practically it would be interesting to understand what types of motivation or incentives can affect users' behaviour in each role.

In the fourth section, I stressed the importance of the company–customer relationship. In the value co-creation literature, brand influence on customer-innovator decisions receives little attention. A proposition can be made that:

Brand impacts upon users' decisions on joining and contributing to value co-creation activities.

Future studies could test links between described brand variables (trust, attachment, satisfaction) and contribution or entry to communities of user-innovators.

CONCLUSION

This chapter has looked at the value co-creation platforms and covered three main factors that influence a user's contribution within them. I focused on the user perspective on a co-creation platform's acceptance as a tool, its role within value co-creation communities, and the relationship with companies. In each section, I illustrated concepts with a case study taken from external sources. Finally, I looked at possible theoretical propositions that could drive future research. Besides scholars, this chapter should be helpful for community developers, managers, moderators, software designers and tribe leaders, for finding out reasons for and ways to influence user behaviour in value co-creation platforms and other online communities.

ACKNOWLEDGEMENT

This research is funded by the Mare Curie FP7 programme with Grant–238382.

REFERENCES

Abbing, E.R. and Gessel, C. van. (2008). Brand-driven innovation. *Design Management Review*, 19(3), 51–58.

Adar, E. and Huberman, B.A. (2000). Free riding on Gnutella. *First Monday*, 5(10), 8.

Ajzen, I. (1985). From intentions to actions: a theory of planned behavior. In Kuhl, P.D.J. and Beckmann, D.J. (eds), *Action Control: From Cognition to Behavior*. Berlin: Springer, pp. 11–39.

Bagozzi, R.P. (2007). The legacy of the technology acceptance model and a proposal for a paradigm shift. *Journal of the Association for Information Systems*, 8(4), 244–254.

Bagozzi, R.P., Bergami, M., Marzocchi, G.L. and Morandin, G. (2012). Customer–organization relationships: development and test of a theory of extended identities. *Journal of Applied Psychology*, 97(1), 63–76.

Butler, B. (1999). When is a group not a group: an empirical examination of metaphors for online social structure. PhD thesis, Carnegie Mellon University, Pittsburgh, PA.

Cothrel, J. and Williams, R.L. (1999). On-line communities: helping them form and grow. *Journal of Knowledge Management*, 3(1), 54–60.

Davis, F.D. (1989). Perceived usefulness, perceived ease of use, and user acceptance of information technology. *MIS Quarterly*, 13(3), 319.

De Souza, C.S. and Preece, J. (2004). A framework for analyzing and understanding online communities. *Interacting with Computers*, 16(3), 579–610.

Doan, A., Ramakrishnan, R. and Halevy, A.Y. (2011). Crowdsourcing systems on the World-Wide Web. *Communications of the ACM*, 54(4), 86.
Esch, F.-R., Langner, T., Schmitt, B.H. and Geus, P. (2006). Are brands forever? How brand knowledge and relationships affect current and future purchases. *Journal of Product and Brand Management*, 15(2), 98–105.
Farzan, R., Dimicco, J.M. and Brownholtz, B. (2010). Mobilizing lurkers with a targeted task. *International AAAI Conference on Weblogs and Social Media*, pp. 235–238.
Fishbein, M. and Ajzen, I. (1975). *Belief, Attitude, Intention and Behavior: An Introduction to Theory and Research*. Reading, MA: Addison-Wesley.
Fuchs, C., Prandelli, E. and Schreier, M. (2010). The psychological effects of empowerment strategies on consumers' product demand. *Journal of Marketing*, 74(January), 65–79.
Füller, J. (2010). Refining virtual co-creation from a consumer perspective. *California Management Review*, 52(2), 98–123.
Füller, J., Bartl, M., Ernst, H. and Mühlbacher, H. (2004). Community based innovation: a method to utilize the innovative potential of online communities. *Proceedings of the 37th Hawaii International Conference on System Sciences*.
Füller, J., Hutter, K. and Faullant, R. (2011). Why co-creation experience matters? Creative experience and its impact on the quantity and quality of creative contributions. *R&D Management*, 41, 259–273.
Geiger, D., Seedorf, S., Schulze, T., Nickerson, R. and Schader, M. (2011). Managing the crowd: towards a taxonomy of crowdsourcing processes. *Proceedings of the Seventeenth Americas Conference on Information Systems*, pp. 1–11. Detroit, MI.
Godin, S. (2008). *Tribes*. New York: Portfolio Hardcover.
Karjalainen, T.-M. and Snelders, D. (2010). Designing visual recognition for the brand. *Journal of Product Innovation Management*, 27(1), 6–22.
Keller, K.L. (1993). Conceptualizing, measuring, and managing customer-based brand equity. *Journal of Marketing*, 57(1), 1.
Kiron, D. (2012). What sells CEOs on social networking. *MIT Sloan Management Review*, 53(3), 1–6.
Lakhani, K.R. and Von Hippel, E. (2003). How open source software works: 'free' user-to-user assistance. *Research Policy*, 32(6), 923–943.
Lakhani, K. and Wolf, R.G. (2003). Why hackers do what they do: understanding motivation and effort in free/open source software projects. SSRN Electronic.
Lindman, J., Rajala, R. and Rossi, M. (2010). FLOSS-induced changes in the software business: insights from the pioneers. ICSOB.
Ma, M. and Agarwal, R. (2007). Through a glass darkly: information technology design, identity verification, and knowledge contribution in online communities. *Information Systems Research*, 18(1), 42–67.
McKee, J. (2005). What is community?. http://www.communityguy.com/2005/02/28/what-is-community/ (accessed 22 November 2014).
Mockus, A., Fielding, R.T. and Herbsleb, J.D. (2002). Two case studies of open source software development: Apache and Mozilla. *ACM Transactions on Software Engineering and Methodology*, 11(3), 309–346.
Nambisan, Satish (2002). Designing virtual customer environments for new product development: toward a theory. *Academy of Management Review*, 27(3), 392–413.

Nambisan, S. and Nambisan, P. (2008). How to profit from a better 'virtual customer environment'. *MIT Sloan Management Review*, 49(3), 53–61.
Pavlou, P.A. (2003). Consumer acceptance of electronic commerce: integrating trust and risk with the technology acceptance model. *International Journal of Electronic Commerce*, 7(3), 101–134.
Piller, F., Schubert, P., Koch, M. and Möslein, K. (2005). Overcoming mass confusion: collaborative customer co-design in online communities. *Journal of Computer-Mediated Communication*, 10(4), http://jcmc.indiana.edu/vol10/issue4/piller.html.
Pomery, E.A., Gibbons, F.X., Reis-Bergan, M. and Gerrard, M. (2009). From willingness to intention: experience moderates the shift from reactive to reasoned behavior. *Personality and Social Psychology Bulletin*, 35(7), 894–908.
Preece, J., Nonnecke, B. and Andrews, D. (2004). The top five reasons for lurking: improving community experiences for everyone. *Computers in Human Behavior*, 20(2), 201–223.
Sawhney, M., Verona, G. and Prandelli, E. (2005). Collaborating to create: the Internet as a platform for customer engagement in product innovation. *Journal of Interactive Marketing*, 19(4), 4–17.
Schau, H.J., Muñiz Jr., A.M. and Arnould, E.J. (2009). How brand community practices create value. *Journal of Marketing*, 73(September), 30–51.
Scott, S.G. and Lane, V.R. (2000). A stakeholder approach to organizational identity. *Academy of Management Review*, 25(1), 43–62.
Shneiderman, B. (2000). Creating creativity: user interfaces for supporting innovation. *ACM Transactions on Computer–Human Interaction*, 7(1), 114–138.
Wellman, B. (1979). The community question: the intimate networks of East Yorkers. *American Journal of Sociology*, 84(5), 1201–1231.

16. Taming the waves of adversity: exploring the multidimensional construct of organizational resilience

Andreas Alexiou

INTRODUCTION

Life is bound to transition and transformation. Day becomes night, death becomes birth, the caterpillar becomes the butterfly. However, alongside linear, gradual and typically predicted change, organizations – ever increasingly so – find themselves facing episodes of disruptive change, usually originating from sources that managers 'don't know that they don't know'. Rapid economic, technological, social and political changes weave a web of effects that generate epistemological (imperfection of knowledge) and ontological (natural variation) uncertainty (Walker et al., 2003) that guarantees that 'accidents' and big failures will occur probably more frequently than managers would like to admit (Starbuck, 2009). However, it is not possible for organizations to plan for every disaster that could conceivably affect them (Mitroff and Alpaslan, 2003). What is of the essence in such cases is that organizations have developed the capability to react and overcome adversity by the means of buffering the damage, creatively adapting and, ultimately, capitalizing on the new reality once it manifests itself. Such a capacity is reflected in the construct of organizational resilience.

Organization theory has examined the maladaptive or pathological cycles of behaviour in response to adversity both at an individual as well as a collective level (threat-rigidity; Staw et al., 1981) and argued on how radical unanticipated change leads to response rigidities by restricting information processing (for example, by narrowing the field of attention) and constricting control (for example, by centralization of authority, increased formalization). However, while there exist cases of organizations that have successfully dealt with exogenous shocks (for example, a shooting and stand-off in a business school; Powley, 2009) or survive ongoing strain in dynamic environments (for example, the interesting case of criminal organizations; Ayling, 2009), there is limited theory on the mechanism that enables organizations to escape a deterministic fate and thrive upon adversity.

Despite the recent popularity of the concept of resilience among practitioners' discussions and literature, scholarly literature remains fragmented and the construct inadequately theorized (Sutcliffe and Vogus, 2003). I believe that the elusiveness of the concept to a great extent lies in its multidimensionality and meta-nature. Resilience is an emergent capacity that lies at the intersection of active and passive capabilities of organizations, across organizational levels of analysis. As a second-order capability, or what Collis (1994) labels a meta-capability, resilience has to do not only with possessing particular resources or capabilities in order to cope with unexpected change, but also having the ability to learn, generate and deploy them when needed.

In this chapter I identify three phases in the circle of disruption: the incubation phase, the impact phase and the enhanced equilibrium phase. Three corresponding sets of capabilities get enacted during these phases in order to secure a resilient response to adversity. These capabilities provide a roadmap for exploring the construct of resilience and are by no means exhaustive. They are the product of literature synthesis from fields as diverse as psychology, ecology and organization theory, and aim at highlighting how diverse and multidimensional is the phenomenon of resilience. To be more specific, the first phase represents the period of time when the threat has not yet materialized; however the mindful awareness of organization members and the attention capability of the management can allow for sensing early ripples and anomalies in the environment that could moderate the surprise factor and even allow for a timely response. During the second phase where disruption is fully manifested, attention is directed to buffering of the damage and preventing disintegration at different levels. Any threat has an immediate negative impact on individuals, inducing psychological stress, anxiety and negative emotions. It also erodes group cohesiveness and cripples the formal structures or communication channels of the organization. At this stage, relational resources, transformational leadership and structural flexibility can minimize the impact of such effects. Finally, during the last phase, the organization needs to reach a new enhanced state of equilibrium, past the point of simple recovery. Capitalizing on adversity is the quintessence of resilience and requires the ability to learn from crisis and innovate at a strategic level.

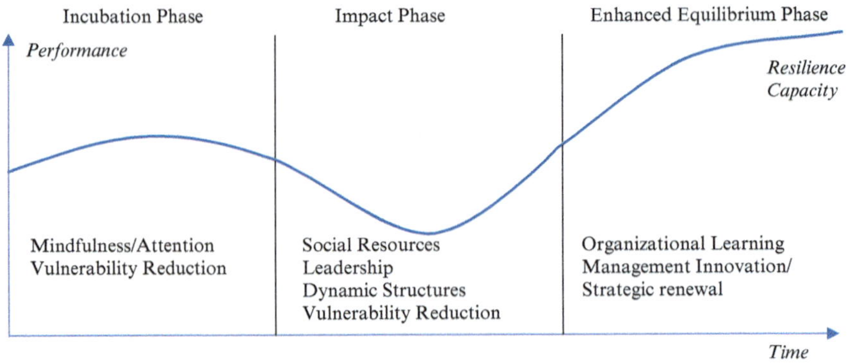

Figure 16.1 The phases of the disruption cycle and the capabilities that invoke a resilient response

RESILIENCE: A CONCEPTUAL MEDLEY

The construct of resilience has a rich history in fields outside organization and management studies. Organizational resilience builds on and expands existing literature in the areas of ecology, sociology and psychology in order to bear on a simple but important challenge that modern organizations face. This section will attempt to flesh out the dimensions and properties of organizational resilience by synthesizing literature primarily of two major scientific fields that dominate the resilience discussion in relevance to organizations: psychology and ecology.

Ecological Resilience

Traditional environmental science viewed the behaviour of natural systems in terms of stability near an equilibrium state with the primary focus set on the system's resistance to disturbance and the speed of return to the single global equilibrium (a definition that springs from traditions of engineering). However Holling (1973) illustrated that even undisturbed natural systems could in fact be in a transient state, and that the existence of multiple stability domains in those systems was actually possible. He illustrated how instabilities can force a system into another regime of behaviour and define functionally different states, without however the movement between states to undermine the structure and diversity of the system. Ecological resilience is therefore defined as the magnitude of disturbance that a system can absorb before it redefines its structure and changes stable states (Gunderson, 2000). Holling redirected the emphasis away from maintaining the efficiency of function, towards maintaining the existence of function, and high variability had become an attribute for existence, adaptability and learning (Folke, 2006).

The adaptive capacity of resilience was gradually brought forward and the dynamic adaptive interplay between sustainability and development with change became more explicit (Folke, 2006). Recent research in social-ecological systems incorporated the properties of adaptation, self-organization and learning alongside the ability to resist adversity. Along these terms, Carpenter et al. (2001) defined resilience as: (1) the amount of disturbance a system can undergo and still remain within the same domain of attraction; (2) the degree to which the system is capable of self-organization; and (3) the degree to which the system can build the capacity to learn and adapt.

Particularly interesting for this analysis is the departure from a more mechanistic view of resilience that has at its epicentre the return to pre-disruption equilibrium, to the idea that multiple different equilibrium states exist for a system, and transformation through adaptation is a viable strategy for retaining the existence of function. The role of learning in the process of adaptation and reconfiguration is also particularly relevant to organizations.

Human Resilience

Resilience in the field of developmental psychopathology made its first appearance by providing an important construct to examine why some at-risk children exposed to negative life events are in fact better able to succeed in their life (Masten et al., 1999). Within this literature resilience is defined as 'a class of phenomena characterized by patterns of positive adaptation in the context of significant adversity or risk' (Masten and Reed, 2002: 74). Positive adaptation accounts for the atypical process inherent to resilience that predicts positive outcomes and growth (benefit) out of circumstances that typically lead to maladjustment (loss).

Early researchers considered resilience a trait; a unique and static characteristic of individuals. But soon empirical evidence demonstrated that factors external to the child (for example, characteristics of their families and wider social environment) could also be contributing to resilience, resulting in the departure from phenomenological descriptions of resilient qualities of individuals, towards a dynamic process of resilience: a disruptive and reintegrative process of acquiring the desired resilient qualities (Richardson, 2002), fluctuating rather than fixed over time (Luthar, 2006), 'dependent upon interactions between individuals and contextual variables' (Metzl, 2009: 113).

A review of the literature reveals a set of antecedents that dominate the discussion on psychological resilience. These factors are typically related to: (1) how individuals make sense of the situation; (2) creative competence; (3) the role of values; and (4) the role of social relationships. Sense-making in the case of resilience primarily has to do with finding meaning and purpose (e.g. Coutu, 2002; Fine, 1991; Luthar, 2006); creative competence reflects the skills of improvisation and ingenuity that allow for finding solutions and surviving with whatever means are available at the time; values play the role of an anchor or reference point that an individual holds on to in order to cope with the overwhelming effect of dramatic change; social relationships through compassion and solidarity create a net of support that helps people to get back on their feet, make sense of the new reality and assist their psychological development.

The discussion on psychological resilience brings forward the element of positive adaptation instead of mere survival, and the dynamic nature of resilience as a capability that can be developed and enhanced over time by manipulating its internal variables through interaction with the environment. Organizations as human systems can benefit from some of the same protective factors that shield resilient individuals – for example, creativity, social resources and sense-making – and can learn to shape and enhance such qualities over time.

Organizational Resilience

Resilience, recently introduced to organization studies, is often defined as a capacity to cope with unanticipated dangers and bounce-back (e.g. Wildavsky, 1988). Based on the previous discussion, I consider the adoption of a wider definition of organizational resilience: one that includes the dynamic nature of resilience and the capacity to sense and capitalize on the opportunities that change offers in order to further develop. I define organizational resilience as the dynamic capacity of an organization to: (1) be mindfully aware of its environment; (2) absorb the negative impact of disruption; and (3) positively adjust and capitalize on adversity. In the next section I will elaborate on some core antecedents of these three dimensions of resilience.

The first dimension represents the relatively passive phase that takes place before the disruption materializes and includes the capabilities for sensing change and threats, and developing an integrative approach to risk management. The second dimension corresponds to the second phase of resilience activation that takes place once the disruption occurs, and portrays the ability of an organization to buffer the initial damage and shock primarily through liquid or dynamic structures and by utilizing the strong interpersonal relationships among its members. The last phase portrays the ability of an organization to capitalize on adversity and grow. During this period the organization needs to be able to learn from crisis, and allow for innovation at different levels (management, strategy, business processes).

ANTECEDENTS OF ORGANIZATIONAL RESILIENCE

Incubation Phase

While it goes without saying that in the case of some rare, dramatic events (for example, a terrorist attack) it is almost impossible for organizations to anticipate, when it comes to disruptions in the business cycle, for instance as a result of the dynamism of the environment (for example technological innovation), it is often the case that early signals do herald the upcoming events. Unfortunately, organizations often lack the ability to sense and interpret such signals, and there are many reasons for that. One typical example is the case of technological disruption challenging some deeply held beliefs within a firm, resulting in an identity crisis (Tripsas, 1997).

Organizational awareness is a capability that assumes the role of a protective factor against unexpected change when active during times of equilibrium. Recent research highlights the critical role of managerial attention (e.g. Maula et al., 2013) and the interrelation between processes of perception and cognition (mindfulness; Weick and Sutcliffe, 2006) in coping with exogenous disruption or unanticipated events. Empirical evidence has illustrated how top managers can shape a firm's strategy and handle disruptive events by 'interpreting changes in the external environment and formulating (or at least orchestrating the formulation of) appropriate and timely responses' (Maula et al., 2013: 15). Attention can be better understood as a variety of interconnected processes with their roots lying deep in the cognitive dimension of human behaviour. As such, it has been used in organization literature to describe different but interrelated mechanisms, processes and outcomes that Ocasio (2010) effectively groups in three ideal types: (1) attentional perspective: referring to the top-down cognitive structures that generate heightened awareness as response to specific stimuli and responses; (2) attentional engagement: referring to the process of intentional, prolonged and concentrated allocation of cognitive resources into a selected set of environmental stimuli; and (3) executive attention and vigilance: referring to the highly important to decision-making abilities of being able to switch from and resume focus between different stimuli, and fixing the attention on one particular stimulus.

All three types of attention are highly interconnected and vital to decision making, strategic action, adaptation and change (Maula et al., 2013; Ocasio, 1997). Attentional engagement in particular can be considered an important indicator of resilience as it signifies the 'quality' of organizational attention (Ocasio, 2010). It entails the mindful processing of information which reflects high attentiveness to context and the capacity to respond to unanticipated cues or signals (Levinthal and Rerup, 2006). Internal and external situation monitoring and reporting routines (Seville, 2008) that enhance the situational awareness of the organization, as well as information technology (IT) infrastructure that enriches action repertoires and engages actors in 'more extensive search processes, increased collaboration, and more careful contemplation of issues and alternative courses of action' (Valorinta, 2009: 991), are expected to further enhance the collective mindfulness of organizations.

Overall, mindfulness in the context of organizational resilience involves the passive element of awareness but also the active element of reflection, the combination of which enhances insight and the process of interpreting events constructively while experiencing a situation openly and holistically (Bishop et al., 2006; Brown and Ryan, 2003; Weick et al., 1999). Its role in developing a capacity for resilience is critical in the way that it: (1) fosters the capability of discovering and managing unexpected events (Weick et al., 1999); and (2) enhances the capacity of learning from crisis (Levinthal and Rerup, 2006).

Embedded in the managerial efforts to sense and predict upcoming turbulence exists a strategic focus on vulnerability reduction through the active assessment, monitoring and treatment of risk. Such risk management approaches can never wholly develop a capacity for resilience, however the positive effect of proactive vulnerability reduction to the overall robustness of a system cannot be ignored. Vulnerability analysis includes a set of processes that are embedded in the operation of an organization, and determine which areas and threats should be considered part of a catastrophic risk analysis by using frequency data or on the basis of expert judgements, scenarios and subjective probabilities (Haimes, 1998). One very important aspect of such an approach is the realization of the interconnectedness of risks and the continuous, proactive and systematic process of understanding, managing and communicating risk across organizational boundaries.

Impact Phase

When equilibrium gets punctuated by disruption, focus gets shifted towards overcoming the rigidities that arise and reversing the disintegrating effect that it conjures. A threat situation can have a detrimental effect on different levels, such as individual psychology, team cohesiveness and organizational processes (Staw et al., 1981). Social resources, leadership and structural flexibility represent the most important organizational reflexes during a period of impact.

One of the detrimental effects that adversity has on people is the activation of negative emotions such as psychological stress and anxiety. Emotions are tightly interwoven with cognition and motivation (Lazarus, 1991), are a determinant of human behaviour and relationships, and play an important role in the process of adaptation and change (Huy, 1999, 2002). Research in psychology has shown how negative emotions narrow individuals' scope of attention (Easterbrook, 1959) and thought–action repertoires (Fredrickson and Branigan, 2005), and call forth specific action tendencies (for example flight, resignation). Managing negative emotions and enabling principles such as solidarity and compassion is extremely important for avoiding maladaptive adjustments to change (Avey et al., 2008; Dutton et al., 2006).

Initial empirical evidence shows that during a crisis, resilience is activated through the people's ability to alter social structure, extend compassion and act in ways that foster caring and supportive relationships, and enlarging information inputs by drawing on their social capital (e.g. Powley, 2009). Literature in developmental psychology, adult coping and post-traumatic growth (Tedeschi and Calhoun, 2004) has consistently illustrated that social support and strong human relationships can contribute to the positive adjustment of individuals after exposure to adversity.

The role of leadership during this process, while largely overlooked (Luthans and Avolio, 2003), appears to be important (Harland et al., 2005). Leaders can influence people and direct action by engaging in symbolic actions or creating powerful symbols that can influence people at a cognitive and emotional level (Dutton et al., 2006). At the same time they should have the ability to convert crisis into a developmental challenge (Bass, 1990), assist the meaning-making process of organization members as well as provide 'intellectual stimulation to promote subordinates' thoughtful, creative, adaptive solutions to stressful conditions, rather than hasty, defensive maladaptive ones' (p. 690). As resources and time can be limited during times of crisis, the ability to improvise a solution without proper or obvious tools is a core element of resilience (Coutu, 2002). While individuals with such capabilities (bricoleurs) can be found among the ranks of every organization, it is important that they are granted expanded decision-making authority and access to resources in order to assume such a role (Mallak, 1998). Therefore appropriate improvisation established through tools, rules and routines augments contingent structuring mechanisms, leading to increased responsiveness and contributing to overall resilience (Bigley and Roberts, 2001).

The same degree of flexibility and responsiveness however must be also displayed at an organizational level, primarily through dynamism in terms of structure and the strategic use of information systems. Structural flexibility is dependent 'on the management's capabilities to adapt the organization structure, and its decision and communication processes, to suit changing conditions in an evolutionary way' (Volberda, 1996: 362). Examples of internal structural flexibility include changes in organizational responsibilities, alterations in control systems, and the existence of virtual role systems (Volberda, 1996; Weick, 1993). Enterprise systems on the other hand, while often considered the backbone of modern organizations, can become a double-edged sword when it comes to flexibility. Their mission is to facilitate the seamless integration and exchange of data between the various departments of an organization; however if not managed properly they can introduce rigidities and decreased flexibility due to increased control, a response to power differentials introduced by the centrality of the configuration and the concentration of power in the hands of selected individuals (Ignatiadis and Nandhakumar, 2006).

Enhanced Equilibrium Phase

Once the new reality starts to crystallize the organization enters a new phase of equilibrium where in the case of positive adaptation it signifies a phase of opportunity and growth. In order to turn the disruption into opportunity however the organization needs to be capable of integrating the lessons learned and innovating on a strategic level. The quintessence of resilience is not the mere ability to survive but rather to transform through change, and grow through learning. Learning from crisis is a fundamental dimension of resilience; however, evidence shows that organizations hardly ever do so (Smith and Elliott, 2007). In order to capitalize on the newly formed reality, organizations need the ability to learn from their own failures, and hopefully, the failures of others. Many things can go wrong in this process. First, an organization could deny that it fell victim to an internal vulnerability by seeing the failure as having idiosyncratic and largely exogenous causes (Starbuck, 2009). Denial inhibits renewal and, according to Hamel and Välikangas (2003), managers can counteract this by collecting first-hand information and experience on the sites that first confront change, and making sure that units that possess the required mindfulness to interpret events are not censored. Second, an organization could engage in first-order learning by evaluating and reconsidering some of its processes and practices, although the danger of failing to question the true nature of the problem and engage in deeper cultural or strategy change always lingers (Smith and Elliott, 2007).

The second step in this phase is the realization that strategy eventually decays. This is an important prerequisite for capitalizing on the newly formed reality and sustaining competitive advantage. Through rapid-resource redeployment and strategic variety, organizations can accommodate radical change in their environments (Hamel and Välikangas, 2003). Strategic flexibility reflects a set of capabilities related to the reconfiguration of the organization's goal, vision and mission, and interpretation of the environment that is necessary when the organization faces disruptive or unfamiliar changes that have far-reaching consequences and require timely response (Volberda, 1996). The lack of previous experience of similar circumstances however might involve a holistic reconfiguration or renewal of plans, strategies, resources or products. New values and norms are necessary for the phase of renewal, and as past experience may not be relevant, the ability to innovate at a managerial level would allow the organization to introduce new practices, processes or structures that would further organizational goals (Birkinshaw et al., 2008).

CONCLUSION

Organization scholars have recently picked up on the extensive literature on ecological, sociological and human resilience as well as related psychological phenomena such as hardiness (Kobasa et al., 1982) or invulnerability (Anthony, 1987) that reflect the ability of certain individuals to positively adjust and survive trauma, in an attempt to explain how organizations cope with potential threats, and what distinguishes organizations that survive such adverse circumstances from those that fail (Sutcliffe and Vogus, 2003). Organizational resilience represents a broaden-and-build approach to managing radical change instead of a deterministic and, in essence, mechanistic view of organizational survival in response to disruption.

Creativity and improvisation (e.g. Bigley and Roberts, 2001; Lengnick-Hall et al., 2011; Mallak, 1998), sense-making (e.g. Coutu, 2002; Jacelon, 1997; Weick, 1993), strong values (Coutu, 2002; Sutcliffe and Vogus, 2003) and organizational assets such as knowledge management systems, clear communication channels and so on are some of the factors that have been identified to foster organizational resilience. However as Zellars et al. (2011: 1) observe in a recent review of the literature, 'much remains unknown about the ability to build resilience capacity at work'. Given the dearth of empirical research in this area (Vogus and Sutcliffe,

2007) the literature remains fragmented and many mechanisms behind the capacity for resilience remain unexplored. I argue that it is helpful to deem organization resilience as a meta-capability, a capacity that involves the nurturing and management of distinct first-order capabilities. Consequently, this chapter has attempted to provide a working definition and flesh out the construct's dimensions and outcomes as well as the human-related, operational and strategic underpinnings of its antecedents.

Resilient organizations are not a mere collection of resilient individuals (Ill and Orr, 1998). Future research should look in more detail at how resilience is embedded in individual-level knowledge, skills and abilities, how it is enabled by organizational routines and processes, and how it gets reinforced by social interaction and human relationships. A major challenge in particular is in drawing the link between individual and organizational-level resilience (Youssef, 2004; Zellars et al., 2011). The role of networks among organizational members as well as the relationships of the wider ecosystem in building a capacity for resilience is another fruitful area for research. Given the ever-increasing dynamism of the business environment of most industries – let alone organizations operating in traditionally dynamic environments like that of emerging technologies – and the countless contingencies that our modern socio-economic reality forces on firms, it is imperative that we understand how organizations can positively adjust under conditions of radical change and disruption and emerge stronger and wiser. By answering this, future research will be able to contribute towards meeting one of the most pressing needs of modern organizations.

REFERENCES

Anthony, E.J. 1987. Children at high risk for psychosis growing up successfully. In Cohler B.J. (ed.), *The Invulnerable Child*. New York: Guildford Press, pp. 147–184.

Avey, J.B., Wernsing, T.S. and Luthans, F. 2008. Can positive employees help positive organizational change? Impact of psychological capital and emotions on relevant attitudes and behaviors. *Journal of Applied Behavioral Science* 44(1): 48–70.

Ayling, J. 2009. Criminal organizations and resilience. *International Journal of Law, Crime and Justice*. 37(4): 182–196.

Bass, B. 1990. *Handbook of Leadership*, 3rd edn. New York: Free Press.

Bigley, G. and Roberts, K. 2001. The incident command system: high-reliability organizing for complex and volatile task environments. *Academy of Management Journal* 44(6): 1281–1299.

Birkinshaw, J., Hamel, G. and Mol, M. 2008. Management innovation. *Academy of Management Review* 33(4): 825–845.

Bishop, S.R., Lau, M., Shapiro, S., Carlson, L., Anderson, N.D., Carmody,

J., Segal, Z.V., Abbey, S., Speca, M., Velting, D. and Devins, G. 2006. Mindfulness: a proposed operational definition. *Clinical Psychology: Science and Practice* 11(3): 230–241.

Brown, K.W. and Ryan, R.M. 2003. The benefits of being present: mindfulness and its role in psychological well-being. *Journal of Personality and Social Psychology* 84(4): 822–848.

Carpenter, S., Walker, B., Anderies, J.M. and Abel, N. 2001. From metaphor to measurement: resilience of what to what? *Ecosystems* 4(8): 765–781.

Collis, D. 1994. Research note: how valuable are organizational capabilities? *Strategic Management Journal* 15: 143–152.

Coutu, D. 2002. How resilience works. *Harvard Business Review* 80(5): 46–55.

Dutton, J., Worline, M., Frost, P.J. and Lilius Jacoba, M. 2006. Explaining compassion organizing. *Administrative Science Quarterly* 51(1): 59–96.

Easterbrook, J.A. 1959. The effect of emotion on cue utilization and the organization of behavior. *Psychological Review* 66(3): 183–201.

Fine, S. 1991. Resilience and human adaptability: who rises above adversity? *American Journal of Occupational Therapy* 45(6): 493–503.

Folke, C. 2006. Resilience: the emergence of a perspective for social–ecological systems analyses. *Global Environmental Change* 16(3): 253–267.

Fredrickson, B.L. and Branigan, C. 2005. Positive emotions broaden the scope of attention and thought–action repertoires. *Cognition and Emotion* 19(3): 313–332.

Gunderson, L. 2000. Ecological resilience – in theory and application. *Annual Review of Ecology and Systematics* 31: 425–439.

Haimes, Y. 1998. *Risk Modeling, Assessment and Management*. New York: John Wiley.

Hamel, G. and Välikangas, L. 2003. The quest for resilience. *Harvard Business Review* 81(9): 52–63, 131.

Harland, L., Harrison, W., Jones, J.R. and Reiter-Palmon, R. 2005. Leadership behaviors and subordinate resilience. *Journal of Leadership and Organizational Studies* 11(2), 2–14.

Holling, C. 1973. Resilience and stability of ecological systems. *Annual Review of Ecology and Systematics* 4: 1–23.

Huy, Q. 1999. Emotional capability, emotional intelligence, and radical change. *Academy of Management Review* 24(2): 325–345.

Huy, Q. 2002. Emotional balancing of organizational continuity and radical change: the contribution of middle managers. *Administrative Science Quarterly* 47(1): 31–69.

Ignatiadis, I. and Nandhakumar, J. 2006. The impact of enterprise systems on organizational resilience. *Journal of Information Technology* 22(1): 36–43.

Ill, J.F.H. and Orr, J.E. 1998. Assessing behaviors that create resilient organizations. *Employment Relations Today* 24(4): 29–39.

Jacelon, C.S. 1997. The trait and process of resilience. *Journal of Advanced Nursing* 25(1): 123–129.

Kobasa, S.C., Maddi, S.R. and Kahn, S. 1982. Hardiness and health: a prospective study. *Journal of Personality and Social Psychology* 42(1): 168–177.

Lazarus, R.S. 1991. Progress on a cognitive–motivational–relational theory of emotion. *American Psychologist* 46(8): 819–34.

Lengnick-Hall, C.A., Beck, T.E. and Lengnick-Hall, M.L. 2011. Developing a capacity for organizational resilience through strategic human resource management. *Human Resource Management Review* 21(3): 243–255.

Levinthal, D. and Rerup, C. 2006. Crossing an apparent chasm: bridging mindful and less-mindful perspectives on organizational learning. *Organization Science* 17(4): 502–513.
Luthans, F. and Avolio, B. 2003. Authentic leadership development. In Cameron, K.S., Dutton, J.E. and Quinn, R.W. (eds), *Positive Organizational Scholarship: Foundations of a New Discipline*. San Fransisco, CA: Berrett-Koehler, pp. 241–258.
Luthar, S. 2006. Resilience in development: a synthesis of research across five decades. In Cicchetti, D. and Cohen, D.J. (eds), *Developmental Psychopathology: Risk, Disorder, and Adaptation*, Vol. 3. New York: Wiley, pp. 739–795.
Mallak, L. 1998. Putting organizational resilience to work. *Industrial Management* 40(6): 8–13.
Masten, A.S. and Reed, M.J. 2002. Resilience in development. In Snyder, C.R. and Lopez, S.J. (eds), *Handbook of Positive Psychology*. London: Oxford University Press, pp. 74–78.
Masten, A.S. Hubbard, J.J., Gest, S.D., Tellegen, A., Garmezy, N. and Ramirez, M. 1999. Competence in the context of adversity: pathways to resilience and maladaptation from childhood to late adolescence. *Development and Psychopathology* 11(1): 143–69.
Maula, M., Keil, T. and Zahra, S. 2013. Top management's attention to discontinuous technological change: corporate venture capital as an alert mechanism. *Organization Science* 24(3): 926–947.
Metzl, E.S. 2009. The role of creative thinking in resilience after Hurricane Katrina. *Psychology of Aesthetics, Creativity, and the Arts* 3(2): 112–123.
Mitroff, I.I. and Alpaslan, M.C. 2003. Preparing for evil. *Harvard Business Review* 81(4): 109–115.
Ocasio, W. 1997. Towards an attention-based view of the firm. *Strategic Management Journal* 18(Summer Special Issue): 187–206.
Ocasio, W. 2010. Attention to attention. *Organization Science* 22(5): 1286–1296.
Powley, E.H. 2009. Reclaiming resilience and safety: resilience activation in the critical period of crisis. *Human Relations* 62(9): 1289–1326.
Richardson, G.E. 2002. The metatheory of resilience and resiliency. *Journal of Clinical Psychology* 58(3): 307–321.
Seville, E. 2008. Resilience: great concept but what does it mean? Council on Competitiveness – Risk Intelligence and Resilience Workshop, Wilmington, USA, November.
Smith, D. and Elliott, D. 2007. Exploring the barriers to learning from crisis: organizational learning and crisis. *Management Learning* 38(5): 519–538.
Starbuck, W. 2009. Perspective – cognitive reactions to rare events: perceptions, uncertainty, and learning. *Organization Science* 20(5): 925–937.
Staw, B., Sandelands, L. and Dutton, J. 1981. Threat rigidity effects in organizational behavior: a multilevel analysis. *Administrative Science Quarterly* 26(4): 501–524.
Sutcliffe, K.M. and Vogus, T.J. 2003. Organizing for resilience. In Cameron, K.S., Dutton, I.E. and Quinn, R.E. (eds). *Positive Organizational Scholarship: Foundations of a New Discipline*. San Francisco, CA: Berrett- Koehler, pp. 94–110.
Tedeschi, R. and Calhoun, L. 2004. Posttraumatic growth: conceptual foundations and empirical evidence. *Psychological Inquiry* 27(1): 1–18.
Tripsas, M. 1997. Unraveling the process of creative destruction: complementary assets and incumbent survival in the typesetter industry. *Strategic Management Journal* 18(S1): 119–142.

Valorinta, M. 2009. Information technology and mindfulness in organizations. *Industrial and Corporate Change* 18(5): 963–997.
Vogus, T.J. and Sutcliffe, K.M. 2007. Organizational resilience: towards a theory and research agenda. *2007 IEEE International Conference on Systems, Man and Cybernetics*, pp. 3418–3422.
Volberda, H.W. 1996. Toward the flexible form: how to remain vital in hypercompetitive environments. *Organization Science* 7(4): 359–374.
Walker, W., Harremoës, P., Rotmans, J., van der Sluijs, J.P., van Asselt, M.B., Janssen, P. and Krayer von Krauss, M.P. 2003. Defining uncertainty: a conceptual basis for uncertainty management in model-based decision support. *Integrated Assessment* 4(1): 5–17.
Weick, K.E. 1993. The collapse of sensemaking in organizations: the Mann Gulch disaster. *Administrative Science Quarterly* 38(4): 628.
Weick, K. and Sutcliffe, K. 2006. Mindfulness and the quality of organizational attention. *Organization Science* 17(4): 514–524.
Weick, K.E., Sutcliffe, K.M. and Obstfeld, D. 1999. Organizing for high reliability: process of collective mindfulness. In Sutton, R.S. and Staw, B. (eds), *Research in Organizational Behavior*. Stanford, CA: JAI Press, pp. 81–123.
Wildavsky, A. 1988. *Searching for Safety*. New Brunswick, NJ: Transaction Books.
Youssef, C.M. 2004. Resiliency development of organizations, leaders and employees: multi-level theory building and individual-level, path-analytical empirical testing. ETD collection for University of Nebraska – Lincoln.
Zellars, K., Justice, L. and Beck, T. 2011. Resilience: new paths for building and sustaining individual and organizational capacity. *Research in Occupational Stress and Well-being* 9: 1–37.

17. From Web 2.0 to Social CRM: the place and value

Pradeep Durgam

INTRODUCTION: THE INTERACTIVE SOCIETY AND THE LACK OF SOCIAL EXPERIMENTATION

Since 2000, contemporary information and communication technology (ICT) as an emerging phenomenon has suggested several enhancements and provisions for the traditional customer relationship management (CRM) systems that have already been deployed, and for those on the verge of being implemented. The sole reason behind this strategy has been to better organize customers and proficiently manage them (acquisition, retention, and so on) (Rodriguez and Honeycutt, 2011; Xiong et al., 2011; Gebert et al., 2002). CRM systems were devised in order to streamline the processes related to marketing, sales, services and various other activities which involved customers. Therefore, the CRM system, a part of the enterprise's CRM strategy, is the core element for maintaining and managing customer information. For years, CRM systems have been counted as one of the many technological facilitators for simulating and revitalizing CRM strategies. The main aim of CRM strategy is to try and administer customers' behaviour and transfigure customer relationships into profitability. It also involves collecting customer data and monitoring customer transactions to determine customers' thought process (Feiz et al., 2011; Nambisan and Baron, 2007). For instance, understanding of the customer, and process efficiency with the help of aligned CRM systems, increases the purchasing level for the customer and decreases the cost of customer retention for the firm (Lemon et al., 2002; Xu and Walton, 2005; Verhoef, 2003; Jayashree et al., 2011; Zineldin, 2006). But that was not always the case. There was a time when CRM strategy was embedded or encompassed within the CRM system. It was after the advent of Web 2.0 innovation as an emerging Internet technology that CRM systems started becoming 'dependent' on CRM strategy, which changes constantly due to varying customer needs (Faase et al., 2011; Li et al., 2012). As CRM systems provide one-way interaction that flows only from the organization to its customers (knowledge flow is one-directional), it is important to note the role of Web 2.0 in transforming the manner in which individuals and firms interact and collaborate via the Internet today (two-way collaboration). Communication, which was previously restricted to two individuals, has now broadened to a wider audience with the use of Internet technology via Web 2.0, through blogs, wikis, communities, news, e-mails, online television, websites and the firm's interactive platform for its customers (existing and new), to name a few. The advancement of Web 2.0 has encouraged and persuaded this digital environment to innovate and produce modern tools and interactive platforms (LinkedIn, Facebook, Twitter, and so on). Web 2.0 also empowers cultures and societies to operate, access, familiarize and generate knowledge in quality and quantity more rapidly at minimal costs and suggest vast opportunities for boosting business and economic sustainability. The use of Web 2.0 has facilitated amplified social participation, self-profiling (Ryals and Knox, 2001; Sigala, 2008), social networking, reinforcing democracy (Levy, 2009; Barsky and Purdon, 2006; Alavi et al., 2011), vying in a global business environment, eliminating obstacles to modernization, and has overall revolutionized the once static society to create an interactive society now. Considering a social ecosystem that involves interaction and collaboration between various actors and elements, there is a purpose for the alliance and association for a desired outcome that involves feedback, reviews, product/service information and similar brand interest, amongst others. To date, CRM systems have been researched as a technical entity related to a particular firm, or as a case for failure and success (Table 17.1); and Web 2.0 has been analysed individually as an advancement in Internet technology, but not much has been explored about the accommodation of Web technologies in the existing CRM strategies. Web 2.0 provokes social intervention and social experimentation that can enhance the roadmap planned under the CRM strategy of a firm. There have been very few articles that have integrated both the concepts of Web 2.0 and CRM and branded it as 'Social CRM' or SCRM, as there is little knowledge or expertise that can create a Social CRM environment. The technology–organization–environment (TOE) framework that Oliveira et al. (2011) developed in the 1990s presents a perfect platform upon which to base the inclusion of SCRM in CRM strategy, because both of these discuss and classify three important facets – namely the technological context, the organizational context and environmental context – within an enterprise that impacts upon the process of adopting and implementing a technological innovation. While in a technological context, the SCRM phenomenon emerging from Web 2.0 represents both the internal and external technologies relevant to the firm; the explicit and tacit knowledge present within

Table 17.1 CRM failure

Authors	CRM failures
Rigby et al. (2002)	1. Implementing CRM before creating a customer strategy. 2. Rolling out CRM before changing your organization to match. 3. Assuming that more CRM technology is better. 4. Stalking, not wooing, customers.
Mithas et al. (2005)	The propensity of firms to avoid the important data transformation and convergence processes including all transactions, interactions, and networked touch points.
Dong and Zhu (2006)	Lack of business value derived from CRM investment; 41% of the firms with CRM projects were either experiencing significant difficulties or close to failure.
Stringfellow and Bowen (2004)	55% of CRM projects are not expected to produce results. A lack of commitment to the CRM system, its associated actors and customer focus leads to the highlighted failures.
Frow et al. (2011)	Failure due to tactical issues, including quality of data, project management and technological skills, and strategic aspects of CRM implementation.
Zablah et al. (2005)	CRM implementation initiatives have ended in failure; failure rate ranges from 35% to 75%.
Bolton and Tarasi (2007)	The collection of the data does not imply the existence of useful information that will be disseminated and acted upon appropriately.

a firm, the scope (products and services) and the size of the firm and the structure of management can be categorized under the organizational context. The firm that manages its business with its consumers, competitors and political elements can be then classified under the environmental context. The TOE framework offers a useful analytical framework that can support investigation of the adoption and integration of different types of technological innovation such as SCRM concepts originating from Web 2.0 and the CRM domain. I refer to this framework as it has a concrete theoretical basis, consistent historical empirical support and the prospect of application to technological innovation domains. This research also mentions some articles on SCRM as references supporting the chapter. The chapter starts by focusing on CRM strategy, CRM systems, Web 2.0, Semantic Web and its associated tools, and moves on to explain the concept of SCRM and discuss the place and value SCRM currently has and may create in the future.

SWITCHING GEARS WITHIN STRATEGIC CRM

Customer relationship management or CRM has been defined in numerous contexts, as and when CRM as a strategy emerged. Although researchers, scholars, marketing experts and champions in various customer-oriented professions have delved into this constantly developing, important topic and recognized it in different ways, the main crux has always revolved around strategy, people, technology, capabilities and processes (Kevork and Vrechopoulos, 2009; Chen and Popovich, 2003; Xu et al., 2002; Zablah et al., 2004). Viewing the classic definition of traditional CRM theorized by Paul Greenberg (2009), where he states that 'CRM is a philosophy and a business strategy supported by a system and a technology designed to improve human interactions in a business environment', firms are recognizing that their customers have different economic significance, and are thereby successively modifying their customer strategy by including communication appropriately (Reinartz et al., 2004). Firms are also constantly shifting from a product- or service-centric approach to a customer-centric approach. Increasingly, organizations are highlighting their CRM strategy, emphasizing the customer-oriented processes where the elements of Web technology are unified into the CRM systems, facilitating the business process related to marketing, sales and services (Reinhold and Alt, 2012; Torggler, 2009; Deng et al., 2009; Osarenkhoe and Bennani, 2007; Frow and Payne, 2009). Payne and Frow underline Boulding et al.'s (2005) thoughts that 'strategy' resides at the core of any successful CRM. Fan and Ku (2010), Campbell (2003) and Sawhney et al. (2005) argue that an organization venturing into CRM initiatives should evaluate the growth opportunities that are available in the business ecosystem and choose the customer relationships that are suitable in nature for the planned customer segments.

For a successful CRM, a customer first should be categorized as a crucial stakeholder for the future success of the firm, rather than regarding them merely as a consumer. A successful CRM also takes into account an organization's current situation in the market and its customers, around which the strategy is planned (Greenberg, 2009; Ata and Toker, 2012; Boulding et al., 2005). A business strategy should be the first step to decide how a customer strategy needs to be established, followed by forecasting the evolution over time (Woodcock et al., 2011; Ang, 2011; Greenberg, 2010a; Tollin, 2002). While a business strategy process can be initiated with an analysis and vocalization of a firm's foresight, specifically when it relates to CRM followed by the investigation of the industry and its competitive ecosystem (Payne and Frow, 2005; Frow and Payne, 2009), a customer strategy involves exploring the current and probable customer base and classifying which forms of customer segmentation are the most suitable (Zahaya et al., 2004; Pavicic et al., 2011; Sofianti et al., 2010; Kumar, 2008; Reinhold and Alt, 2012). Segment granularity plays an important role, involving assessments of a suitable macro, micro or one-to-one segmentation approach (Rubin, 1997). Many authors underline the capability of shifting to one-to-one marketing from a mass-market environment by utilizing the essential economic features of the Internet that can empower a rooted level of segmentation granularity (Peppers and Rogers, 1993; Tseng and Piller, 2003; Wells et al., 1999; Piller et al., 2004; Pitta, 1998).

Having said that, it is crucial to focus on a few important aspects that must be infused in CRM strategic initiatives, namely meaningful advancement in services by integrating satisfaction, information and communication technology that involves loyalty and advocacy, aspects related to people implicating culture transformation, new interactions between individuals and groups, the realization of changing expectation and customer behaviour, and personal agendas (Foss et al., 2008; Anshari and Almunawar, 2011). Following which, if the CRM strategy is enhancing the profitability and quality of the product and service including the realistic price strategy, then it can be said that the firm is moving in the right direction and holding a strong position in the market (Zineldin, 2006). That explains why CRM strategy necessitates a cross-functional integration of processes, people and technology (CRM systems and Web 2.0).

THE PROCESS LINKAGE

Process is one of the important pillars of CRM strategy. Its role is crucial, as CRM is a cross-functional, customer-driven, technology-integrated business process management strategy that exploits relationships and embodies the entire organization for facilitating marketing, sales, service and other customer-oriented processes. That is why a customer-centric organization impeccably integrates business processes and marketing to handle customers and react to market pressures. Business process management assists the formation of the logical cores of CRM, which are (to name a few): lifetime value, customer profitability, retention, relationship marketing and satisfaction (Sahaf et al., 2011; Kumar et al., 2010; Osarenkhoe and Bennani, 2007; Roberts et al., 2005; Gummesson, 2004;

Kim and Lee, 2007). CRM can thus be portrayed as a blend of business process and technology (Faase et al., 2011) that aims to understand a company's customer from the standpoint of who they are, what they do and what they are like (Chen and Popovich, 2003). It is an endless activity that necessitates the restructuring of central business processes, beginning from the customer perspective to customer feedback. In general, firms tend to incorporate their central business processes, associate interrelated activities and eliminate the processes that do not benefit or add value to their customers. And for customer-focused firms aiming at customer processes, there is a need for 'knowledge' of a significant degree, because they deliver knowledge that customers request, process the knowledge received from their customers, and possess knowledge about their customers (Fan and Ku, 2010). CRM processes need transactional data that are automatically gathered and stockpiled in relational databases, and also a significant amount of knowledge from other sources (Greenberg, 2009; Bueren et al., 2004; Liao et al., 2010). Also, marketing, sales and service being the main business functions have a great level of direct customer interaction and operate in a knowledge-intensive environment. Bueren et al. (2005) rightly indicate that a CRM business process engages direct customer contact, exchanges information and services between enterprise and customer, and processes customer knowledge to fulfil the goals of relationship marketing. The customer initiates processes with the aim of receiving information and services and also at the same time the enterprise initiates processes with the aim of delivering information and services to customers. A CRM process consistently compels changes throughout the firm and with the whole-hearted consent and commitment of top management (Almotairi, 2008; Chen and Hung, 2010; Coltman, 2007; Lin and Lee, 2006; Ernst, 2002). Kumar et al. (2004) regard CRM as the process of realizing and retaining an ongoing relationship with its customers across multiple customer touch points via distinct and tailored treatment of individual customers created on their possible responses to the substituted marketing programmes, so that the involvement of each customer towards the whole profitability of the firm is maximized. Boulding et al. (2005) decipher the scope of CRM as an encompassing strategy, acquisition and dissemination of customer knowledge to appropriate stakeholders, intelligent use of data and technology, management of the dual creation of value, development of appropriate (long-term) relationships with specific customers and/or customer groups, and the integration of processes across the many areas of the firm and across the network of firms that collaborate to generate customer value, thereby reiterating the importance of processes in CRM.

POWER OF PEOPLE IS PIVOTAL

While technology and business processes are both crucial to fruitful CRM initiatives, it is the people within the organization that continue to remain the focal point of customer relationships (Goldenberg, 2000). Implementation of relationship activities in a firm develops the social and structural bonds which integrate people, organizational systems and processes together. Investing in people is the most important task because internal relationships are as important as external relationships. Execution of a relationship alignment is derived from the people in the firm who understand the goals set and meet the essential standards (Donaldson and O'Toole, 2007). Constructing, altering and operating a CRM solution comprises many important actors, who include systems experts, business analysts, backroom operations specialists, managers who use customized reports to fine-tune sales, marketing and customer service strategies, and the frontline sales and service agents who are accountable for recording data that the CRM initiative needs to further generate rich insights (Roberts et al., 2005). Donaldson and O'Toole mention the core dimensions of relationship strategy implementation as being organizational structure, team-based structure, the staff and the style that managers use to define the outcome of an implemented strategy by setting up relational systems (for example, sales service processes, supply chain management system, relationship performance scorecards and the order fulfilment system). But that is one side of the coin. The other, as Boulding et al. (2005) highlight, is that little attention is given to people issues and debates in an organization despite the fact that employees are a fundamental part in the delivery of CRM activities. Add to that concerns from the consumers' or customers' point of view, and what you have is a cause for worry. Customers believe that firms use CRM with a strategic agenda, which sows the seed of mistrust in their minds. They are then cautious while observing the firm's behaviour and prefer to retain the power to strategically initiate interactions with the firm (Yang et al., 2010; Lin and Lee, 2006). The real question from the consumers' viewpoint is whether the firm will use the gathered information and data, and whether it will help the consumer in return (Boulding et al., 2005). It is therefore important to note that while technological processes and CRM systems are critical for CRM strategic initiatives, without proper human interaction with these processes and systems, the return on investment (ROI) for a firm is at risk (Al-mashari and Zairi, 2000).

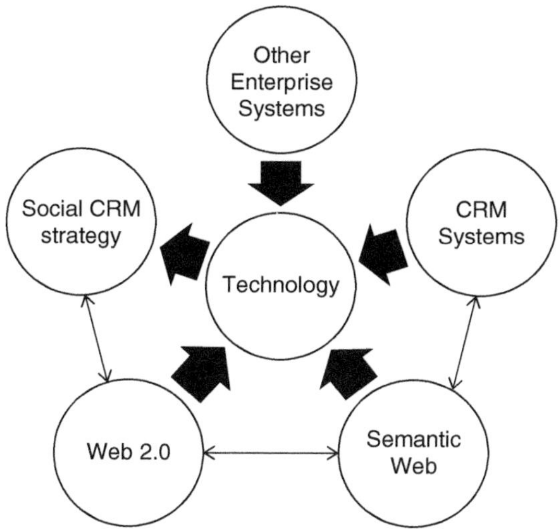

Figure 17.1 Elements of technology

TECHNOLOGY ESSENTIALS

CRM Systems as Business Enablers

From people to technology: a lethal and indispensable combination. The socio-technical system can be deemed as a purposeful collaborative system or a social work system. It comprises the collectiveness of the social aspect (people), the ecosystem and the technological element (see Figure 17.1) (Heller, 1997). In order to reap the benefits from the collaborative architecture, the technology and people have to be combined within a system, as they cannot be maximized separately. The use of a socio-technical system could be a joint optimization technique to be employed towards a determined objective. Socio-technical systems take the notion of an 'open system' for analysing, describing, designing and managing the environmental complexity and competition (Appelbaum, 1997). Working on relationship-building agendas can be reinforced by technologies via CRM systems that permit the firms to gain insights into the behaviour of their customers and produce crucial information about those customers. Interfunctional assimilation of processes, people, operations and marketing know-how within the organization are enabled through CRM systems and applications (Choy et al., 2004). The prominence of technology in assisting CRM strategic initiatives is represented by the challenges, which define that notion. CRM is outlined as the association of business strategies and business processes to create customer loyalty and profitability that is enabled by technology (Rigby et al., 2003). CRM systems generally depict a single point of all data linked with customer processes and the functional aspect to manage them (Reinhold and Alt, 2011). They are a composite assimilation of hardware and software applications that involves a detailed analysis of the firm's business processes (Bose, 2002). For firms moving forward to adopt CRM applications in order to improve customer satisfaction, it is worthwhile to focus on understanding and fulfilling the needs of customers by choosing an appropriate CRM technology and developing a CRM attitude on customer-facing business processes (Stone, 2009; Ali and Alshawi, 2004; Stefanou et al., 2003).

Web 2.0: The Showstopper

Technology took a new leap with Web 2.0. O'Reilly coined and introduced Web 2.0 as a network platform, spanning across all connected digital devices and idiosyncratic channels illustrated by user participation, openness, ease of collaboration and sharing and network effects (O'Reilly, 2007). Web 2.0 technologies offer the capacity to partake in multidirectional communications giving the liberty and opportunity for interaction, engagement and collaboration with the known and the unknown stakeholders (Black and Jacobs, 2010; Kamel Boulos and Wheelert, 2007). Social networking sites (SNSs), information aggregator sites, blogs and wikis, amongst others, can be categorized under Web 2.0 technologies, which is altering the dynamics of current businesses with the help of knowledgeable consumers and their collective intelligence. Firms have been innovating and creating new business models such as the crowdsourcing concept, a community-based model, for exploiting the knowledge of customers, employees and partners. Constant development of Web 2.0 technologies helps to recognize new products and services for consumers. But that was not always the case. Faase (Faase et al., 2011) complements Reinhard's (2009) statement which says that before the evolution of Web 2.0, the Web was not considered as a competitor and a threat to broadcasting, since Web 1.0 was just an information and transaction medium, whereas now Web 2.0 is used as an advertising and entertainment medium. Faase et al. (2011) use the deductive comparison approach and classify the seven principles and the fundamental aspects of Web 2.0. Through the Web 2.0 platform, the traditional knowledge management with the centralized knowledge repository has transformed into a more interactive conversational approach (Lee and Lan, 2007). Hossain and Aydin (2011) take the Web 2.0 dynamics a bit further by presenting the participation in virtual communities to practise effective multicultural learning, skills and self-educating with no or minimum cost, thereby highlighting the breadth of its reach and use.

Semantic Web or Web 3.0 and More

Tim Berners-Lee, the director of the World Wide Web consortium (W3C) and prime architect of the Semantic Web, said: 'The goal of semantic Web is to be a unifying system that will (like the Web for human communication) be as un-restraining as possible so that the complexity of reality can be described' (Lee, 1999). With the Semantic Web, it will be easy to recognize a variety of tools and applications that are difficult to handle in the framework of the current Web (Li and Horrocks, 2004). It delivers a response to the demands and requests for the existing and evolving volumes of data within the online communities to a phenomenon represented by information sharing, cooperation and collaboration (Capuano et al., 2010). According to Berners-Lee, the Semantic Web is not a new Web but a natural extension of the World Wide Web; it has the potential to revolutionize the way we look at the oceans and, in effect, allows us to protect our endangered environment and dependent fragile ecosystems (Le, 2008). Taking it a step further, the Semantic Web initiative, or what is touted to be Web 3.0, brings non-human content, which includes services, applications, bots and other automated components to the consumers (Kobielus, 2007). Although the touted development is laudable, it must be pointed out that the foresight precedes the intensification of Web 2.0 and has not accounted for Web 2.0 completely. The current market dynamism has thus forced the Semantic Web community to give Web 2.0 a closer look, particularly its applications and tools (Ankolekar et al., 2007).

THE INEVITABLE ROLE OF SCRM IN CRM STRATEGY

Amidst the co-existence of all components of CRM strategy (Figure 17.2), voices are getting louder to include Social CRM as one of its main pillars. After all, Social CRM is an innovative concept that combines social media technologies with CRM strategy, CRM systems and other CRM initiatives in order to enhance customer rendezvous (Greenberg, 2009). Paul Greenberg (2009) conceptualizes SCRM in a classic definition by stating that: 'Social CRM is a philosophy and a business strategy supported by technology platforms, processes and social characteristics, designed to engage the customer in a collaborative conversation to provide mutually beneficial value in a trusted and transparent business environment'.

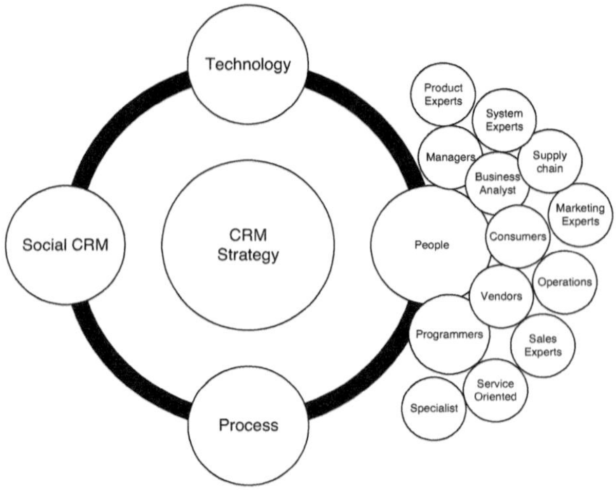

Figure 17.2 Elements of CRM strategy

This collaboration can be garnered through SCRM activities, which are constructed around users, user-generated shared content and the constant usage of Web 2.0. While privacy can be a bone of contention in this collaboration process, the value of SCRM activities in CRM strategy cannot be denied, if the motive relates to the firm's business. For organizations that use SCRM or Web 2.0 approaches (Table 17.2), knowledge generated from the content of the conversations may be helpful within a business for understanding customer segmentation, to establish proper communication and to rightly comprehend opinions, expectations and relations. One particularly fascinating and potentially explosive opportunity is that the customer or a consumer as a social element creates new prospects and insights through the actionable knowledge brought via different Web 2.0 platforms in real time. The process of SCRM focuses to bring business value to firms, altering the outputs of the strategy development process into active agendas by extracting and delivering value. This social phenomenon can determine the value that the company can provide to its customers, establish the value that the company can obtain from its customers, and efficiently manage the value exchange, which involves a procedure to co-produce, co-create and increase the lifetime value of its desired customer segments. Regrettably, there are no SCRM products or social softwares available.

Baird and Parasnis (2011) state that the implementation of a Social CRM programme is still in its early stages in many companies, and execution is unreliable with ROI concerns. Currently SCRM is being derived

Table 17.2 Social CRM initiatives

SCRM facilitated firms	Challenges	SCRM constituents and its functionalities
MTS	Negative sentiment on the Web. Listen and monitor online conversations. Customer engagement with unhappy customers. Predict and manage situation with potential of full-blown crisis for the brand.	Engaging online communities, customer care executives, management and online properties of the brand. Social media monitoring tool to track online conversations. Aggressively engage unhappy customers by offering them speedy solutions.
Best Buy	Customer engagement, providing support, customer advocacy, providing recommendations across multiple channels, in a scalable and manageable way.	Social and community team, monitoring blogosphere, social online community, Twelpforce.
Fiskars	Boost specialty store sales, customer engagements, product innovations and developments.	Blog, fiskateers.com, live events, message board, online tools.
Finnair	Service improvement, addressing negative feedback.	Finnair Blog, quality hunters, travel blog, basic online tools.
Telefonica	Customer satisfaction, customer engagement.	Online community, forum, crowdsourcing of ideas, customer participation, hilfe.o2online.de.
Starbucks	New products, innovation, reach.	Starbucks Gossip, My Starbucks Idea, social media monitoring, online communities, idea generators, mystarbucksidea.force.com.
Procter & Gamble	Customer connect, innovation.	Connect and develop, vocal point, being girl.
Dell	Support, addressing issues, product and service innovation.	Forum, communities, connect.dell.com, user-to-user interaction.

Source: Internet data.

from various Web 2.0 services that have been mildly integrated in a CRM environment. And given the popularity of the Internet, a collaborative ecosystem is being created by involving SCRM as one of the main pillars of CRM strategy. To copiously recognize and use the power of social media to enable firms to associate with their customers, they need to amalgamate platforms that use Social CRM to reinvent the customer relationship (Pavicic et al., 2011). So, while blogs, forums, communities, and so on have been a part of the CRM strategy, the important question is whether we should categorize them as elements of SCRM. Because if these Internet-dependent and open information portals can be grouped into one category (that is, under the umbrella of a social component that can contribute, induce knowledge sharing and fill the gap that the CRM as a strategy is facing to be successful), then it is about time that SCRM should be the core of any CRM strategy.

SCRM is a new strategy and a system that assimilates Web 2.0 and the influence of online communities in parallel with traditional CRM systems, for inspiring customers to play an important part alongside the firms in formulating decisions that have an impact on creating sensible conversations and high-quality relationships between firms and customers (Woodcock et al., 2011). It is clear that SCRM is not just a technology or a system as it includes several components that need to be taken into consideration in order to offer value for the organization and its customers (Reinhold and Alt, 2012, 2011; Faase et al., 2011). That then means that networking, collaboration and participation through SCRM systems (Figure 17.3) must be a standard and not an expectation (Gurau, 2003; Askool and Nakata, 2010b). SCRM is the connection of social data, wherever it may reside, with existing customer records or customer databases which empowers firms to provide new forms of consumer insight and appropriate context (Woodcock et al., 2011). To the best of my knowledge and research, there is little or limited literature presented (Table 17.3) that

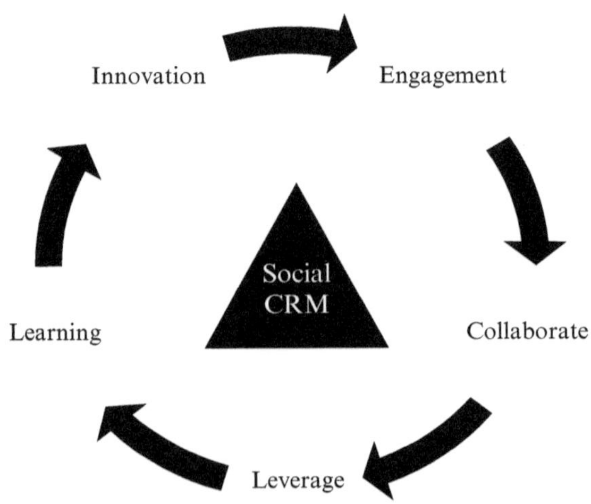

Figure 17.3 Framework for SCRM strategy

Table 17.3 SCRM literature

Authors	Describing SCRM
Reinhold and Alt (2011)	Investigates the requirements of SCRM and the role of analytical SCRM in integrated SCRM systems, analyses available solutions with their shortcomings and develops a blueprint of an analytical SCRM system.
Almunawar and Anshari (2012)	Lays the foundation for developing a system for managing customer relationships using Web 2.0 technology and Social CRM in the health care industry.
Faase et al. (2011)	Defines social CRM and to create an understanding of the concept.
Mohan et al. (2008)	Discussion on how social networking helps organizations to increase their customer life cycle and to acquire new customers by developing a conceptual architecture merging the CRM System with Social Network Engine and Web 2.0.
Askool and Nakata (2010a)	Describes a set of factors believed to influence the adoption of SCRM and provides a conceptual model of SCRM usage and acceptance followed by a discussion.
Deng et al. (2009)	Defines the key components of Social CRM application that aims at finding the potential needs of customers and translating them to sale opportunities by integrating outside information sources and leveraging Web 2.0 concepts of collaboration and community based on service-oriented architecture (SOA).
Ang (2011)	Defining whether SCRM is a really good social media strategy.
Zhang (2011)	Investigates the optimal level of social software for a firm which maximizes its transactional benefits through managing a consumer knowledge system. The system consists of a customer knowledge base and social software that dynamically affects the customer knowledge base through customers' interactions.
Li et al. (2012)	Literature reviews of social network service, social system of enterprise and collaborative product development (CPD). Suggestions on system architecture and function of enterprise social collaboration platform. Describes product catalogue sharing and synchronization from enterprise resource planing (ERP) to its dealer or customer. A deduction of the development and application of a social collaboration platform.
Woodcock et al. (2011)	Understanding SCRM as a business strategy.
Baird and Parasnis (2011)	Reinventing the customer relation via Social CRM.
Greenberg (2010b)	How CRM 2.0 impacts upon customer insights. Incorporation of social tools and strategies with traditional operational functions.
Greenberg (2009)	Book focusing on Social CRM.

specifically concentrates on Social CRM. I have thus collated the available research writings and findings on traditional CRM, Web 2.0 and Semantic Web to highlight the importance and need for SCRM to be included in CRM strategy.

DISCUSSION AND CONCLUSION

The constant evolution of Web 2.0 has given rise to Social CRM with a mature collaborative channel and knowledge source for both firms and their customers. This chapter focuses on the core elements for SCRM and its deployment, where stakeholders within the firm and outside are interconnected. It exemplifies that SCRM revolves around CRM processes by enhancing the CRM strategy through social platforms. That is also because the customer information and product/service knowledge emerging from SCRM can complement CRM strategic initiatives and help in satisfying the customer's needs. It is the company's response to the customer's ownership of the conversation. Nevertheless, there exists bewilderment about how SCRM initiatives can be incorporated and implemented, and what are the infrastructure essentials to construct it. SCRM as an emerging concept has been around for some time but is still in its nascent stage in terms of use and benefits (Greenberg, 2010b). For now, it is certain that Social CRM can be described as an extension to traditional CRM, but not a 'replacement' for traditional CRM. Wahlberg et al. (2009) label it as an extension of collaborative CRM, describing it as one of the main pillars of traditional CRM (Iriana and Buttle, 2006; Askool and Nakata, 2010a). SCRM can be recognized as a separate entity that concentrates on customer engagement, generating comprehensive and effective customer knowledge rather than just customer transactional data. There is a definite need for deep integration of SCRM capability into the functions and processes of the organization for further automation. This chapter presents a few of the many SCRM cases that are emerging even though the concept of SCRM is in its nascent and testing stage. This research tries to provoke and motivate readers to investigate SCRM further and find the missing constituents that could make SCRM close to perfect and find a permanent place in a firm's CRM strategy. Future research could include more relevant case studies on SCRM strategies and tools and highlight its capabilities and limitations.

ACKNOWLEDGEMENTS

I would like to thank the Management of Emerging Technologies for Economic Impact (ManETEI) network (with Grant agreement no.: 238382) that offers a research-led training programme to create a rigorous collaborative research agenda centred on the multifaceted phenomenon of managing emergent technologies for maximum economic and societal impact. It will have advanced capacity building and career development that will benefit the early-career and experienced researchers needed to ensure Europe becomes a leading knowledge economy driven by its unique European research area (ERA). The network has identified arguably Europe's most important opportunity for its future management researchers, managers and policy makers.

I would also like to convey my sincere thanks to Professor Matti Rossi and Dr Risto Rajala for guiding and supporting me in my research.

REFERENCES

Alavi, S., Ahuja, V. and Medury, Y. (2011). ECRM using online communities. *Journal of Marketing Management*, 10(1), 83–96.

Ali, M. and Alshawi, S. (2004). Developing a structurational approach for electronic customer relationship management (ECRM). *European and Mediterranean Conference on Information Systems*, Tunis, pp. 25–27.

Al-mashari, M. and Zairi, M. (2000). The effective application of SAPR/3: a proposed model of the effective application of SAPR/3: a proposed model of best practice. Logistics Information Management Emerald Article.

Almotairi, M. (2008). CRM success factors taxonomy. European and Mediterranean Conference on Information Systems.

Almunawar, M.N. and Anshari, M. (2012). Improving customer service in healthcare with CRM 2.0. arXiv preprint arXiv:1204.3685.

Ang, L. (2011). Is SCRM really a good social media strategy? *Journal of Database Marketing and Customer Strategy Management*, 18(3), 149–153.

Ankolekar, A., Krötzsch, M., Tran, T. and Vrandecic, D. (2007). The two cultures. ACM WWW. http://ids.snu.ac.kr/w/images/1/1a/sc21.pdf.

Anshari, M. and Almunawar, M.N. (2011). Evaluating CRM implementation in healthcare organization. *2011 International Conference on Economics and Business Information*, Vol. 9, pp. 30–34.

Appelbaum, S.H. (1997). Socio-technical systems theory: an intervention strategy for organizational development. *Management Decision*, 35(6), 452–463.

Askool, S. and Nakata, K. (2010a). A conceptual model for acceptance of social CRM systems based on a scoping study. *AI and Society*, 26(3), 205–220.

Askool, S.S. and Nakata, K. (2010b). Scoping study to identify factors influencing the acceptance of social CRM. http://ieeexplore.ieee.org.libproxy.aalto.fi/xpl/articleDetails.jsp?arnumber=5492888.

Ata, Z. and Toker, A. (2012). The effect of customer relationship management adoption in business-to-business markets. *Journal of Business and Industrial Marketing*, 27(6), 497–507.

Baird, C.H. and Parasnis, G. (2011). From social media to Social CRM: reinventing the customer relationship. *Strategy and Leadership*, 39(5), 30–37.

Barsky, E. and Purdon, M. (2006). Introducing Web 2.0: social networking and social bookmarking for health librarians. *Journal of the Canadian Health Libraries Association*, 27(3), 65–67.

Black, S. and Jacobs, J. (2010). Using Web 2.0 to improve software quality. *Proceedings of the 1st Workshop on Web 2.0 for Software Engineering – Web2SE '10*, p. 6.

Bolton, N.R. and Tarasi, O.C. (2007). Managing customer relationships. *Review of Marketing Research*. http://dx.doi.org/10.1108/S1548-6435(2007)0000003005.

Bose, R. (2002). Customer relationship management: key components for IT success. *Industrial Management and Data Systems*, 102(2), 89–97.

Boulding, W., Staelin, R., Ehret, M. and Johnston, W.J. (2005). Management roadmap: what is known, potential pitfalls. *Journal of Marketing*, 69(October), 155–166.

Bueren, A., Schierholz, R., Kolbe, L. and Brenner, W. (2004). Customer knowledge management – improving performance of customer relationship management with knowledge management. *46th Hawaii International Conference on System Sciences*, Vol. 7, pp. 70172b–70172b.

Bueren, A., Schierholz, R., Kolbe, L.M. and Brenner, W. (2005). Improving performance of customer-processes with knowledge management. *Business Process Management Journal*, 11(5), 573–588.

Campbell, J.A. (2003). Creating customer knowledge competence: managing customer relationship management programs strategically. *Industrial Marketing Management*, 32(5), 375–383.

Capuano, N., Gaeta, M., Orciuoli, F. and Ritrovato, P. (2010). Semantic Web Fostering Enterprise 2.0. *2010 International Conference on Complex, Intelligent and Software Intensive Systems*, pp. 1087–1092.

Chen, C.-J. and Hung, S.-W. (2010). To give or to receive? Factors influencing members' knowledge sharing and community promotion in professional virtual communities. *Information and Management*, 47(4), 226–236.

Chen, I.J. and Popovich, K. (2003). Understanding CRM: people, process and technology. *Business Process Management Journal*, 9(5), 672–688.

Choy, K.L., Lee, W.B., Lau, H.C.W., So, S.C.K. and Lo, V. (2004). An enterprise collaborative management system: a case study of supplier selection in new product development. *International Journal of Technology Management*, 28(2), 206.

Coltman, T. (2007). Why build a customer relationship management capability? *Journal of Strategic Information Systems*, 16(3), 301–320.

Deng, X.L., Zhang, L., Wang, B. and Wu, B. (2009). Implementation and Research of Social CRM Tool in Mobile BOSS Based on Complex Network. *1st International Conference on Information Science and Engineering*, pp. 870–873.

Donaldson, B. and O'Toole, T. (2007). *Strategic Market Relationships: From Strategy to Implementation*. Chichester: Wiley.

Dong, S. and Zhu, K. (2006). The business value of CRM systems: productivity, profitability, and time lag. In *Workshop on Information Systems and Economy* (WISE 2006), Northwestern University, Evanston, IL, pp. 1–5.

Ernst, H. (2002). Success factors of new product development: a review of the empirical literature. *International Journal of Management Reviews*, 4(1), 1–40.
Faase, R., Helms, R. and Spruit, M. (2011). Web 2.0 in the CRM domain: defining social CRM. *International Journal of Electronic Customer Relationship Management*, 5(1), 1.
Fan, Y.-W. and Ku, E. (2010). Customer focus, service process fit and customer relationship management profitability: the effect of knowledge sharing. *Service Industries Journal*, 30(2), 203–223.
Feiz, S., Zeinab, D., Khalifah, B. and Ghotbabadi, A.R. (2011). Customer relationship management in organizations. International Conference on Management (ICM) Penang, Malaysia.
Foss, B., Stone, M. and Ekinci, Y. (2008). What makes for CRM system success – or failure? *Journal of Database Marketing and Customer Strategy Management*, 15(2), 68–78.
Frow, P.E. and Payne, A.F. (2009). Customer relationship management: a strategic perspective. *Journal of Business Marketing Management*, 3(1), 7–27.
Frow, P., Payne, A., Wilkinson, I.F. and Young, L. (2011). Customer management and CRM: addressing the dark side. *Journal of Services Marketing*, 25(2), 79–89.
Gebert, H., Geib, M., Kolbe, L. and Riempp, G. (2002). Towards customer knowledge management : integrating customer relationship management and knowledge management concepts. *Second International Conference on Electronic Business*, Taipei, pp. 10–13.
Goldenberg, B. (2000). What is CRM? What is an e-customer? Why you need them now. *DCI Customer Relationship Management Conference*. Boston, MA: DCI CRM, pp. 27–29.
Greenberg, P. (2009). *CRM at the Speed of Light: Social CRM 2.0 Strategies, Tools, and Techniques for Engaging Your Customers*. New York: McGraw-Hill Osborne Media.
Greenberg, P. (2010a). The impact of CRM 2.0 on customer insight. *Journal of Business and Industrial Marketing*, 25(6), 410–419.
Greenberg, P. (2010b). The impact of CRM 2.0 on customer insight. *Journal of Business*, 6(April), 410–419.
Gummesson, E. (2004). Return on relationships (ROR): the value of relationship marketing and CRM in business-to-business contexts. *Journal of Business and Industrial Marketing*, 19(2), 136–148.
Gurau, C. (2003). Tailoring e-service quality through CRM. *Managing Service Quality*, 13(6), 520–531.
Heller, F. (1997). Sociotechnology and the environment. *Human Relations*, 50(5), 605–624.
Hossain, M.M. and Aydin, H. (2011). A Web 2.0-based collaborative model for multicultural education. *Multicultural Education and Technology Journal*, 5(2), 116–128.
Iriana, R. and Buttle, F. (2006). Strategic, operational, and analytical customer relationship management: attributes and measures. *Journal of Relationship Marketing*, 5(4), 23–42.
Jayashree, S., Shojaee, S. and Pahlavanzadeh, S. (2011). A critical analysis of customer relationship management from strategic perspective. *2010 International Conference on E-business, Management and Economics*, Vol. 3, pp. 340–345.
Kamel Boulos, M.N. and Wheelert, S. (2007). The emerging Web 2.0 social

software: an enabling suite of sociable technologies in health and health care education. *Health Information and Libraries Journal*, 24(1), 2–23.
Kevork, E.K. and Vrechopoulos, A.P. (2009). CRM literature: conceptual and functional insights by keyword analysis. *Marketing Intelligence and Planning*, 27(1), 48–85.
Kim, E. and Lee, B. (2007). An economic analysis of customer selection and leveraging strategies in a market where network externalities exist. *Decision Support Systems*, 44(1), 124–134.
Kobielus, J. (2007). The Semantic Web: meaning and SOA. *Business Communication Review*, 37(10), 30–36.
Kumar, A. (2008). From mass customization to mass personalization: a strategic transformation. *International Journal of Flexible Manufacturing Systems*, 19(4), 533–547.
Kumar, V., Aksoy, L., Donkers, B., Venkatesan, R., Wiesel, T. and Tillmanns, S. (2010). Undervalued or overvalued customers: capturing total customer engagement value. *Journal of Service Research*, 13(3), 297–310.
Kumar, V., Ramani, G. and Bohling, T. (2004). Customer lifetime value approaches and best practice applications. *Journal of Interactive Marketing*, 18(3), 60–72.
Le, S. (2008). Semantic web holds promises for ocean observing needs. In *Oceans 2008*, Quebec City, 15–18 September, pp. 1–9.
Lee, M.R. and Lan, Y. (2007). From Web 2.0 to conversational knowledge management: towards collaborative intelligence. *Journal of Entrepreneurship Research*, 2(2), 47–62.
Lee, T.B. (1999). *Weaving the Web*. San Francisco: Harper.
Lemon, K.N., White, T.B. and Winer, R.S. (2002). Dynamic customer relationship management: incorporating future considerations into the service retention decision. *Journal of Marketing*, 66(1), 1–14.
Levy, M. (2009). Web 2.0 implications on knowledge management. *Journal of Knowledge Management*, 13(1), 120–134.
Li, L. and Horrocks, I. (2004). A software framework for matchmaking based on semantic web technology. *International Journal of Electronic Commerce*, 8(4), 39–60.
Li, M., Chen, G., Zhang, Z. and Fu, Y. (2012). A social collaboration platform for enterprise social networking. In *Proceedings of the 2012 IEEE 16th International Conference on Computer Supported Cooperative Work in Design*, pp. 671–677.
Liao, S., Chen, Y.-J. and Deng, M. (2010). Mining customer knowledge for tourism new product development and customer relationship management. *Expert Systems with Applications*, 37(6), 4212–4223.
Lin, H.-F. and Lee, G.-G. (2006). Effects of socio-technical factors on organizational intention to encourage knowledge sharing. *Management Decision*, 44(1), 74–88.
Mithas, S., Krishnan, M.S. and Fornell, C. (2005). Why do customer relationship management applications affect customer satisfaction? *Journal of Marketing*, 69(4), 201–209.
Mohan, S., Choi, E. and Min, D. (2008). Conceptual modeling of enterprise application system using social networking and Web 2.0 'Social CRM System'. In *International Conference on Convergence and Hybrid Information Technology 2008*, pp. 237–244.
Nambisan, S. and Baron, R.A. (2007). Interactions in virtual customer

environment: implications for product support and CRM. *Journal of Interactive Marketing*, 21(2), 42–62.

O'Reilly, T. (2007). What is Web 2: design patterns and business models for the next generation of software. *Communications and Strategies*, 65(1), 17–37.

Oliveira, T., Martins, M.F. and Lisboa, U.N. De (2011). Literature review of information technology adoption models at firm level. *Electronic Journal Information Systems Evaluation*, 14(1), 110–121.

Osarenkhoe, A. and Bennani, A.-E. (2007). An exploratory study of implementation of customer relationship management strategy. *Business Process Management Journal*, 13(1), 139–164.

Pavicic, J., Alfirevic, N. and Znidar, K. (2011). Customer knowledge management: toward social CRM. *International Journal of Management Cases*, 13(3), 203–209.

Payne, A. and Frow, P. (2005). A strategic framework for CRM. *Journal of Marketing*, 2429, 167–176.

Peppers, D. and Rogers, M. (1993). *The One to One Future: Building Relationship One Customer at a Time*. New York: Doubleday.

Piller, F.T., Moeslein, K. and Stotko, C.M. (2004). Does mass customization pay? An economic approach to evaluate customer integration. *Production Planning and Control*, 15(4), 435–444.

Pitta, D.A. (1998). Marketing one-to-one and its dependence on knowledge discovery in databases. *Journal of Consumer Marketing*, 15(5), 468–480.

Reinartz, W., Krafft, M. and Hoyer, W.D. (2004). The customer relationship management process: its measurement and impact on performance. *Journal of Marketing Management*, 41(August), 293–305.

Reinhard, C.D. (2009). Discourse swings in understanding audiences: case studies on Hollywood's cooptation of audience activity(s) as emergent discourse. Paper presented at International Communication Association conference, Chicago, IL, 21–25 May. *International Communication Association*, pp. 1–29.

Reinhold, O. and Alt, R. (2011). Analytical Social CRM: concept and tool support. *Media*, 226–241, http://aisel.aisnet.org/bled2011/50/.

Reinhold, O. and Alt, R. (2012). Social customer relationship management: state of the art and learnings from current projects. In *25th Bled eConference eDependability: Reliable and Trustworthy eStructures, eProcesses, eOperations and eServices for the Future*, pp. 155–169.

Rigby, B.D.K., Reichheld, F. and Dawson, C. (2003). Winning customer loyalty is the key to a winning CRM strategy. *Ivey Business Journal*, 519, 2.

Rigby, D.K., Reichheld, F.F. and Schefter, P. (2002). Avoid the four perils of CRM. *Harvard Business Review*, 80(2), 101–106, 108–109, 130; available at http://www.ncbi.nlm.nih.gov/pubmed/11894676.

Roberts, M. Lou, Liu, R.R. and Hazard, K. (2005). Strategy, technology and organisational alignment: key components of CRM success. *Journal of Database Marketing and Customer Strategy Management*, 12(4), 315–326.

Rodriguez, M. and Honeycutt, E.D. (2011). Customer relationship management (CRM)'s impact on B to B sales professionals' collaboration and sales performance. *Journal of Business-to-Business Marketing*, 18(4), 335–356.

Rubin, M. (1997). Creating customer-oriented companies. *Prism*, 4, 5–27.

Ryals, L. and Knox, S. (2001). Cross-functional issues in the implementation of relationship marketing through customer relationship management (CRM). *European Management Journal*, 19(5), 534–42.

Sahaf, M.A., Qureshi, H.I. and Khan, R.A. (2011). The science and art of

customer relationship management: a blend of business processes and technology solutions. *African Journal of Business Management*, 5(13), 5057–5064.
Sawhney, M., Verona, G. and Prandelli, E. (2005). Collaborating to create: the internet as a platform for customer engagement in product innovation. *Journal of Interactive Marketing*, 19(4), 1–15.
Sigala, M. (2008). Developing and implementing an eCRM 2.0 strategy: usage and readiness of Greek tourism firms. *Information and Communication Technologies in Tourism 2008*, pp. 463–474. Proceedings of Fifteenth International Conference on Information and Communication Technologies in Travel and Tourism, Innsbruck, Austria, 23–25 January.
Sofianti, T.D., Suryadi, K., Govindaraju, R. and Prihartono, B. (2010). Customer knowledge co-creation process in new product development. *Proceedings of the World Congress on Engineering*, Vol. I.
Stefanou, C.J., Sarmaniotis, C. and Stafyla, A. (2003). CRM and customer-centric knowledge management: an empirical research. *Business Process Management Journal*, 9(5), 617–634.
Stone, M. (2009). Staying customer-focused and trusted: Web 2.0 and Customer 2.0 in financial services. *Journal of Database Marketing and Customer Strategy Management*, 16(2), 101–131.
Stringfellow, A. and Bowen, D.E. (2004). CRM: profiting from needs. *Business Horizons*, October, 45–52.
Tollin, K. (2002). Customization as a business strategy – a barrier to customer integration in product development? *Total Quality Management*, 13(4), 427–439.
Torggler, M. (2009). The functionality and usage of CRM systems. *Environment*, 30(41), 47.
Tseng, M. and Piller, T. (2003). *The Customer Centric Enterprise-Advances in Mass Customization and Personalization*. New York: Springer.
Verhoef, P.C. (2003). Understanding the effect of efforts on customer retention and customer share development. *Journal of Marketing*, 67(4), 30–45.
Wahlberg, O., Strandberg, C., Sundberg, H. and Sandberg, K.W. (2009). Trends, topics and under-researched areas in CRM research. *International Journal of Public Information Systems*, 3, 191–208.
Wells, J.D., Fuerst, W.L. and Choobineh, J. (1999). Managing information technology (IT) for one-to-one customer interaction. *Information and Management*, 35(1), 53–62.
Woodcock, N., Green, A. and Starkey, M. (2011). Social CRM as a business strategy. *Journal of Database Marketing and Customer Strategy Management*, 18(1), 50–64.
Xiong, L., Qiu-yue, H. and Chun-li, J.I. (2011). Creating competitive advantage through customer knowledge management in customer relationship management. In *2011 International Conference on E-Business and E-Government (ICEE)*, pp. 1–4.
Xu, M. and Walton, J. (2005). Gaining customer knowledge through analytical CRM. *Industrial Management and Data Systems*, 105(7), 955–971.
Xu, Y., Yen, D.C., Lin, B. and Chou, D.C. (2002). Adopting customer relationship management technology. *Industrial Management and Data Systems*, 102(8), 442–452.
Yang, C.-W., Fang, S.-C. and Lin, J.L. (2010). Organisational knowledge creation strategies: a conceptual framework. *International Journal of Information Management*, 30(3), 231–238.

Zablah, A.R., Bellenger, D.N. and Johnston, W.J. (2004). An evaluation of divergent perspectives on customer relationship management: towards a common understanding of an emerging phenomenon. *Industrial Marketing Management*, 33, 475–489.

Zablah, A.R., Bellenger, D.N. and Johnston, W.J. (2005). CRM implementation gaps. *Journal of Personal Selling and Sales Management*, 24(4), 279–295.

Zahaya, D., Griffin, A. and Frederick, E. (2004). Sources, uses, and forms of data in the new product development process. *Industrial Marketing Management*, 33(7), 657–666.

Zhang, Z. (J.). (2011). Customer knowledge management and the strategies of social software. *Business Process Management Journal*, 17(1), 82–106.

Zineldin, M. (2006). The royalty of loyalty: CRM, quality and retention. *Journal of Consumer Marketing*, 23(7), 430–437.

Index